Employee Green Behavior

M. Y. Yusliza • Douglas W. S. Renwick
Jing Yi Yong • Nurul Liyana Mohd Kamil
Editors

Employee Green Behavior

Transforming HR for a Sustainable Tomorrow

Springer

Editors
M. Y. Yusliza
Faculty of Business, Economics and Social Development
Universiti Malaysia Terengganu
Kuala Nerus, Terengganu, Malaysia

Douglas W. S. Renwick
Department of HRM, Nottingham Business School
Nottingham Trent University
Nottingham, UK

Jing Yi Yong
School of Management and Marketing
Faculty of Business and Law
Taylor's University
Subang Jaya, Selangor, Malaysia

Nurul Liyana Mohd Kamil
Faculty of Business and Economics
Universiti Malaya
Kuala Lumpur, Malaysia

ISBN 978-981-95-1421-2 ISBN 978-981-95-1422-9 (eBook)
https://doi.org/10.1007/978-981-95-1422-9

This Springer imprint is published by the registered company Springer Nature Singapore Pte Ltd.
The registered company address is: 152 Beach Road, #21-01/04 Gateway East, Singapore 189721, Singapore

Acknowledgements

- Anonymous reviewers
- Editors
- Publisher
- Faculty Special Research Grant (UMG006I-2023)
- Fundamental Research Grant Scheme (FRGS) from the Ministry of Higher Education, Malaysia (FRGS/1/2019/SS03/UMT/02/3)
- Fundamental Research Grant Scheme (FRGS) from the Ministry of Higher Education, Malaysia (FRGS/1/2022/SS01/TAYLOR/02/3)
- Fundamental Research Grant Scheme (FRGS) from the Ministry of Higher Education Malaysia (FRGS/1/2022/SS01/TAYLOR/02/4).
- Talent and Publications Enhancement Research Grant (TAPERG) from the Universiti Malaysia Terengganu, Malaysia (TAPERG/2023/UMT/2682)

Contents

About the Editors

M. Y. Yusliza is a distinguished academic and researcher from Universiti Malaysia Terengganu in the field of business and management, widely recognized for her pioneering work in sustainable human resource management (HRM). Her research has significantly advanced understanding in areas such as green HRM and employee green behavior. More recently, her scholarly pursuits have expanded into emerging domains including artificial intelligence in HRM and Green HRM, blockchain applications, and the integration of HRM within the circular economy. With over 125 scholarly articles published in prestigious Web of Science and Scopus-indexed journals, Professor Dr. Yusliza has made a profound impact on the academic community. Her work has garnered more than 4700–citations on Scopus, affirming her status as a thought leader in HRM, organizational behavior, and sustainable management practices.

Professor Dr. Yusliza's contributions have been widely recognized. In 2022, she received the Icon Academia Award from the Malaysian Ministry of Higher Education and the Academic Leadership Award from Universiti Malaysia Terengganu, honoring her leadership in teaching, research, and international collaboration. By 2025, she is celebrated as one of Malaysia's foremost scholars in business and management. Her groundbreaking research earned her a place on Stanford University's list of the world's top 2% scientists for four consecutive years (2021–2024), a testament to her global academic influence. Additionally, she has been ranked among the top 0.05% of scholars worldwide in Human Resource and Resource Management by ScholarGPS, reflecting her exceptional research productivity and scholarly impact.

Douglas W. S. Renwick is Associate Professor of Sustainable Work at Nottingham Business School, UK, and Senior Research Fellow at Rennes School of Business, France, EU, where he undertakes research in green HRM, sustainable workforce management, green values and culture, and green jobs. Doug has published over 38 journal articles, 7 edited journal collections, 7 books, 23 chapters, 27 conference papers, symposia and workshops, and 16 case studies, discussion, and other papers (115 works overall).

His green HRM and sustainability publications appear in the German Journal of HRM (2011), International Journal of Management Reviews (2013), The International Journal of HRM (2016), Journal of Environmental Management (2018), Business Strategy and the Environment (2022) and Organization & Environment, USA (2022, 2024). His books examine green HRM research (Routledge, 2020), UK green jobs (Palgrave MacMillan, 2024), green HRM in the global south (Springer Nature, 2024), and employee green behavior (Springer Nature, 2024), plus a new contracted book with Palgrave MacMillan (2027). Doug is a co-editor of a journal special issue on the circular economy and sustainable development for Business & Society (2027). His work informs both British government (2024) and United Nations policy (2019) and continues in his role as advisor to the German government on green skills development (Paderborn University, from 2024), and the UK government on green skills education (in UK Parliament Post 711, 2024). Dr. Renwick's work also informs practice, as seen in his thought leadership on green jobs for HR Magazine (2024).

Doug has won competitive external research funding of £408,000, is ranked in the: top 1% on Scholar GPS (2024) in HRM; top 3% for HRM productivity; top 0.38% for HR impact; top 2% for HR quality; and top 6.5% for all academic fields, plus in the top 2% of all global scientists by Scopus/Elsevier in 2023 and 2024. He won Emerald's Citation of Excellence award and an Outstanding Paper prize in the Emerald Literati Awards for his green HRM work (in 2016 & 2023); achieved 10,000 reads of his work on ResearchGate (2024); and 3000 citations to his green HRM paper on Google Scholar (2025). Dr. Renwick is an editorial Board Member of The International Journal of HRM, Employee Relations, the German Journal of HRM and Sustainability Letters (Netherlands, EU), and an invited Academic Council Member for Harvard Business Review.

Douglas W.S. Renwick, Associate Professor in Sustainable Work, Nottingham Business School, Nottingham Trent University, United Kingdom. https://www.ntu.ac.uk/staff-profiles/business/douglas-renwick.

Jing Yi Yong is a Senior Lecturer in the School of Management and Marketing at Taylor's University. She earned her Doctorate degree from Universiti Sains Malaysia, where she delved deep into her passion for exploring the antecedents and outcomes of green HRM, unearthing invaluable insights into sustainable organizational practices. She has been published in esteemed scholarly journals such as the *Journal of Cleaner Production, Business Strategy and Environment, Benchmarking: An International Journal, Industrial and Commercial Training, Journal of Management Development*, and the *International Journal of Engineering & Technology.* Dr. Yong's contributions extend beyond publication, as she actively participates as a reviewer for international journals and conferences. Her research interests span a wide spectrum, including green HRM, intellectual capital, environmental, social, and governance practices, sustainability, and family business dynamics.

Nurul Liyana Mohd Kamil is a senior lecturer in HRM and organizational behavior at the Faculty of Business and Economics, Universiti Malaya, Malaysia. She was a visiting fellow in the Department of Management of the London School of Economics and Political Science (LSE), the UK. Her research interests include leadership, organizational climate, voice, turnover intentions, employee well-being, job performance, and green climate. She is a membership officer at the Asia Pacific Academy for Psychosocial Factors at Work. Her research and writing have appeared in ISI and Scopus journals including *Journal of Occupational and Organisational Psychology, Leadership and Organisational Development*, *Non-Profit Management and Leadership*, among others. Her work also has been featured in books and book chapters published by Springer and Palgrave.

From Culture to Action: Unveiling the Path to Ecological Behavior Through Green HRM

M. Y. Yusliza, Juhari Noor Faezah, T. Ramayah, and Muhammad Hamza Qummar

Abstract This study investigates the influence of green culture (GC) on employee ecological behavior (EEB), with green human resource management (HRM) serving as a mediating factor, framed within the Ability–Motivation–Opportunity (AMO) theory. Utilizing a quantitative approach, data were gathered from 308 academics across five higher education institutions in Malaysia and analyzed through partial least-squares structural equation modeling (PLS-SEM). The findings indicate that while green culture has a positive effect on green HRM, it does not directly impact EEB. Instead, green HRM fully mediates the relationship between green culture and EEB. Notable insights include the inconsistent role of green culture in promoting EEB, attributed to ineffective green practices, as well as the indirect influence of green culture on employee behaviors through established organizational norms. The study emphasizes the necessity of authentic investments in green HRM to effectively motivate pro-environmental behaviors among employees. The conclusion addresses the implications of the findings, acknowledges limitations, and suggests avenues for future research.

M. Y. Yusliza · M. H. Qummar
Faculty of Business, Economics and Social Development, Universiti Malaysia Terengganu, Kuala Nerus, Terengganu, Malaysia
e-mail: yusliza@umt.edu.my; Hamzaqummar786@gmail.com

J. Noor Faezah (✉)
Faculty of Economics and Management, Universiti Kebangsaan Malaysia, Bangi, Selangor, Malaysia
e-mail: faezahjuhari@ukm.edu.my

T. Ramayah
School of Management, Universiti Sains Malaysia, Penang, Malaysia
e-mail: ramayah@usm.my

M. Y. Yusliza et al. (eds.), *Employee Green Behavior*,
https://doi.org/10.1007/978-981-95-1422-9_1

1 Introduction

The primary objective of sustainable development (SD) is to ensure a viable future for both humanity and the planet. In pursuit of this goal, regulatory bodies at national and international levels have endorsed relevant guidelines. In response, governments and organizations worldwide have been implementing effective strategies to address the escalating environmental degradation (Yang et al., 2023). Several governments have adopted sustainable policies, while organizations are increasingly fostering the environment-friendly practices in the workplace (Mousa & Othman, 2020) by embracing eco-friendly initiatives and environmental management systems (EMS) (Anwar et al., 2020).

The effectiveness of eco-friendly initiatives is significantly influenced by the ecological behavior of employees (Nisar et al., 2021). Employee ecological behavior (EEB) is crucial for achieving optimal environmental performance and addressing issues related to environmental degradation (Raza & Khan, 2022; Mohamed et al., 2020). Researchers have emphasized the necessity for individuals to enhance their efforts in safeguarding the natural environment (Yusliza et al., 2020). EEB involves responsible actions that promote environmental sustainability, such as waste reduction and resource recycling (Farooq et al., 2021). Given the increasing environmental challenges stemming from human actions, organizations are increasingly promoting EEB (Shah et al., 2023) and should establish comprehensive policies to embed sustainability across all operations and procedures (Saeed et al., 2019).

These actions aim to alleviate the adverse effects on the environment resulting from human actions (Stern, 2000) and promote an environmentally sustainable workplace (Saifulina & Carballo-Penela, 2017). Therefore, the concept of employee ecological behavior (EEB) holds significant importance for organizations across various sectors, including tertiary education (Yusliza et al., 2020). However, recent EEB research has predominantly focused on corporate sectors rather than the educational domain (Anwar et al., 2020). Studies have explored EEB in industries such as hospitality (He et al., 2021), manufacturing (Bodrud-Doza et al., 2020), and small- and medium-sized enterprises (SMEs) (Khan et al., 2024). There is still a notable research gap in the area of employee ecological behavior (EEB), especially concerning the emerging focus on sustainability within higher education institutions (Dumitru et al., 2016). Furthermore, environmental management practices in Asian countries facing environmental challenges have not received adequate attention in the literature (Renwick et al., 2013). Therefore, there is a need for EEB investigations in developing countries like Malaysia, particularly within higher education institutions (Fawehinmi et al., 2020; Anwar et al., 2020). Moreover, Malaysia, as an emerging economy, grapples with significant environmental challenges. For instance, Malaysia exhibits an annual carbon dioxide (CO_2) emission growth rate exceeding 6%, which is close to China's annual growth rate of 7.42% (Anwar et al., 2020).

Several organizations have revised their policies and strategies to promote environmental sustainability. However, these endeavors alone are insufficient. To foster EEB, organizations must undergo a cultural transformation (Gurlek & Tuna, 2018).

Without cultivating a culture that aligns with a sustainability-oriented approach, organizations engaging in environmental initiatives may not realize the benefits of their environmental management investments (Yeşiltaş et al., 2022). A GC can enhance the successful adoption of eco-friendly practices and behaviors by encouraging individuals to follow established cultural norms and values (Gürlek & Çemberci, 2020). Scholars argue that organizational culture significantly influences the HRM system (Schneider, 1988; Gürlek & Çemberci, 2020).

From this perspective, fostering environmentally conscious cultural norms within an organization can support the implementation of green human resource management (HRM) practices. The integration of environmental management with HRM practices is often termed green HRM (Renwick et al., 2013). Green HRM involves leveraging HRM policies, principles, and strategies to promote the sustainable utilization of resources and mitigate organizational losses resulting from environmental challenges (Ercantan & Eyupoglu, 2022). Research conducted by Al-Zawahreh et al. (2019) indicates that higher education institutions (HEIs) often function similarly to business organizations, contributing to carbon emissions and environmental degradation (Tangwanichagapong et al., 2017). For example, research by Naderipour et al. (2021) indicated that university activities, particularly air travel, were significant contributors to greenhouse gas emissions, accounting for 34.74%, 19.37%, and 18.19% of total emissions from electricity consumption. Therefore, the adoption of green HRM practices in the educational sector is crucial for advancing environmental sustainability.

The concept of organizational GC can be understood as a set of core beliefs related to environmental management and concerns within an organization (Gurlek & Tuna, 2018). As a relatively new concept, its exact definition remains ambiguous, reflecting the nascent stage of research in this area (Afum et al., 2020; Al-Swidi et al., 2021). The effects of green organizational culture on EEB and the possible mediating role of green HRM in higher education contexts have not yet been thoroughly explored. The investigation involved the evaluation of 308 academic professionals from five universities in Malaysia. The primary objective was to analyze how organizational GC influences ecological behavior and the potential mediating impact of green HRM in Malaysian higher education institutions. The study focused on addressing the following research questions (RQ): (1) Does GC affect EEB? (2) Does GC affect green HRM? (3) Does green HRM affect EEB? And (4) Does green HRM mediate the relationship between GC?

This research aims to make significant contributions to the existing literature in several ways. Firstly, it extends the current body of knowledge by demonstrating that GC fosters EEB, an area that has been underexplored in prior research except for a few select studies (Al-Swidi et al., 2021). This contribution is particularly valuable in highlighting the significance of this relationship. Secondly, this study represents a pioneering effort in a developing nation, emphasizing the crucial role of individual employees in environmental protection, an aspect that has received limited attention, particularly relevant to the higher education sector (Anwar et al., 2020). Therefore, it contributes to the existing literature by demonstrating the mediating role of green HRM in the relationship between GC and EEB.

2 Literature Review and Hypotheses Development

2.1 Theoretical Foundation

This study is grounded in the Ability–Motivation–Opportunity (AMO) hypothesis, which serves as a key framework for examining the relationship between HRM and performance. The AMO theory is frequently applied in research focused on individual-level outcomes in HRM contexts (Jiang et al., 2012; Bos-Nehles et al., 2013). Previously, this theory has been utilized concerning commitment (Gong et al., 2009), service provision (Chuang & Liao, 2010), and green practices (Rayner & Morgan, 2018). Yong et al. (2020) encouraged the application of the AMO theory in the context of adopting green HRM practices. Cheema and Javed (2017) provided empirical evidence supporting the relevance of the AMO model by investigating how employee engagement, senior management involvement, and organizational motivation influence sustainable environmental performance. Additionally, Yu et al. (2019) explored the implications of green HRM in China through the lens of the AMO framework.

Abilities are based on "recruiting, selection, and training and development programs" (Al-swidi et al., 2021), which guarantee that potential employees have the relevant skills, essential information, values, and attitudes necessary to accomplish specific duties and to help the organization achieve its goals (Sibian & Ispas, 2021). Next, the term "motivation" refers to workers' drive to accomplish, which can be boosted by either extrinsic or inherent inspiration (Sibian & Ispas, 2021). To motivate employees, a wide range of incentives can be provided, such as rewards based on group or personal performance (Kehoe & Wright, 2013), job security, job promotion (Jiang et al., 2012), and awards (Yusliza et al., 2019). The third aspect is opportunity, defined by the presence (or the absence) of facilities and communication systems that facilitate product recycling and environmentally friendly behaviors. According to Jerónimo et al. (2020), employees are more likely to participate in environmental protection-related activities if they perceive the company's commitment to those efforts as being favorable or if they are given the opportunity.

2.2 Green Culture and Employee Ecological Behavior

A green culture in an organization indicates that the employees care deeply about the environment and demonstrate this through their values, beliefs, and activities (Roscoe et al., 2019). Employees who embrace these green principles are more likely to operate in environmentally friendly ways. This culture enables the firm to make any changes that are required to address environmental challenges (Aggarwal & Agarwala, 2025). The way employees behave and think is shaped by the organization's culture, structure, and identity (Lasrado & Zakaria, 2020). Hence, creating GC within an organization can effectively encourage employees to engage in environmentally friendly behaviors (Pham et al., 2018). In such an environment, employees receive signals that PEBs are not only expected but also valued and rewarded by the organization (Norton et al., 2014). When organizations communicate their green practices and strategies, employees perceive this as support and encouragement for environmental values, which, in turn, enhances their engagement in PEBs (Norton et al., 2017; Ramus & Steger, 2000).

Prior studies have shown that a GC is positively associated with environmentally friendly behaviors (Pham et al., 2018; Abadiyah et al., 2020). Lasrado and Zakaria (2020) emphasize that an organization's identity, culture, and structure significantly shape employees' attitudes and behaviors. When a specific culture is established in the workplace, it encourages employees to align their actions with the values that culture embodies (Gurlek & Tuna, 2018; Al-swidi et al., 2021). By fostering a GC, organizations can effectively bridge the gap and encourage employees to adopt simple, eco-friendly behaviors (Azhar & Yang, 2022). According to Mirahsani et al. (2023), organizations can foster environmentally friendly behaviors among their employees by implementing policies that promote a green identity and by creating supportive working conditions. When employees perceive the organization's support and feel a connection to its green initiatives, they are more likely to engage in pro-environmental behaviors. Thus, by cultivating a GC, organizations can indirectly encourage their employees to adopt such behaviors (Noor Faezah et al., 2024). Based on this theoretical framework, the following hypothesis was proposed:

Hypothesis 1 (H1) Green culture has a positive effect on employee ecological behavior.

2.3 *Green Culture and Green Human Resource Management*

A green culture emerges when employees integrate environmentally friendly practices into their daily work routines, prioritizing ecological considerations over mere profit maximization to reduce the negative impacts of their organization's operations on the environment (Aggarwal & Agarwala, 2025). This pro-environmental mindset, expressed through the values, beliefs, and behaviors of organizational members, gradually evolves into habitual actions that define the organization's GC (Hooi et al., 2021). GC is expected to grow as organizational members collaborate on environmental initiatives and foster workplace green behavior (Hooi et al., 2012). Thus, as GC increases organizational members' understanding of the value of green practices, this will motivate them to be more demanding about implementing those practices through green HRM (Gurlek & Tuna, 2018).

This can also be achieved through environmentally focused hiring practices, incentives, performance development, and training (Luu, 2019). For instance, in organizations with a GC, upper management encourages employees to raise awareness about environmental issues and engage in discussions surrounding them. This collaborative approach fosters a shared sense of responsibility for environmental protection, thereby enhancing the GC within the organization. These initiatives support the organization's environmental goals by familiarizing new employees with the organization's eco-friendly culture. Additionally, green training programs help instill pro-environmental values among employees (Aggarwal & Agarwala, 2025). Research by Porter et al. (2016) indicates that the adoption of sustainable practices contributes to the development of a green organizational culture. Similarly, Roscoe et al. (2019) found that collaborative employees work to addresses the environmental challenges, it promotes the development of a pro-environmental culture within the organization. Based on this understanding, we propose the following hypothesis:

Hypothesis 2 (H2) Green culture has a positive effect on green HRM.

2.4 Green Human Resource Management and Employee Ecological Behavior

Green HRM focuses on hiring individuals who align with the organization's environmental values and objectives. It develops training programs aimed at enhancing employees' environmental knowledge, skills, awareness, and attitudes (Cheema et al., 2020). Additionally, green HRM integrates environmental considerations into performance evaluations, compensation management, and employee empowerment (Sancho et al., 2018). Through these environmentally focused management practices, green HRM can effectively assess and influence both task-related and voluntary green behavior among employees. Key components of green HRM, such as environmental training, involvement, and performance, are essential for preparing employees with the necessary knowledge, skills, abilities, motivation, and opportunities to foster ecological behavior (Noor Faezah et al., 2024). Engaging employees in the implementation of green initiatives and providing them with relevant training can significantly enhance their knowledge and capabilities, making them more psychologically prepared to participate in green behavior (Chaudhary, 2020). Furthermore, when employees are recognized and rewarded for their green actions, they are more likely to be motivated to engage in extra-role behavior, including PEBs. Consequently, it is suggested that green HRM activities can positively influence employee PEBs within the workplace (Dumont et al., 2017).

Previous research has established a significant relationship between green HRM and ecological behavior (Ababneh, 2021; Aboramadan, 2022; Chaudhary, 2020; Chen et al., 2021; Ercantan, 2022; Noor Faezah et al., 2024). For instance, Saeed et al. (2019) found that green HRM practices positively influence pro-environmental behavior among employees across various industries in Pakistan. These practices enhance environmental awareness among employees and help shape their attitudes toward sustainability in both their personal and professional lives (Saeed et al., 2019). Furthermore, green HRM encourages environmentally friendly activities by involving employees in green initiatives (Cincera & Krajhanzl, 2013). Consequently, this study utilizes the AMO framework to explore how green HRM practices impact behavioral outcomes within the workforce. In summary, it is anticipated that academic staff who experience green HRM practices will demonstrate more environmentally conscious behavior. Based on this discussion, we propose the following hypothesis:

Hypothesis 3 (H3) Green HRM has a positive effect on employee ecological behavior.

2.5 Mediating Effect of Green Human Resource Management

Over time, a GC within an organization is likely to develop as members encourage one another in their environmental initiatives and actively promote sustainability in the workplace (Hooi et al., 2021). In environments where a strong GC is established, employees tend to feel motivated to engage in eco-friendly behaviors and adopt sustainable practices (Beksultanova et al., 2022). These can include actions such as reducing paper usage, turning off lights when not in use, and utilizing recyclable materials (Aggarwal & Agarwala, 2025). After developing green HRM, firms are expected to communicate the organization's environmental preservation philosophy and regard to their staff. Once employees recognize the importance of green HRM, GC acts as a key mechanism by shaping shared values, norms, and practices that guide and reinforce ecological behavior. It is apparent that GC and perceived green HRM will catalyze employee involvement in pro-environmental initiatives and nurture favorable green behaviors in tackling environmental issues (Obeidat et al., 2020). The study of Yeşiltaş et al. (2022) showed that GC has a mediating effect on green HRM and EEB. As Gurlek and Tuna (2018) explained, culture exerts some pressure on people to behave in certain ways. Thus, when GC is adopted in an organization, members are pressured to demonstrate green behavior.

The implementation of green HRM practices within an organization plays a crucial role in enhancing employees' environmental awareness. This increased awareness can significantly shape the beliefs and values of employees, fostering a GC within the organization (Roscoe et al., 2019). A strong GC has the potential to change employee behaviors and emphasizes the importance of adopting sustainable practices in the workplace (Aggarwal & Agarwala, 2025). Green HRM, which reflects the human resources department's commitment to environmental sustainability, plays a key role in fostering eco-friendly attitudes and behaviors among employees (Dumont et al., 2017; Paille et al., 2014; Yusliza et al., 2017). This, in turn, encourages the integration of a GC throughout the organization. As declared in the AMO theory (Appelbaum et al., 2000), increasing employees' abilities, motivation, and opportunities through green HRM efforts will stimulate positive employee behavior. So, in theory, it is plausible that, through the mediating effect of green HRM, GC would result in green behavior among employees. Thus, the current research seeks to propose that:

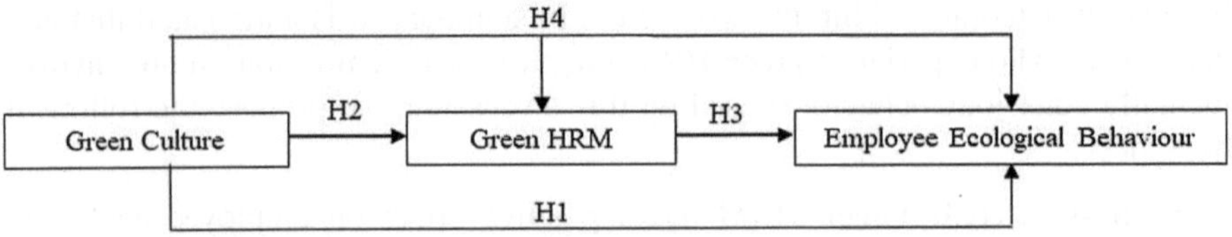

Fig. 1 Theoretical framework. Source: Author's own work

Hypothesis 4 (H4) Green HRM positively mediates the relationship between green culture and employee ecological behavior.

The proposed theoretical framework summarizing these hypothesized relationships is illustrated in Fig. 1.

3 Methodology

3.1 Sample

Before distributing the survey, the researcher obtained official approval from the registrars of each university to facilitate data collection. The registrars were informed of the research objectives and scope, and permission was sought for the data gathering process. To ensure confidentiality, the researcher assured the registrars that individual responses would remain anonymous and offered to share general findings as well as any specific results pertinent to their respective universities. The email questionnaire was sent to all five public research universities, with the sample being sourced from each university's official website. The email contained a comprehensive explanation of the study's purpose and included a hyperlink that directed participants to the Google Forms survey. Participants were made aware that their involvement was voluntary and confidential. To improve the response rate, follow-up emails were sent monthly to remind participants about the survey and encourage them to complete it.

3.2 Ethical Considerations

Universiti Malaysia Terengganu (UMT) arranged approval for the study. Prior to data collection or subject enrollment, informed consent was obtained from all individuals involved. Subject enrollment involves a systematic process to ensure ethical recruitment and inclusion of participants. In this study, academic staff from Malaysian HEIs who meet the inclusion criteria will be identified and invited through appropriate channels, such as emails, to participate. Comprehensive information about the study's purpose, objectives, procedures, potential risks, and benefits will be provided to ensure informed consent. Participation will be voluntary, with confidentiality assured, and participants will retain the right to withdraw at any time without consequences. Signing an informed consent form will confirm their agreement to participate, after which they will receive detailed instructions for completing surveys.

3.3 Participants and Procedure

The survey data were gathered from academic staff at Malaysian Higher Education Institutions (HEIs) between May and July 2021 using a structured questionnaire. Given the challenges of reaching the entire target population, convenience sampling was utilized to select participants. This cross-sectional study employed non-probability convenience sampling, resulting in the distribution of 650 questionnaires through an online survey platform.

The questionnaire assessed GC, green HRM, and EEB. Participants were briefed on the research objectives and assured of the confidentiality of their responses through an accompanying cover letter. The questionnaire was designed to be brief and user-friendly, taking approximately 10–15 min to complete. It consisted of two sections: the first section focused on the research variables, while the second section gathered demographic information.

Out of the 650 questionnaires distributed, a total of 315 responses were received, although seven were excluded from the analysis due to the presence of multiple outliers. Consequently, 308 questionnaires were considered valid for statistical analysis, yielding a response rate of 47.38%. The demographic breakdown of the respondents revealed a predominance of male participants, who constituted 73.1% of the sample, while female participants made up 26.9%. The age group with the highest representation among respondents was 25–30 years, accounting for 44.7%. This was followed by individuals over 40 years old at 28.4%, those aged 36–40 at 21.6%, and a smaller proportion of 5.3% comprising participants aged 31–35. Regarding work experience, the majority (53.8%) had 1–5 years of experience, while 30.8% were more experienced, with 11–15 years in the field. Approximately 14% had 6–10 years of experience, and only 1% reported having over 15 years of work experience.

4 Data Analysis

The model was tested using SmartPLS version 4 (Ringle et al., 2022), which is advantageous as it does not require the assumption of normality where data are usually not normally distributed.

4.1 Measurement Model

We followed the suggestion of Hair et al. (2022) and assessed the measurement model followed by the structural model. As shown in Table 1, all the loadings were >0.5, the composite reliability (CR) was higher than 0.7, and the average variance extracted (AVE) were higher than 0.5, thus the measurement items are valid and reliable (Ramayah et al., 2018).

To test whether the measures were distinct, we used the HTMT ratio (Franke & Sarstedt, 2019). As shown in Table 2, all the HTM ratios were equal to or lower than 0.85, indicating that the discriminant validity of the measures was acceptable.

Table 1 Measurement model

Variable	Item	Loading	CR	AVE
Employee ecological behavior	EEB4	0.760	0.815	0.527
	EEB5	0.664		
	EEB6	0.812		
	EEB7	0.656		
Green commitment	GC1	0.902	0.941	0.799
	GC2	0.938		
	GC3	0.909		
	GC4	0.823		
Green HRM	GHRM1	0.776	0.905	0.613
	GHRM2	0.810		
	GHRM3	0.807		
	GHRM4	0.777		
	GHRM5	0.769		
	GHRM6	0.757		

Note: EEB1, EEB2, and EEB3 were deleted due to low loadings
Source: Author's own work

Table 2 Discriminant validity

Variable	EEB	GC	Green HRM
Employee ecological behavior			
Green commitment	0.211		
Green HRM	0.469	0.557	

Source: Author's own work

4.2 *Structural Model*

Following Hair et al. (2022) and Becker et al. (2023), we reported the path coefficients, standard errors, t-values, and p-values for the structural model using a 10,000-sample resample bootstrapping procedure. Table 3 shows the summary of the criteria we have used to test the hypotheses developed.

The R^2 for green HRM ($R^2 = 0.266$) and EEB ($R^2 = 0.144$) shows that GC can explain 26.6% of the variance in green HRM, while GC and green HRM together can explain 14.4% of the variance in EEB. GC ($\beta = -0.028$, $p > 0.05$) was not significant, while green HRM ($\beta = 0.394$, $p < 0.01$) was positively related to EEB. Thus, H3 was supported, while H1 was not supported. GC ($\beta = 0.516$, $p < 0.01$) was positively related to EEB, which supports H2.

Finally, following Preacher and Hayes (2004, 2008), we bootstrapped the indirect effect to test for mediation. The mediation effect of GC ➔ green HRM ➔ EEB ($\beta = 0.203$, $p < 0.01$) was significant, with no 0 straddling the lower and upper limits confirming a positive mediation, which supports H4 of this study.

We also tested the predictive power of our model using PLS-Predict (Shmueli et al., 2019) with a tenfold procedure, and based on Table 4, all the errors of the PLS model were lower than those of the LM model; thus, we can conclude that our model has strong predictive power.

5 Discussion

This finding is inconsistent with previous research that emphasizes the significance of GC in fostering ecological behavior (H1). One potential explanation for this discrepancy could be the inadequate implementation of green practices within higher education institutions (HEIs). Prior studies have highlighted the importance of GC in shaping employees' perspectives to align with pro-environmental initiatives in organizations (Ashikali & Groeneveld, 2015). A strong GC offers employees clear indicators of expected behaviors that support environmental values, which may be recognized or rewarded by the organization (Norton et al., 2014). Furthermore, employees are more motivated to engage in PEBs when they perceive organizational support and encouragement that aligns with their own environmental values (Ramus & Steger, 2000).

The findings indicate a positive relationship between GC and green HRM (H2). When a university integrates GC into its daily operations, it is more likely to establish a strong green HRM policy. The adoption of green practices and policies raises environmental awareness among employees. Additionally, an organizational culture that emphasizes adaptability enables employees to identify and assess adaptive traits that can enhance their competitive edge (Sharma et al., 2021). This suggests that employees in organizations with a strong GC internalize these values and norms, often reflecting them in their socially responsible behaviors beyond the workplace.

Table 3 Hypothesis testing

Hypothesis	Relationship	Standardized beta (β)	Standard deviation (Std. Dev.)	t-value	p-value	PCI LL	PCI UL	f^2
H1	GC ➔ EEB	−0.028	0.071	0.398	0.345	−0.141	0.094	0.001
H2	GC ➔ Green HRM	0.516	0.042	12.330	$p < 0.001$	0.437	0.578	0.363
H3	Green HRM ➔ EEB	0.394	0.063	6.295	$p < 0.001$	0.276	0.482	0.133
H4	GC ➔ Green HRM ➔ EEB	0.203	0.038	5.383	$p < 0.001$	0.129	0.277	

Source: Author's own work

Table 4 PLS-predict

MV	Q^2predict	PLS-SEM_RMSE	LM_RMSE	PLS-LM_RMSE
EEB4	0.027	1.612	1.624	−0.012
EEB5	0.003	1.893	1.907	−0.014
EEB6	0.006	1.359	1.372	−0.013
EEB7	0.005	1.412	1.429	−0.017

Source: Author's own work

Next, the findings also support H3, showing that green HRM has a positive and significant effect on employees' EEB. These results imply that employees are more likely to be environmentally friendly in the workplace and, more significantly, would willingly participate in the company's green activities if they were aware of the advantages of utilizing green practices and their consequences. This is attributed to the fact that organizations genuinely committed to green HRM tend to set out more true efforts in recruiting, training, and motivating their employees about green initiatives and practices compared to those that do not prioritize these efforts. Similar results have been observed in prior studies, which show that organizations often promote green behavior to contribute to a common goal (Ababneh, 2021; Aboramadan, 2022; Chaudhary, 2020; Farrukh et al., 2022; Norton et al., 2017; Saeed et al., 2019). Thus, the current research result has justified how organizations may influence employees' PEB by providing a green HRM-enriched environment. Green HRM practices would thus have a favorable impact on employees' environmentally friendly behavior and increase the effectiveness of organizations' environmental productivity program (Ansari et al., 2021).

Given the results obtained, it is concluded that GC indirectly impacts EEB through perceived green HRM (H4). Although their investigation was conducted in the context of the hotel industry instead of HEIs, Yeşiltaş et al. (2022) also came to the same agreement that there is a mediating effect of green HRM on GC and EEB. Implementing GC in universities strengthens the practices of green HRM, which stimulates employees to engage in ecological behavior. Since employees who demonstrate pro-environmental behavior are not officially rewarded by the HR department, these favorable behaviors are the results of a green climate rather than being directly impacted by green HRM practices.

5.1 Theoretical Contributions

This study presents a unique theoretical argument exploring the relationship between green HRM and EEB by empirically testing the factor of GC. Additionally, the study employed the AMO theory. This enriches the literature regarding this theory in the context of HEIs and fulfills the research gap. The utilization of this theory proved that when academic staff are equipped with the required ability, motivation, and ample opportunities to engage in green practices, they will have the necessary knowledge and drive to participate in environmental protection. This deduction is consistent with research by Saeed et al. (2019), who stated that the successful implementation of green HRM practices depends on the improvement of the environmental knowledge of employees.

5.2 Practical Implications

This study underscores the critical role of green HRM and GC in promoting ecological behavior within HEIs. To foster sustainable practices, HR departments should integrate green initiatives into core HRM functions, including training, performance appraisal, recruitment, and employee engagement. Green recruitment strategies can attract environmentally conscious candidates, while green training enhances employees' ecological knowledge and skills, empowering them to adopt pro-environmental behaviors in the workplace. Based on the importance of green HRM practices in this study, it is suggested that HR professionals in higher education put green HRM practices as the top agenda in their effort to develop a green campus. To further integrate GC into green HRM practices, the process of green recruitment, training and development, green performance evaluation, and rewarding green policies should be adopted. Consequently, improved green performance can be seen not only in these organizations but also in the community.

Institutions should create a supportive GC by aligning organizational norms and policies with sustainability goals. Top management plays a vital role in this transformation, as their commitment signals the importance of green practices to employees. Offering workshops, seminars, and awareness campaigns can amplify the impact of green HRM and create a cohesive, sustainability-driven workplace. In this context, it is essential for organizations not only to adopt green practices but also to encourage their employees to engage in environmentally friendly behaviors (Abbas, 2020). Achieving sustainability goals is not feasible without fostering a green mindset among the workforce. Therefore, top management should prioritize educating employees on the significance and advantages of GC through workshops and seminars (AlSuwaidi et al., 2021). This approach will help instill socially and environmentally responsible behaviors among employees. These practical steps enable HEIs to build environmentally responsible campuses, improve organizational effectiveness, and contribute to broader societal sustainability goals. Policymakers and university leaders can leverage these findings to design actionable strategies that enhance ecological behavior and align HEIs with global sustainability agendas.

5.3 Limitation and Future Research

This study acknowledges several limitations that suggest avenues for future research. A primary limitation is that the data were collected from Malaysia, a developing country in Asia, which may restrict the generalizability of the findings. To enhance the validity of this concept, further research should be conducted in other emerging or developed regions. The current study utilized a quantitative approach to achieve its objectives; however, future investigations could benefit from employing a mixed-methods design to better understand the factors that promote environmental and social responsibility among employees. Additionally, subsequent studies could explore the influence of leadership and motivational factors on employees' responsible behaviors within society.

This study is constrained by its reliance on cross-sectional data. To better understand the long-term impacts of implementing GC and green HRM on ecological behavior, future research should adopt a longitudinal approach. Such studies would enable researchers to capture the evolving effects of GC over time and explore the causal relationships between GC, green HRM, and ecological behavior.

Another limitation of this study is the use of convenience sampling methods for participant recruitment. As a non-probability sampling technique, it does not guarantee a representative sample of the broader population. While this approach is frequently used to examine theoretical relationships (Norton et al., 2017), it limits the ability to statistically generalize the findings to the entire population. Moreover, considering that this is a "cross-sectional and quantitative" study, future works should conduct "longitudinal and qualitative" studies to provide new insights into the ecological behavior framework.

Acknowledgments This research was funded by the Fundamental Research Grant Scheme (FRGS) provided by the Ministry of Higher Education in Malaysia (FRGS/1/2019/SS03/UMT/02/3).

References

Abadiyah, R., Eliyana, A., & Sridadi, A. R. (2020). Motivation, leadership, supply chain management toward employee green behavior with organizational culture as a mediator variable. *International Journal of Supply Chain Management, 9*(3), 981–989.

Abbas, J. (2020). Impact of total quality management on corporate green performance through the mediating role of corporate social responsibility. *Journal of Cleaner Production, 242*, 118458. https://doi.org/10.1016/j.jclepro.2019.118458

Aboramadan, M. (2022). The effect of green HRM on employee green behaviors in higher education: The mediating mechanism of green work engagement. *International Journal of Organizational Analysis., 30*(1), 7–23. https://doi.org/10.1108/IJOA-05-2020-2190

Afum, E., Agyabeng-Mensah, Y., & Owusu, J. A. (2020). Translating environmental management practices into improved environmental performance via green organizational culture: Insight from Ghanaian manufacturing SMEs. *Journal of Supply Chain Management Systems, 9*(1), 31–49.

Aggarwal, P., & Agarwala, T. (2025). Green organizational culture: An exploration of dimensions. *Global Business Review, 26*(4), 24.

Al-Swidi, A. K., Gelaidan, H. M., & Saleh, R. M. (2021). The joint impact of green human resource management, leadership and organizational culture on employees' green behaviour and organisational environmental performance. *Journal of Cleaner Production, 316*, 128112. https://doi.org/10.1016/j.jclepro.2021.128112

Al-Zawahreh, A., Al-Jaradat, M., & Khasawneh, S. (2019). Green management practices in higher education: The status of sustainable leadership. *Tertiary Education and Management, 25*(1), 53–63. https://doi.org/10.1007/s11233-018-09014-9

Ansari, N. Y., Farrukh, M., & Raza, A. (2021). Green human resource management and employees pro-environmental behaviours: Examining the underlying mechanism. *Corporate Social Responsibility and Environmental Management, 28*(1), 229–238. https://doi.org/10.1002/csr.2044

Anwar, N., Nik Mahmood, N. H., Yusliza, M. Y., Ramayah, T., Noor Faezah, J., & Khalid, W. (2020). Green human resource management for organisational citizenship behaviour towards the environment and environmental performance on a university campus. *Journal of Cleaner Production, 256*, 120401. https://doi.org/10.1016/j.jclepro.2020.120401

Appelbaum, E., Bailey, T., Berg, P., Kalleberg, A. L., & Bailey, T. A. (2000). *Manufacturing advantage: Why high-performance work systems pay off.* Cornell University Press.

Ashikali, T., & Groeneveld, S. (2015). Diversity management in public organizations and its effect on employees' affective commitment: The role of transformational leadership and the inclusiveness of the organizational culture. *Review of Public Personnel Administration, 35*(2), 146–168. https://doi.org/10.1177/0734371X13511088

Azhar, A., & Yang, K. (2022). Examining the influence of transformational leadership and green culture on pro-environmental behaviors: Empirical evidence from Florida City governments. *Review of Public Personnel Administration, 42*(4), 738–759. https://doi.org/10.1177/0734371x211027347

Ababneh, O. M. A. (2021). How do green HRM practices affect employees' green behaviors? The role of employee engagement and personality attributes. *Journal of Environmental Planning and Management, 64*(7), 1204-1226.

AlSuwaidi, M., Eid, R., & Agag, G. (2021). Understanding the link between CSR and employee green behaviour. *Journal of Hospitality and Tourism Management, 46*, 50-61.

Becker, J.-M., Cheah, J.-H., Gholamzade, R., Sarstedt, M., & Ringle, C. M. (2023). PLS-SEM's most wanted guidance. *International Journal of Contemporary Hospitality Management, 35*(1), 321–346. https://doi.org/10.1108/ijchm-04-2022-0474

Beksultanova, A. I., Ibragimova, P. A., & Sarkarova, D. S. (2022). Green, social bonds in the context of the pandemic. In *International scientific and practical conference "Sustainable development of environment after Covid-19" (SDEC 2021)* (pp. 320–324). Atlantis Press.

Bodrud-Doza, M., Bahlman, L., Islam, A. R. M. T., Rahman, M. M., & Shammi, M. (2020). Psychosocial and socio-economic crisis in Bangladesh due to COVID-19 pandemic: A perception-based assessment. *Frontiers in Public Health, 8*. https://doi.org/10.3389/fpubh.2020.00341

Bos-Nehles, A. C., Van Riemsdijk, M. J., & Kees Looise, J. (2013). Employee perceptions of line management performance: Applying the AMO theory to explain the effectiveness of line managers' HRM implementation. *Human Resource Management, 52*(6), 861–877. https://doi.org/10.1002/hrm.21578

Chaudhary, R. (2020). Green human resource management and employee green behavior: An empirical analysis. *Corporate Social Responsibility and Environmental Management, 27*(2), 630–641. https://doi.org/10.1002/csr.1827

Cheema, S., Afsar, B., Al-Ghazali, B. M., & Maqsoom, A. (2020). How employee's perceived corporate social responsibility affects employee's pro-environmental behaviour? The influence of organizational identification, corporate entrepreneurship, and environmental consciousness. *Corporate Social Responsibility and Environmental Management, 27*(2), 616–629. https://doi.org/10.1002/csr.1826

Cheema, S., & Javed, F. (2017). The effects of corporate social responsibility toward green human resource management: The mediating role of sustainable environment. *Cogent Business & Management, 4*(1), 1310012. https://doi.org/10.1080/23311975.2017.1310012

Chen, S., Jiang, W., Li, X., & Gao, H. (2021). Effect of employees' perceived green HRM on their workplace green behaviors in oil and mining industries: Based on cognitive-affective system theory. *International Journal of Environmental Research and Public Health, 18*(8). https://doi.org/10.3390/ijerph18084056

Chuang, C. H., & Liao, H. U. I. (2010). Strategic human resource management in service context: Taking care of business by taking care of employees and customers. *Personnel Psychology, 63*(1), 153–196. https://doi.org/10.1111/j.1744-6570.2009.01165.x

Cincera, J., & Krajhanzl, J. (2013). Eco-schools: What factors influence pupils' action competence for pro-environmental behaviour? *Journal of Cleaner Production, 61*, 117–121. https://doi.org/10.1016/j.jclepro.2013.06.030

Dumitru, A., De Gregorio, E., Bonnes, M., Bonaiuto, M., Carrus, G., Garcia-Mira, R., & Maricchiolo, F. (2016). Low carbon energy behaviors in the workplace: A qualitative study in Italy and Spain. *Energy Research & Social Science, 13*, 49–59. https://doi.org/10.1016/j.erss.2015.12.005

Dumont, J., Deng, X., & Shen, J. (2017). Effects of green HRM practices on employee workplace green behavior: The role of psychological green climate and employee green values. *Human Resource Management, 56*(4), 613–627. https://doi.org/10.1002/hrm.21792

Ercantan, O. (2022). How do green human resource management practices encourage employees to engage in green behavior? Perceptions of university students as prospective employees. *Sustainability, 14*(1718). https://doi.org/10.3390/su14031718

Ercantan, O., & Eyupoglu, S. (2022). How do green human resource management practices encourage employees to engage in green behavior? Perceptions of university students as prospective employees. *Sustainability, 14*(3), 1718.

Farooq, K., Yusliza, M. Y., Wahyuningtyas, R., Haque, A. U., Muhammad, Z., & Saputra, J. (2021). Exploring challenges and solutions in performing employee ecological behaviour for a sustainable workplace. *Sustainability, 13*(17), 9665.

Farrukh, M., Ansari, N., Raza, A., Wu, Y., & Wang, H. (2022). Fostering employee's pro-environmental behavior through green transformational leadership, green human resource management and environmental knowledge. *Technological Forecasting and Social Change, 179*, 121643. https://doi.org/10.1016/j.techfore.2022.121643

Fawehinmi, O., Muhammad, Z., Noor Faezah, J., Mohamad, Z., & Yusliza, M. Y. (2020). Assessing the green behaviour of academics. *International Journal of Manpower, 41*(7), 879–900. https://doi.org/10.1108/ijm-07-2019-0347

Franke, G., & Sarstedt, M. (2019). Heuristics versus statistics in discriminant validity testing: A comparison of four procedures. *Internet Research, 29*(3), 430–447. https://doi.org/10.1108/intr-12-2017-0515

Gong, Y., Xin, K. R., Chang, S., & Law, K. S. (2009). Human resources management and firm performance: The differential role of managerial affective and continuance commitment. *Journal of Applied Psychology, 94*(1), 263–275. https://doi.org/10.1037/a0013116

Gürlek, M., & Çemberci, M. (2020). Understanding the relationships among knowledge-oriented leadership, knowledge management capacity, innovation performance and organizational performance. *Kybernetes, 49*(11), 2819–2846. https://doi.org/10.1108/k-09-2019-0632

Gurlek, M., & Tuna, M. (2018). Reinforcing competitive advantage through green organizational culture and green innovation. *Service Industries Journal, 38*(7–8), 467–491. https://doi.org/10.1080/02642069.2017.1402889

Hair, J. F., Thomas, G., Hult, M., Ringle, C. M., & Sarstedt, M. (2022). *A primer on partial least squares structural equation modeling* (3rd ed.). Sage.

He, J., Zhang, H., & Morrison, A. M. (2021). Being sustainable: The three-way interactive effects of CSR, green human resource management, and responsible leadership on employee green behavior and task performance. *Corporate Social Responsibility and Environmental Management, 28*(3), 1043–1054. https://doi.org/10.1002/csr.2104

Hooi, K. K., Hassan, F., & Mat, M. C. (2012). An exploratory study of readiness and development of green university framework in Malaysia. *Procedia - Social and Behavioral Sciences, 50*, 525–536. https://doi.org/10.1016/j.sbspro.2012.08.056

Hooi, L. W., Liu, M. S., & Lin, J. J. (2021). Green human resource management and green organizational citizenship behavior: do green culture and green values matter?. *International Journal of Manpower, 43*(3), 763-785. https://doi.org/10.1108/IJM-05-2020-0247

Jerónimo, H. M., Henriques, P. L., de Lacerda, T. C., da Silva, F. P., & Vieira, P. R. (2020). Going green and sustainable: The influence of green HR practices on the organizational rationale for sustainability. *Journal of Business Research, 112*, 413–421. https://doi.org/10.1016/j.jbusres.2019.11.036

Jiang, K., Lepak, D. P., Hu, J., & Baer, J. C. (2012). How does human resource management influence organizational outcomes? A meta-analytic investigation of mediating mechanisms. *Academy of Management Journal, 55*(6), 1264–1294. https://doi.org/10.5465/amj.2011.0088

Khan, S. A. R., Sheikh, A. A., & Tahir, M. S. (2024). Corporate social responsibility—An antidote for sustainable business performance: Interconnecting role of digital technologies, employee eco-behavior, and tax avoidance. *Environmental Science and Pollution Research, 31*(3), 4365–4383.

Kehoe, R. R., & Wright, P. M. (2013). The impact of high-performance human resource practices on employees' attitudes and behaviors. *Journal of management, 39*(2), 366-391. https://doi.org/10.1177/0149206310365901

Lasrado, F., & Zakaria, N. (2020). Go green! Exploring the organizational factors that influence self-initiated green behavior in The United Arab Emirates. *Asia Pacific Journal of Management, 37*(3), 823–850. https://doi.org/10.1007/s10490-019-09665-1

Luu, T. T. (2019). Green human resource practices and organizational citizenship behavior for the environment: The roles of collective green crafting and environmentally specific servant leadership. *Journal of Sustainable Tourism, 27*(8), 1167–1196. https://doi.org/10.1080/09669582.2019.1601731

Mirahsani, N., Azizan, O., Shahriari, M., Rexhepi, G., & Najmi, A. (2023). Green culture toward employee green behavior; the mediation roles of perceived support and green identity. *Environment, Development and Sustainability, 26*, 16149–16172. https://doi.org/10.1007/s10668-023-03291-0

Mohamed, N. H., Noor, Z. Z., & Sing, C. L. I. (2020). Environmental sustainability of universities: Critical review of best initiatives and operational practices. In *Green engineering for campus sustainability* (pp. 5–17). Springer. https://doi.org/10.1007/978-981-13-7260-5_2

Mousa, S. K., & Othman, M. (2020). The impact of green human resource management practices on sustainable performance in healthcare organisations: A conceptual framework. *Journal of Cleaner Production, 243*, 118595. https://doi.org/10.1016/j.jclepro.2019.118595

Naderipour, A., Chelliapan, S., Abdul-Malek, Z., Arshad, R. N., Ashokkumar, V., Kamyab, H., & Tavalaei, J. (2021). Assessment of carbon footprint from transportation, electricity, water, and waste generation: Towards utilisation of renewable energy sources. *Clean Technologies and Environmental Policy, 23*(1), 183–201. https://doi.org/10.1007/s10098-020-02017-4

Nisar, Q. A., Haider, S., Ali, F., Jamshed, S., Ryu, K., & Gill, S. S. (2021). Green human resource management practices and environmental performance in Malaysian green hotels: The role of green intellectual capital and pro-environmental behavior. *Journal of Cleaner Production, 311*, 127504. https://doi.org/10.1016/j.jclepro.2021.127504

Noor Faezah, J., Yusliza, M. Y., Chaudhary, R., Ramayah, T., & Fawehinmi, O. (2024). Green human resource management and employee ecological behaviour: The mediating role of green commitment. *Social Responsibility Journal, 20*(6), 1157–1176. https://doi.org/10.1108/SRJ-07-2023-0399

Norton, T. A., Ashkanasy, N. M., Zacher, H., & Parker, S. L. (2017). Bridging the gap between green behavioral intentions and employee green behavior: The role of green psychological climate. *Journal of Organizational Behavior, 38*(7), 996–1015. https://doi.org/10.1002/job.2178

Norton, T. A., Zacher, H., & Ashkanasy, N. M. (2014). Organisational sustainability policies and employee green behaviour: The mediating role of work climate perceptions. *Journal of Environmental Psychology, 38*, 49–54. https://doi.org/10.1016/j.jenvp.2013.12.008

Obeidat, S. M., Al Bakri, A. A., & Elbanna, S. (2020). Leveraging "Green" Human Resource practices to enable environmental and organizational performance: Evidence from the Qatari Oil and Gas Industry. *Journal of Business Ethics, 164*(2), 371–388. https://doi.org/10.1007/s10551-018-4075-z

Paille, P., Boiral, O., & Chen, Y. (2013). Linking environmental management practices and organizational citizenship behaviour for the environment: A social exchange perspective. *International Journal Human Resource Management, 24*, 3552–3575. https://doi.org/10.1080/09585192.2013.777934

Pham, N. T., Nguyen, L. H. L., Vo, N., Phan, Q. P. T., & Tučková, Z. (2018). Enhancing the organizational citizenship behavior for the environment: The roles of green training and organizational culture. *Management & Marketing, 13*(4), 1174–1189. https://doi.org/10.2478/mmcks-2018-0030

Porter, T. H., Gallagher, V. C., & Lawong, D. (2016). The greening of organizational culture: Revisited fifteen years later. *American Journal of Business, 31*(4), 206–226. https://doi.org/10.1108/AJB-04-2016-0011

Preacher, K. J., & Hayes, A. F. (2004). SPSS and SAS procedures for estimating indirect effects in simple mediation models. *Behavior Research Methods, Instruments, & Computers, 36*(4), 717–731. https://doi.org/10.3758/bf03206553

Preacher, K. J., & Hayes, A. F. (2008). Asymptotic and resampling strategies for assessing and comparing indirect effects in multiple mediator models. *Behavior Research Methods, 40*(3), 879–891. https://doi.org/10.3758/brm.40.3.879

Ramayah, T., Cheah, J., Chuah, F., Ting, H., & Memon, M. A. (2018). *Partial least squares structural equation modeling (PLS-SEM) using SmartPLS 3.0: An updated guide and practical guide to statistical analysis* (2nd ed.). Pearson.

Ramus, C. A., & Steger, U. (2000). The roles of supervisory support behaviors and environmental policy in employee "Ecoinitiatives" at leading-edge European companies. *Academy of Management Journal, 43*(4), 605–626. https://doi.org/10.5465/1556357

Rayner, J., & Morgan, D. (2018). An empirical study of 'green' workplace behaviors: Ability, motivation, and opportunity. *Asia Pacific Journal of Human Resource, 56*(1), 56–78. https://doi.org/10.1111/1744-7941.12151

Raza, S. A., & Khan, K. A. (2022). Impact of green human resource practices on hotel environmental performance: The moderating effect of environmental knowledge and individual green values. *International Journal of Contemporary Hospitality Management, 34*(6), 2154–2175. https://doi.org/10.1108/ijchm-05-2021-0553

Renwick, D. W. S., Maguire, S., & Redman, T. (2013). Green human resource management: A review and research agenda. *International Journal of Management Reviews, 15*(1), 1–14. https://doi.org/10.1111/j.1468-2370.2011.00328.x

Ringle, C. M., Wende, S., & Becker, J.-M. (2022). *SmartPLS 4*. SmartPLS. Retrieved from https://www.smartpls.com

Roscoe, S., Chong, T., Jabbour, C. J. C., & Subramanian, N. (2019). Green human resource management and the enablers of green organisational culture: Enhancing a firm's environmental performance for sustainable development. *Business Strategy and the Environment, 28*(5), 737–749. https://doi.org/10.1002/bse.2277

Saeed, B. B., Khan, I., Afsar, B., Hafeez, S., Tahir, M., & Afridi, M. A. (2019). Promoting employee's proenvironmental behavior through green human resource management practices. *Corporate Social Responsibility and Environmental Management, 26*(2), 424–438. https://doi.org/10.1002/csr.1694

Saifulina, N., & Carballo-Penela, A. (2017). Promoting sustainable development at an organizational level: An analysis of the drivers of workplace environmentally friendly behaviour of employees. *Sustainable Development, 25*(4), 299–310. https://doi.org/10.1002/sd.1654

Sancho, M. P. L., Herrera Madueño, J., Larran Jorge, M., & Martínez-Martínez, D. (2018). Understanding the link between socially responsible human resource management and competitive performance in SMEs. *Personnel Review, 47*(6), 1211–1243. https://doi.org/10.1108/pr-05-2017-0165

Schneider, S. C. (1988). National vs. corporate culture: Implications for human resource management. *Human Resource Management, 27*(2), 231–246. https://doi.org/10.1002/hrm.3930270207

Shah, S. H. A., Fahlevi, M., Al-Ghazali, B. M., Bhatti, S., Aljuaid, M., Aman, N., & Hasan, F. (2023). The impact of perceived CSR on employees' pro-environmental behaviors: The mediating effects of environmental consciousness and environmental commitment. *Sustainability, 15*(5), 4350. https://doi.org/10.3390/su15054350

Shmueli, G., Ting, H., Hair, J. F., Sarstedt, M., Ringle, C. M., Vaithilingam, S., & Cheah, J.-H. (2019). Predictive model assessment in PLS-SEM: Guidelines for using PLSpredict. *European Journal of Marketing, 53*(11), 2322–2347. https://doi.org/10.1108/ejm-02-2019-0189

Sibian, A. R., & Ispas, A. (2021). An approach to applying the ability-motivation-opportunity theory to identify the driving factors of green employee behavior in the hotel industry. *Sustainability (Switzerland), 13*(9), 4659. https://doi.org/10.3390/su13094659

Stern, P. C. (2000). New environmental theories: Toward a coherent theory of environmentally significant behavior. *Journal of Social Issues, 56*(3), 407–424.

Sharma, S., Singh, K., & Arya, V. (2021). Innovation and employee turnover in biotechnology companies: rethinking the role of strategic human resource management. *World Review of Entrepreneurship, Management and Sustainable Development, 17*(2-3), 291-305. https://doi.org/10.1504/WREMSD.2021.114435

Tangwanichagapong, S., Mohanty, B., Nitivattananon, V., & Visvanathan, C. (2017). Greening of a campus through waste management initiatives. *International Journal of Sustainability in Higher Education, 18*(2), 203–217. https://doi.org/10.1108/ijshe-10-2015-0175

Yang, M., Rooney, D. W., Chen, L., Wang, J., Yap, P.-S., Fawzy, S., Msigwa, G., & Osman, A. I. (2023). Circular economy strategies for combating climate change and other environmental issues. *Environmental Chemistry Letters, 21*(1), 55–80. https://doi.org/10.1007/s10311-022-01499-6

Yeşiltaş, M., Gürlek, M., & Kenar, G. (2022). Organizational green culture and green employee behavior: Differences between green and non-green hotels. *Journal of Cleaner Production, 343*, 131051. https://doi.org/10.1016/j.jclepro.2022.131051

Yusliza, M.-Y., Jabbour, C. J. C., & Othman, N. Z. (2017). Deciphering the implementation of green human resource management in an emerging economy. *Journal of Management Development, 36*(10), 1230–1246. https://doi.org/10.1108/jmd-01-2017-0027

Yusliza, M. Y., Norazmi, N. A., Jabbour, C. J. C., Fernando, Y., Fawehinmi, O., & Seles, B. M. R. P. (2019). Top management commitment, corporate social responsibility and green human resource management: A Malaysian study. *Benchmarking, 26*(6), 2051–2078. https://doi.org/10.1108/BIJ-09-2018-0283

Yusliza, M. Y., Amirudin, A., Rahadi, R. A., Athirah, N. A. N. S., Ramayah, T., Muhammad, Z., Mas, F. D., Massaro, M., Saputra, J., & Mokhlis, S. (2020). An investigation of pro-environmental behaviour and sustainable development in Malaysia. *Sustainability, 12*(17), 7083. https://doi.org/10.3390/su12177083

Yong, J. Y., Yusliza, M. Y., Ramayah, T., Chiappetta Jabbour, C. J., Sehnem, S., & Mani, V. (2020). Pathways towards sustainability in manufacturing organizations: Empirical evidence on the role of green human resource management [Article]. *Business Strategy and the Environment, 29*(1), 212-228. https://doi.org/10.1002/bse.2359

Exploring Ethics and Employee Green Behavior

Douglas W. S. Renwick and Rohan Crichton

Abstract Building on an earlier work regarding employee green behavior (EGB) (see Renwick et al., The limitations of individual analysis and alternative framing of employee green behavior, In: Green human resource management and sustainability: a view from the global south countries, Springer, 2024)), this chapter seeks to integrate the literature streams of EGB and business ethics. Findings from our exploratory review reveal that ethical issues surrounding EGBs center on issues of deontology and the extent to which such actions comply with basic moral rules and considerations like honesty, fairness, and justice. We advocate that all organizational stakeholders should consider the ethical issues involved in EGB to improve their operational legitimacy.

1 Introduction

Ethical issues within human resource management (HRM) currently arise from some employers not accepting responsibility for staff illnesses and deaths arising from workplace-related accidents involving toxic leaks and chemical spills. Such high-profile incidents include those experienced at Union Carbide (Bhopal, India), Chernobyl (Ukraine), British Petroleum (Texas, USA), and the Fukushima Daiichi (Japan), among others. Here, present orthodoxy assumes that organizations or their senior managers are responsible for such accidents and may be prosecuted under corporate manslaughter charges in some countries, e.g., the United Kingdom (Ministry of Justice, 2007). However, this link may not be logical, as such

D. W. S. Renwick (✉)
Department of HRM, Nottingham Business School, Nottingham Trent University, Nottingham, UK
e-mail: douglas.renwick@ntu.ac.uk

R. Crichton
Reh School of Business, Institute for a Sustainable Environment, Clarkson University, Potsdam, NY, USA
e-mail: rchrichto@clarkson.edu

M. Y. Yusliza et al. (eds.), *Employee Green Behavior*,
https://doi.org/10.1007/978-981-95-1422-9_2

circumstances arguably also appear due to governments and regulators not enacting or enforcing relevant legislation to protect workplace staff, or individual workers unfortunately "lapsing on the job" and contributing to them too (Kagan et al., 2003). Additionally, HR managers may need to take additional ownership of workplace environmental accidents, as HR acting as a "business partner" (Ulrich, 1997; Ulrich & Brockbank, 2005) arguably asks HR staff to put the profit-based desires of "the business" before more externally green-focused ecological goals. Indeed, an ethical approach to HRM seems needed as current "strategic" or "best practice" HRM models appear to place less emphasis on ethical dimensions and the safeguarding of employee welfare (Winstanley & Woodall, 2006, p. 6, in Wilton, 2013, p. 69), meaning that green safety and health interventions may change due to market conditions and their perceived impact on organizational profits.

Ergo, a key ethical issue emerging is the extent to which organizational staff are put under pressure to deliver organizational goals which compromise employee safety and health at work, and the unintended consequences for other staff arising from such actions, i.e., physical harm to such staff. As such, HR managers and HRM itself (e.g., HR strategies, policies, procedures, processes, and practices) could be seen as a barrier to achieving wider ethical and societal green goals, which focus on reducing ecological degradation, rather than facilitating their achievement. While the relevant literature focuses on going green via ethical HRM (Manjula & Parvathy, 2019), green HRM, ethical leadership, and ethical behaviors (Islam et al., 2021), ethical leadership as a predictor of green HRM (Hossain et al., 2024), and links between HRM and eco-friendly culture (Tirno et al., 2023), plus effects of green HRM practices on employee in-role behavior (Li et al., 2023), few studies critically and explicitly detail individual green HRM workplace practices ethically—which is a research gap that we aim to fill herein. Our key theoretical contribution lies in using deontology to assess whether such workplace green HRM practices satisfy moral rules and considerations such as honesty, fairness, and justice, while our practical contributions include interrogating how HR practitioners can make "moral sense" of the ethical issues arising in green HRM, along the key criteria to do so of virtue ethics, consequentialism, and deontology, and how HR can increase its professionalism in this area by considering the use of an ethical code in green HRM.

2 Background

A growing body of research has focused on international environmental issues and human resource management (HRM) (e.g., Jackson et al., 2011: Wehrmeyer, 1996), which some term "green HRM" (e.g., Renwick et al., 2008, 2013; Renwick et al., 2016). This green HRM literature also includes advocates of voluntary workplace green behaviors (VWGBs) (Kim et al., 2017; Jiang et al., 2022) and a most recent critique of employee green behaviors (EGBs) too (e.g., Renwick et al., 2024). Drawing on both the ethical management of people and human resources works

(Greenwood, 2002, 2012; Crichton & Shrivastava, 2017), and an "entry to exit" conceptualization of all green workplace-based HR processes and practices (Renwick et al., 2013, 2016), we explore ethical questions arising in the field of EGB.

While some researchers have examined environmental sustainability (ES) and morality (e.g., Kim et al., 2017; Crichton et al., 2018; Jackson et al., 2019), their focus is specifically on "eco-advocates," i.e., pro-green employees. As such, progress on investigating ethics and EGB—which apply to nongreen, neutral, and anti-green staff—is still very limited. Thus, although some works link EGB (broadly defined) to social issues such as diversity and inclusion (Renwick, 2018, pp. 182–184; Renwick et al., 2024), we respond to literature calls for more work linking green HRM and social concerns (Aust et al., 2020). To our knowledge, such social concerns include limited works examining ethics and EGB—a literature gap which we seek to reduce. Our aim is therefore to critique the people-related aspects of EGB research from an ethical standpoint. In doing so, we look to surface and problematize such issues to comprehend their inherent tensions.

3 Key Concepts and Issues

3.1 An Integrated Critique of Green HRM, EGB, and Ethics

3.1.1 Recruitment and Selection

While Ehnert (2009) and Albinger and Freeman (2000) detail the role of environmental management (EM) in attracting "green aware" job candidates to EM-focused organizations via recruitment and selection practices (in Renwick et al., 2013, 2016), such activities could appear ethically dubious, as they might only appeal to small numbers of pro-green employers and well-educated graduates who wish to work for such ecologically oriented organizations. These pro-environmental practices may therefore appear as a "minority interest" to other nongreen employers and staff and seem unethical to them, as they could prioritize the needs and wishes of pro-green stakeholders above those of other neutral or anti-green employees. Here, green recruitment and selection (R&S) has the potential to ignore or side-line job applicants who may not have a pro-green orientation or outlook. In essence, a danger exists that a select group of pro-environmental employers, pursuing a "minority green interest" perspective, could undertake ecological R&S activities to attract and select an elite group of graduate staff to work for them. In doing so, such organizations may (in)directly discriminate against other rejected or ignored job candidates who hold different, neutral, or nongreen views, on the basis of their alternative beliefs.

In such circumstances, green selection could therefore be seen as unfair if the green or sustainability "credentials" of staff are being used for hiring purposes (e.g., qualifications, experience, and organizational fit). Here, following Baker and Cooper's (2000, p. 61) concern about psychometric testing in Sweden invading staff privacy, the use of green attitudes and beliefs when selecting staff could become problematic, as doing so involves justifying why some staff are selected and not others, what feedback candidates will gain, and if organizations mention such criteria in any conversations. Additionally, as organizational values can be communicated through the selection process, staff whose ethical values diverge from pro-green organizational values may be encouraged to withdraw from such selection processes. Doing so could mean non, neutral, and anti-green staff being subject to (in)direct social exclusion by some employers in the selection process (cf., Buckley et al., 2001, p. 18; Miceli & Near, 2001, p. 6). In essence, are pro-green job criteria a new barrier to job candidates gaining employment today? If so, are green recruitment and selection practices thus becoming biased and discriminatory ones?

3.1.2 Green Jobs

Studies by Aiman-Smith et al. (2001), Behrend et al. (2009), Bauer and Aiman-Smith (1996), Backhaus et al. (2002), and Dolan (1997) (Renwick et al., 2013, 2016) appear to build a solid picture of contemporary pro-green employment—of "green jobs"—existing among current and future graduates (in Renwick et al., 2013, 2016). However, the notion of green work and employment may be a fad or fashion among such students, as it could open up a gap between the espousal of green employment emerging in future years and the relative, real-life dearth of green jobs existing today. For example, in the United Kingdom, a disconnect exists between Van Riel's (2019) estimates of global green job growth and more modest UK government figures on actual green jobs created, i.e., no net gains in UK green jobs (see Renwick, 2024). Ergo, placing the hopes and dreams of gaining a green job into the minds of somewhat impressionable young graduates could appear unethical, as doing so may arguably inaccurately raise their expectations to an unrealistic level regarding the ability of employers to provide such jobs. Green organizational goals and targets may thus not always be realizable for employers to deliver in practice here due to the regulatory, cost, skills, passporting, and training issues involved for employers. Ergo, are key stakeholders being somewhat dishonest regarding future green job opportunities?

A growing number of self-help guides exist on how to find a green job (e.g., Parks & Helmer, 2009; Cassio & Rush, 2009; Llewellyn & Golden, 2008), and some personnel specifications detail green job aspects (including interview protocols to probe on the environmental knowledge, values, and beliefs of potential applicants) (in Renwick et al., 2013, 2016). Here, following Provis (2009), ecologically focused employers could argue on ethical grounds that staff with higher green-related qualifications deserve to be hired more than staff without such attributes, particularly in the "non-polluter, green industries," where such qualifications seem relevant to the tasks being undertaken. However, developing these pro-environmental policies may not seem ethical, as indirect discrimination might arise against applicants whose minds are "unsure" about green issues (e.g., nongreens/neutrals), or those who display mutual and/or multiple commitments to business advancement and environmental concerns (i.e., pro-greens). As such, following Alder and Gilbert (2006), concerns may arise on the fairness and justice organizational staff receive surrounding the formulation and enactment of green job hiring practices. Furthermore, while comparative evidence from the United Kingdom and Japan reveals it is relatively easier for organizations to hire high-quality employees when firms have "good" pro-green reputations (Bansal & Roth, 2000), doing so could arguably be unethical as it implies that organizations may need to hire staff "keen" to practice environmentalism. This situation potentially pressurizes employers to act in odd ways, as it is unclear what happens to employer reputations if they are not inclined to be, or become, pro-green, and what their attractiveness to future job candidates is if such employers choose a "neutral" reputation regarding environmental initiatives too.

3.1.3 Training and Development

The green employee training literature suggests that organizational staff gain "core skills" on how to usefully collect waste data (May & Flannery, 1995) and that advanced sustainability approaches may depend on skill development from employee green training (Brio et al., 2007) (in Renwick et al., 2013, 2016). Yet, such findings seem ethically problematic, as they appear to pressurize employees—and not managers or business owners—to gather and interpret such material(s), meaning that training only seems to be required from organizational staff, meaning is ES training an unjust or unfair burden that only employees carry? Other relevant issues arising regarding green training and development (T&D) (Milliman & Clair, 1996) include the age-old one of ensuring managers release their staff for training days (Wehrmeyer & Vickerstaff, 1996) (in Renwick et al., 2013, 2016). Such events may counter staff cynicism about the importance and relevance of ecological issues, and sustainability training and development programs sometimes being delivered in an authoritarian or "politically correct" way to emphasize green program enforcement (cf., Rees, 1996). Ethical concerns emerging include rationalizing why managers should release staff for green training when they may need workers to do other tasks, and why employees are not allowed to be cynical about ES, to have rights to think critically about green issues and to act independently regarding them. Indeed, Torres (2008) raises the question of what happens regarding the retraining of employees who either have lost, or may lose their jobs, in the so-called "polluter" industries. Here, in consequentialist ethics terms, such staff may benefit from working in greener and safer environments, and on virtue ethics ones, employees may feel that they have rights about, or deserve to work in, "cleaner" organizational environments (cf., Provis, 2009). As such, for green T&D to become more ethical, we may need to understand what specific green training initiatives workers from "polluter" workplaces need, and what other jobs they could usefully occupy in the future.

Indeed, as human resource development (HRD) is increasingly seen to have a role in helping organizations achieve environmental goals, firms may need to decide which staff gain access to green development opportunities (Garavan & McGuire, 2010, p. 488), upon what criteria, why, and who decides this? (cf., Weaver & Trevino, 2001, p. 123). Issues emerging include whether green training opportunities apply to only a select few organizational participants, and if leaders convey ethical messages when delivering such training, daily role modeling behaviors, and an open environment for discussion (cf., Wooten, 2001, pp. 170–172). Here, Ruiz-Palomino et al. (2013, pp. 2189–2190) argue that "pushing" employees to reason cognitively and autonomously about moral virtues is needed in training agendas. As such, pushing staff to be more sustainable may not be acceptable in green training, as it could seem akin to harassment or bullying, an initiative which neutral or anti-green staff may dislike, and a form of coercion forcing organizational participants to engage in green training without their consent (cf., Wooten, 2001, pp. 170–172).

3.1.4 Leadership, Knowledge, and Development

While some business and management (B&M) schools have the potential to deliver enhanced organizational sustainability and increase knowledge to develop future green business leaders (Starkey & Crane, 2003), the pro-environmental motivations of such B&M institutions may be questionable. This may be due to many B&M schools' desire to embrace a corporate social responsibility (CSR) agenda, and arguably appear more acceptable to wider society at large following the 2007/2008 financial crisis; such B&M schools were also arguably "complicit" in encouraging a profit-seeking business agenda before that financial crash occurred (Currie et al., 2010). The green "credentials" of those B&M institutions might therefore seem questionable, and even somewhat "emergent," meaning any B&M school green (re) branding could appear unethical. Additionally, although sustainability is included in some Chinese MBA programs (Fryxell & Lo, 2003), and many "eco-MBAs" also exist globally too (Beyond Grey Pinstripes, 2010) (in Renwick et al., 2013, 2016), it remains unclear how ethically sound enrolling students onto such programs actually is before a significant number of "green jobs" are created worldwide. In sum, is it morally right to encourage students to study ES, eco, and green courses in B&M schools when the prospects of them getting a "green job" currently appear relatively low?

Further, research on green leadership in some United Kingdom and Japanese companies finds single individual staff leaders seeming to champion EM responses as their values appear to drive such decisions forward (Bansal & Roth, 2000) and some Chinese executives embracing green concepts to pursue their own personal values and principles too (Branzei et al., 2004) (in Renwick et al., 2013, 2016). Yet ethical questions arise on the extent to which more junior organizational staff are being led into ES decisions by a small group of pro-green senior managers. Here, it could perhaps be that after Thorpe et al. (2011), a "distributed leadership" strategy might be more ethical to provide evidence of, and build more solid support for, such "green leadership" strategies. Moreover, while self-reports find managers in some emerging markets having a "strong disposition" to take environmental action

(Fryxell & Lo, 2003), it remains unclear if controlling environmental impact is seen as a responsibility for all employees (Boiral, 2002). Ergo, it seems arguably unethical to require the historic "grand challenge" of tackling climate change to be a concern for all workers in all organizations to enact and deliver, as this may be too heavy a burden for such employees to carry (cf., George et al., 2016).

3.1.5 Performance Management and Appraisal

Although some organizations have developed performance indicators for each green risk area (TUSDAC, 2014), difficulties emerge in measuring green performance standards across organizational departments or units and gaining useful environmental data from staff regarding such standards (in Renwick et al., 2013, 2016). Here, questions arise on how ethical it is to measure managers and employees on such green criteria when they are unclear and on the appropriateness of "micromanaging" staff regarding them. After Budd (2004), ethical concerns emerge over the extent to which performance management (PM) schemes using green criteria enable managers to exert "behavioral control" over employees, which may reduce staff autonomy if employee voice processes become restricted. Indeed, the incorporation of green criteria and measurements into performance appraisal (PA) schemes (Wehrmeyer, 1996)—such as assessing staff on the numbers of environmental incidents, amount of recycling, levels of waste management and energy use focus occurring at their desks and workspaces—might place undue "work pressure" on employees to only achieve green goals, or instead see staff engaging in target-driven behaviors to meet them. Moreover, as green elements in PA systems seem limited to plant or divisional managers and executives (Milliman & Clair, 1996), holding senior staff primarily accountable for sustainability performance may appear unethical, as organizations, business owners, pension funds, and unions all seem responsible for it too. For instance, it arguably appeared odd that only a few executives at British Petroleum (BP) were left to publicly "carry the can" for the ecological catastrophe BP produced in the Gulf of Mexico, USA—were no staff, contractors, suppliers, or regulators at fault too?

Further, peer pressure to hit sustainable performance targets, coupled with a "management by blame" culture, might not encourage the mutually supportive and cooperative relationships needed among staff members to deliver green goals (cf., Delbridge & Turnbull, 1992). Here, although negative reinforcements including suspensions, criticisms, and warnings may appear in PM systems to direct staff to make environmental improvements (Chan & Hawkins, 2010), it remains unknown what workers learn from such systems about sustainability in general, and if PM schemes increase employee pro-green behavior. Thus, employees might potentially perceive such managerial actions as a new form of "uncivil" workplace behavior, i.e., a new type of green-related "bullying or harassment" at work. As performance management and appraisal (PMA) appears to heighten managerial control of staff behavior, it raises issues of subjectivity and bias occurring when managers assess staff performance in waste management, energy use, and recycling (in Renwick et al., 2013, 2016). This is because organizations need to ensure procedural fairness for employees in green PMA, scrutinize managerial decisions, and enable staff appeals against them regarding unfair assessments of individual staff green contributions (cf., Wilton, 2013). Thus, following Buckley et al., (2001, p. 18), appraisals may be a process where staff could exaggerate their "green-ness" to impress colleagues and gain a good appraisal score (Renwick et al., 2008). Ethical problems emerging thus concern dealing with multiple peer group, inappropriately aligned and/or culturally unsupportive expectations.

Here, possible unethical behaviors include managers documenting staff green performance ratings which do not represent actual staff performance and managers deliberately keeping green performance factors and expectations ambiguous (cf., Wooten, 2001, pp. 170–172). Further, because of bureaucratic instruments implemented to control and impose behavior, inserting green criteria into appraisal systems may not induce a true internalization of ethical green values in the practice of (moral) virtues, meaning such control and coercive green orientations may be of questionable value too (cf., Ruiz-Palomino et al., 2013, pp. 2188–2190). Indeed, organizational systems could reinforce communicated expectations of ethical green standards and behaviors, e.g., recognition of exemplary, ethical green behavior from the organization's performance appraisal system, but also using performance feedback to apply sanctions and rewards related to ethical green behaviors on staff too (cf., Wells & Schminke, 2001, p. 146).

3.1.6 Pay and Reward Systems

Some organizations develop reward systems to incentivize senior managers in sustainability work (Snape et al., 1994), and research in the United States reveals a positive relationship between CEO total compensation salary and organizational environmental reputation or performance (Berrone & Gomez-Mejia, 2009; Cordeiro & Sarkis, 2008) (in Renwick et al., 2013, 2016; Crichton et al., 2023). Nonetheless, an ethical issue arising concerns the dilemma that sustainability could become a financial or money-based "gravy train" for senior managers to ride, as other findings report managerial pay resulting from, rather than causing, environmental performance (Russo & Harrison, 2005). Here, managers may not have to actively contribute to organizational greening to benefit financially from such an organizational initiative. Fairness issues may therefore arise for organizational members on whether it is ethical for top executives to gain increased pay and bonuses if links between their managerial inputs and green outputs are not clear.

For rewards to employees below senior management levels, a tendency exists to focus on giving workers nonmonetary rewards and recognition for their green suggestions (Govindarajulu & Daily, 2004) (in Renwick et al., 2013, 2016). However, it seems ethically questionable that staff pay and rewards could differ relative to their green inputs—as it is unclear how much control employees have over such inputs. Indeed, this circumstance raises the issue of whether staff without much evidence of producing pro-green initiatives could, or should, receive any increased pay (for managers) and/or nonmonetary rewards (for employees). Moreover, fairness, justice, and equity concerns might emerge across these different staff groups too, because such green pay and reward practices might appear variable, inconsistent, and thus unethical to them.

As organizations face new (un)ethical sustainability situations, shared perceptions of expected behaviors may be driven by ethical practices stimulated and compensated by organizational leaders, which may produce new norms of ethical green behavior (cf., O'Higgins & Kelleher, 2005, pp. 284–286). As green pay and reward is primarily allocated to managers as bonuses (in Renwick et al., 2013, 2016), an emerging issue is the extent to which nonmanagerial staff feel fairly rewarded for green initiatives—an equity concern—as fair reward is a known determinant of staff motivation.

Finally, on the executive compensation front, four issues arise of concern: (1) if promotions and symbolic rewards have their desired impact of changing staff green values and behaviors, (2) whether disciplining staff via organizational reward systems for unethical green behavior is practical and fair, (3) if organizational reward systems encourage unethical acts like advancing business profits ahead of environmental and ecological protection, and (4) whether an absence of accountable reward systems regarding external organizational pollution leads to violations of internal organizational ethical norms (cf., Buckley et al., 2001, p. 20). Thus, while formal and informal rewards could provide powerful reinforcements for green behaviors, they may also demotivate employees through green target-setting and see staff sacrifice more holistic organizational outcomes for performing to clearly stated and highly rewarded, organizational green behaviors instead (cf., Grojean et al., 2004, pp. 229–232).

3.1.7 Employment Relations (ER)

As employee voice initiatives in sustainability controlled by management teams could potentially serve as tools for employee manipulation (cf., Wilton, 2013), assessing whether sustainability initiatives marginalize trade unions, or if unions can use green initiatives to achieve wider goals such as increasing union membership, density, activism and resistance, seems needed (cf., Farnhill, 2018; Hampton, 2015). Here, a business case can be made for the ethical treatment of workers in sustainability by including environmental safety and health issues indirectly impacting on organizational performance through increasing staff engagement and commitment to deliver such organizational goals (cf., Wilton, 2013, p. 11). As such, ethical issues for employers include if ethical decision-making areas pertain to employee rights to have environmentally "good" and safe working conditions, and for employees concern taking responsibility for the green aspects of health and safety for all coworkers too (cf., Wilton, 2013, p. 34).

3.1.8 Employee Involvement

Although some company-wide employee involvement (EI) is seen as critical to achieving successful organizational green outcomes (Bunge et al., 1996; Hanna et al., 2000; Remmen & Lorentzen, 2000) (in Renwick et al., 2013, 2016), it is unclear if it is fair to employers and employees who may lack either the time or resources to deliver such outcomes. Specifically, while findings suggest that EI is historically seen to improve key green performance outcomes such as efficient resource usage (Florida & Davison, 2001), waste management (May & Flannery, 1995), pollution reductions (Denton, 1999; Kitazawa & Sarkis, 2000), and green outcome measures (Brio et al., 2007) (in Renwick et al., 2013, 2016), it is unclear how small firms can deliver such outcomes and may mean that larger firms potentially gain an ethically "unfair" advantage if they can do so. Additionally, EI appears to have achieved impact through tapping into employee tacit green knowledge (Boiral, 2002) by engaging and empowering staff to make suitable green suggestions (Govindarajulu & Daily, 2004) (in Renwick et al., 2013, 2016). Yet after Provis (2009), as contractual employment agreements tend to include "obligations of obedience" for employees, it is unclear if such contracts take precedence over informal ethical obligations from employers to give employees "voice" and "due process" in the formation and operation of green EI schemes. In short, does using EI in organizational green schemes enhance or restrict the use of "employee voice" in companies more widely?

3.1.9 Tacit Knowledge, Employee Empowerment, and Engagement

Although some North American research suggests that tacit green knowledge among production workers appears associated with positive green outcomes (Boiral, 2002; Rothenberg, 2003) (in Renwick et al., 2013, 2016), it is not clear if organizations using such tacit staff knowledge are ethical, or instead merely exploiting employees. Here, empirical data from UK and US companies confirms that increasing employee psychological empowerment seems to generate more staff suggestions for green improvements (Kitazawa & Sarkis, 2000) and that positive links exist between employee empowerment levels in sustainability and environmental performance (Simpson & Samson, 2008) (Renwick et al., 2013, 2016). Despite this circumstance, green employee empowerment initiatives could be considered unethical, as they may be seen as "brainwashing" employees to go green, and in channeling or forcing staff to undertake pro-green behaviors that staff may not have requested, or indeed welcome, especially if such employees have neutral or anti-green beliefs. Here, are environmental empowerment and engagement schemes a liberating force indirectly helping tackle climate breakdown or a new form of "coercive environmentalism" controlling employees in contemporary workplaces?

3.1.10 Supportive Culture and Climate

Changes to corporate culture in organizations toward embedding green values may enhance long-term organizational sustainability, providing employees are well informed about such issues affecting their workplaces (cf., Madsen & Ulhoi, 2001; Stone, 2000; Kitazawa & Sarkis, 2000) (in Renwick et al., 2013, 2016). Indeed, antecedents of supportive ecological cultures also include managerial commitment to, and employee involvement in, such green activities too (Fernandez et al., 2003) (Renwick et al., 2013, 2016). However, it is not known whether it is ethical for organizations to promote such pro-green values if their staff are not informed about, committed to, or involved in organizational ecological schemes, and whether such staff do not have pro-environmental beliefs. Ergo, what happens if organizational staff do not want to embrace organizational green schemes—what career consequences arise for them?

3.1.11 Union Roles

While the British TUSDAC (2014) recommends that UK trade unions negotiate sustainable workplace framework agreements with employers to enhance employee engagement in sustainability, some management teams may see such union actions as being unethical, because (cf., Oates, 1996) their managers may perceive green concerns as "theirs" to pursue, free from any union-led "interference" (Renwick et al., 2013, 2016). As such, questions arise if it is ethical for unions to "push" their role in sustainability, and whether unions are doing this in order to acquire more members—a so-called green union renewal strategy (Hampton, 2015). Here, UK developments in union greening include education programs for "rank and file" union members, joint management-union training programs, and the development of workplace environmental representatives—as "union green representatives" (TUC, 2009). Yet it remains to be seen if lay union members view ecological initiatives positively or instead as a new form of management control to support wider management goals.

Indeed, it may be debatable if it is ethical for UK organizations to allow or deny legal rights for union green representatives to gain time off work to promote sustainability. Here, although the role of unions in green training initiatives seems well developed within the European Union (EU) (Madsen & Ulhoi, 2001; Renwick et al., 2013, 2016), the environmental role of such unions could also appear ethically problematic, as green issues may distract unions and workers from focusing on other employment issues such as job security and pay. In doing so, a sustainability focus potentially asks managers to concentrate less on "bread and butter" issues like pay and terms and conditions in employment relations and more on arguably fashionable ones like green initiatives instead. In sum, is it right for unions to engage with, and promote, pro-green organizational initiatives at the expense of, or risk to, diluting their focus on mainstream union goals? Should the funds of union members be spent in this way?

4 Discussion of Implications

4.1 Toward an Even More Critical View

4.1.1 Irresponsibility

Contemporary management techniques may have a dehumanizing impact on green HRM, such as categorizing employees as pro-, neutral, or anti-green, and represent people as objects in technical and ethically neutral terms. Staff may be exploited as vehicles through which organizations achieve green objectives, and certain staff groups may be excluded from organizational considerations if they do not "fit" pro-green profiles—a new type of social exclusion in organizations (O'Higgins & Kelleher, 2005, p. 284). Thus, it may be important to get green HRM workplace interventions right, as prior research reveals justice perceptions predicting wide-ranging work-related outcomes, including emotions, attitudes, and behaviors, as the consequences of felt injustice occurring in green HRM may seem significant, including "showing counterproductive work behaviors (CWBs) such as organizational deviance, sabotage, or aggression" (Crawshaw et al., 2013, p. 892).

Moreover, some authors argue that "most managers will not be managing ethical questions associated with the organization's response to global warming" and associated attitudes such as "I am not responsible for the environment" (Mellahi et al., 2010, pp. 32–34, 104). As such, we could use postmodern thinking to impose a personal responsibility on us when it comes to making ethical choices. This approach could be used to assess the impact of job losses in the oil industry, and if alternative "green jobs" replace them, they could potentially become more secure and sustainable employment (Van Riel, 2019), which could be seen as an ethical act itself (Mellahi et al., 2010, pp. 32–34, 104).

4.1.2 Preconditions for Ethical Green HRM

Unhindered individualism in organizations by managers and staff pursuing non-green initiatives without notions of a collective good arising from a "social contract" may not produce the "good life" but a nasty one. Here, rather than imposing a green HRM agenda top-down, what seems needed is to involve employees to encourage employee buy-in and commitment, meaning that while the HRM function has been identified as the natural home of organizational consciences," responsibility for business ethics has been handed to other departments (Fisher & Lovell, 2009, p. 413). This is because tensions arising for individual staff regarding ethics in green HRM include those surrounding deontic motives, as individuals face the countervailing motivations of reactance (free behaviors) and deonance (non-free behaviors). In resolving such tensions, some staff may act in ways they perceive as ethical in green HRM but others do not, while withdrawal behaviors and constructive resistance could supply employees with deontic responses to deemed unethical actions by supervisors, without harming others and violating their own ethical and moral green HRM values (Crawshaw et al., 2013, pp. 1895–1897).

4.1.3 Potential Unethical Behaviors

Indeed, an ethical issue arises regarding collusion from internal organizational actors; examples here include: favored groups receiving differential green pay and benefits; an inability to confront green health hazards due to political considerations; not reporting the potentially hazardous anti-green factory conditions; instructing employees that their workplace is safe regarding green safety and health, when it is not (e.g., asbestos-ridden sites); and compromising employee health from increasing factory emissions to enhance organizational productivity (Wooten, 2001, pp. 170–172). We now detail some practitioner implications, limitations, and future research ideas and conclusions arising from our review and critique above.

4.2 Practitioner Implications

A key question for HR professionals and practitioners from our review, critique, and discussion is how they might make "moral sense" of the ethical issues arising in green HRM. Here, some options include practitioners assessing green HRM processes and practices along the criteria of: virtue ethics (moral virtue), consequentialism (consequences of practices), and deontology (observing rules of honesty, fairness, and justice). However, organizations should note that such ethical theories may not always produce simple answers to the "hard" questions involved. For example, an apparent initial simplicity, and then unexpected, eventual complexity involved is seen in the use of an ethical code in green HRM.

The green HRM field may need to focus and increase its professionalization through establishing its core values and constructing a viable ethical green HRM code that applies to the many different areas that comprise it (Wooten, 2001, p. 173). Here, green HRM control systems, formal procedures, and bureaucratic rules may resolve ethical disputes that result from opportunistic green behaviors (Grossman & Schoenfeldt, 2001, p. 61). Yet, as no accepted guidelines exist for resolving green HRM value dilemmas, how managers resolve them defines organizational values and performance and societal values (Acevedo, 2012, p. 201). As such, HRM education and professional associations may need to establish "new" ethical green HRM guidelines and shift HR's focus to establish company green HR firm objectives and capabilities to privilege being for "the other," i.e., fellow humans, in tackling climate change (De Gama et al., 2012, p. 106). However, HR professionals may face pressure to demonstrate their financial bottom-line impact, meaning ethical green HRM actions may involve negative personal and professional consequences for HR practitioners (O'Higgins & Kelleher, 2005, pp. 276–280).

As one senior practitioner we liaised with specifically commented: "I think green HRM could be ethical if organizations introduced an ethical code on it, and asked all staff to sign up to it." While such a code might be feasibly compiled by drawing on best practices from other codes globally, they could become complex and tricky to implement. This is because the key stakeholders involved may wish to input into the terms of reference such codes use, ask who determines such terms, how dividing lines between ethical and unethical green HRM practices are determined, who will introduce a code in action (line managers, HR, or others?), and how far such codes are "pushed" by organizational cultures on workplace staff to conform to them. In essence, the opportunities from using, and problems arising in enacting, ethical codes in green HRM may seem similar to those seen to using ethical codes elsewhere (e.g., as the AoM and CIPD). As such, we might usefully ask, who—which exact, particular stakeholders?—will write an ethics code in green HRM? To date, we have found few answers to this question.

5 Future Directions

5.1 Research Limitations

Our position herein does not claim to exhaust the field of possible ethical issues and concerns arising in EGB, and we readily accept that questions arise on the "truth" of such competing claims which may seem relative to those holding them. We also recognize that many employment contracts are ones employees may have to accept due to their relatively low bargaining power, which could mean that any sustainability-related agreements that staff engage in are not always freely consented to—which is a long-recognized prerequisite for ethical status (cf., Provis, 2009).

5.2 Future Research Ideas

On training design, the malleability of ethical decision-making models on ethical sustainability behaviors such as moral sensitivity, motivation, and character could be empirically investigated. Here, researchers could assess if the context factor of ethical climate reduces employee silence and results in perceived organizational support (POS) by creating a trustful organizational work environment in sustainability (cf., Wang & Hsieh, 2013, pp. 784–787). Further, researchers employing a paired-sample supervisor-subordinate approach might usefully compare responses regarding organizational factors and individual behavior, reliable data on green and sustainability mediators, and if situational and individual variables interact to explain employee ethical green behaviors too (cf., Ruiz-Palomino et al., 2013, pp. 2190–2191).

On gender, age, and religious differences, scholars could investigate whether women receive fewer hours of green ethics training than men (cf., Valentine & Hollingworth, 2014, p. 703), if men hold ethical green values as deep as women appear to (cf., World Bank, 2009; Zelezny et al., 2000), whether younger males have higher ethical environmental concern than younger females seem to possess, and if senior male managers espouse deep(er) green ethical values less than males employed in either staff or line groups appear to state (cf., Wehrmeyer & McNeil, 2000, pp. 211, 214, 220–221). Future research could also investigate whether individual and contextual variables such as age and religion are positively correlated to green ethics too (cf., Segal & Lehrer, 2013, pp. 524–526).

Individual variables could be investigated, such as job type, institution, discipline or country of institutional location, as influences on ethical green expectations, while department chairperson and other administrator perceptions on applying AMCEC standards to professional green ethics education may be a developmental benchmark to gauge ethical green behavior change (cf., Petrick & Scherer, 2005, p. 311). Further, scholars could examine the influence of individuals' cognitive moral development, idealism, and personality (including Machiavellianism and gender) on how they respond to ethical green dilemmas, plus interrogating antecedent conditions, such as cultural environment and personal experiences, as influences on individual perceptions of ethical green problems (cf., O'Higgins & Kelleher, 2005, pp. 282–283). Indeed, research could investigate contextualized ethics and examine the stated HRM discourse and practice through the lived experiences of staff undertaking organizational green HR duties too (cf., De Gama et al., 2012, p. 100).

6 Conclusions

In sum, we advocate a call to action—that HRM stakeholders engage with, assess, and answer the ethical questions and concerns we identify in this exploratory chapter. Perhaps in doing so, such issues may fare a better chance of sound ethical navigation over time.

Academic Contribution This chapter contributes to the literature on ethics and green HRM by addressing the call for more research on it from Renwick (2018, 2020), and extends it by branching out to explore ethics and EGB, which is rarely seen in the literature.

References

Acevedo, A. (2012). Personalist business ethics and humanistic management: Insights from Jacques Maritain. *Journal of Business Ethics, 105*, 197–219. https://doi.org/10.1007/s10551-011-0959-x

Aiman-Smith, L., Bauer, T., & Cable, D. (2001). Are you attracted? Do you intend to pursue. A recruiting policy capturing study. *Journal of Business and Psychology, 16*, 219–237.

Albinger, H. S., & Freeman, S. J. (2000). Corporate social performance and attractiveness as an employer to different job seeking populations. *Journal of Business Ethics, 38*, 243–253.

Alder, G. S., & Gilbert, J. (2006). Achieving ethics and fairness in hiring: Going beyond the law. *Journal of Business Ethics, 68*(4), 449–464.
Aust, I., Matthews, B., & Muller-Camen, M. (2020). Common good HRM: A paradigm shift in sustainable HRM? *Human Resource Management Review, 30*(3), 100705. https://doi.org/10.1016/j.hrmr.2019.100705
Backhaus, K., Stone, B. A., & Heiner, K. (2002). Exploring relationships between corporate social performance and employer attractiveness. *Business and Society, 41*, 292–318.
Baker, B., & Cooper, J. (2000). Occupational testing and psychometric instruments: An ethical perspective. In D. Winstanley & Woodall (Eds.), *Ethical issues in contemporary human resource management*. Palgrave.
Bansal, P., & Roth, K. (2000). Why companies go green: A model of ecological responsiveness. *Academy of Management Journal, 43*, 717–736.
Bauer, T., & Aiman-Smith, L. (1996). Green career choices, the influences of ecological stance on recruiting. *Journal of Business and Psychology, 10*, 445–458.
Behrend, T. S., Baker, B. A., & Thompson, L. F. (2009). Effects of pro-environmental recruiting messages: The role of organizational reputation. *Journal of Business Psychology, 24*, 341–350.
Berrone, P., & Gomez-Mejia, L. R. (2009). Environmental performance and executive compensation: An integrated agency-institutional perspective. *Academy of Management Journal, 52*, 103–126.
Beyond Grey Pinstripes. (2010). *Global 100 list*. Beyond Grey Pinstripes.
Boiral, O. (2002). Tacit knowledge and environmental management. *Long Range Planning, 35*, 291–317.
Branzei, O., Ursacki-Bryant, T. J., Vertinsky, I., & Zhang, W. (2004). The formation of green strategies in Chinese firms: Matching corporate and individual principles. *Strategic Management Journal, 25*, 1075–1095.
Brio, J. A. D., Fernandez, E., & Junquera, B. (2007). Management and employee involvement in achieving an environmental action-based competitive advantage: An empirical study. *The International Journal of Human Resource Management, 18*, 491–522.
Buckley, M. R., Beu, D. S., Frink, D. D., Howard, J. L., Berkson, H., Mobbs, T. A., & Ferris, G. R. (2001). Ethical issues in human resource systems. *Human Resource Management Review, 11*, 11–29.
Budd, J. W. (2004). *Employment with a human face*. Cornell University Press.
Bunge, J., Cohen-Rosenthal, E., & Ruiz-Quintanilla, A. (1996). Employee participation in pollution reduction: Preliminary analysis of the toxic release inventory. *Journal of Cleaner Production, 4*, 9–16.
Cassio, J., & Rush, A. (2009). *Green careers: Choosing work for a sustainable future*. New Society Publishers.
Chan, E., & Hawkins, R. (2010). Attitudes towards EMSs in an international hotel: An exploratory case study. *International Journal of Hospitality Management, 39*, 641–651.
Cordeiro, J. J., & Sarkis, J. (2008). Does explicit contracting effectively link CEO compensation to environmental performance? *Business Strategy and the Environment, 17*, 304–317.
Crawshaw, J. R., Cropanzano, R., Bell, C. M., & Nadisic, T. (2013). Organisational justice: New insights from behavioral ethics. *Human Relations, 66*(7), 885–904.
Crichton, R., & Shrivastava, P. (2017). Sustaining human resource via aesthetic practices. *Journal of Cleaner Production, 153*, 718–726.
Crichton, R., Shrivastava, P., Walker, T., Farhidi, F., Weeratunga, V., & Renwick, D. (2023). Improving executive compensation in the fossil fuel sector to influence green behaviors. *German Journal of Human Resource Management, 37*(3), 199–220.
Crichton, R., Walker, T., & Patel, A. (2018). Slowing climate change: Mitigating poverty and environmental degradation via strategic human resource management and responsible leadership. In *CSR and climate change implications for multinational enterprises* (pp. 37–55). Edward Elgar Publishing.

Currie, G., Knights, D., & Starkey, K. (2010). Introduction: A post-crisis critical reflection on business schools. *British Journal of Management, 21*, S1–S5.

De Gama, N., McKenna, S., & Peticca-Harris, A. (2012). Ethics and HRM: Theoretical and conceptual analysis. An alternative approach to ethical HRM through the discourse and lived experiences of HR professionals. *Journal of Business Ethics, 111*, 97–108.

Delbridge, R., & Turnbull, P. (1992). Human resource maximization: The management of labour under just-in-time manufacturing systems. In P. Blyton & P. Turnbull (Eds.), *Reassessing human resource management* (pp. 56–73). Sage.

Denton, D. K. (1999). Employee involvement, pollution control and pieces to the puzzle. *Environmental Management and Health, 10*, 105–111.

Dolan, K. A. (1997, June 2). *Kinder, Gentler MBAs* (pp. 39–40). Forbes.

Ehnert, I. (2009). *Sustainable human resource management*. Springer.

Farnhill, T. (2018). Union workplace greening - Three case studies. *British Journal of Industrial Relations, 56*(4), 716–743.

Fernandez, E., Junquera, B., & Ordiz, M. (2003). Organizational culture and human resources in the environmental issue. *The International Journal of Human Resource Management, 14*, 634–656.

Fisher, C. M., & Lovell, A. (2009). *Business ethics and values: Individual, corporate and international perspectives* (pp. 1–616). Pearson Education.

Florida, R., & Davison, D. (2001). Gaining from green management: Environmental management systems inside and outside of the factory. *California Management Review, 43*, 64–84.

Fryxell, G. E., & Lo, C. W. H. (2003). The influence of environmental knowledge and values on managerial behaviors on behalf of the environment: An empirical examination of managers in China. *Journal of Business Ethics, 46*, 45–69.

Garavan, T. N., & McGuire, D. (2010). Human resource development and society: Human resource developments role in embedding corporate social responsibility, sustainability, and ethics in organizations. *Advances in Developing Human Resources, 12*, 487–507.

George, G., Howard-Grenville, J. H., Joshi, A., & Tihanyi, L. (2016). Understanding and tackling societal grand challenges through management research. *The Academy of Management Journal, 59*(6), 1880–1895.

Govindarajulu, N., & Daily, B. F. (2004). Motivating employees for environmental improvement. *Industrial Management and Data Systems, 104*, 364–372.

Greenwood, M. (2002). Ethics and HRM: A review and conceptual analysis. *Journal of Business Ethics, 36*, 261–278.

Greenwood, M. (2012). Ethical analyses of HRM: A review and research agenda. *Journal of Business Ethics, 114*(2), 355–366. https://doi.org/10.1007/s10551-012-1354-y

Grojean, M. W., Resick, C. J., Dickson, M. W., & Smith, D. B. (2004). Leaders, values, and organizational climate: Examining leadership strategies for establishing an organizational climate regarding ethics. *Journal of Business Ethics, 55*, 223–241.

Grossman, W., & Schoenfeldt, L. F. (2001). Resolving ethical dilemmas through international human resource management: A transaction cost economics perspective. *Human Resource Management Review, 11*, 55–72.

Hampton, P. (2015). *Workers and trade unions for climate solidarity: Tackling climate change in a neoliberal world* (pp. 1–211). Routledge.

Hanna, M. D., Newman, W. R., & Johnson, P. (2000). Linking operational and environmental improvement through employee involvement. *International Journal of Operations & Production Management, 20*, 148–165.

Hossain, M. I., Teh, B. H., Dorasamy, M., Tabash, M. I., & Ong, T. S. (2024). Ethical leadership, green HRM practices and environmental performance of manufacturing SMEs at Selangor, Malaysia: Moderating role of green technology adoption. In *Economic recovery, consolidation, and sustainable growth* (pp. 85–104). Springer.

Islam, T., Khan, M. N., Ahmed, I., & Mahmood, K. (2021). Promoting in-role and extra-role green behavior through ethical leadership: Mediating role of green HRM and moderating role of individual green values. *International Journal of Manpower, 42*(6), 1102–1123.

Jackson, S. E., Jiang, Y., Shim, H., Budhwar, P., Renwick, D. W. S., Jabbour, C. J. C., de Sousa Jabbour, A. B. L., Tang, G., Muller-Camen, M., Wagner, M., & Kim, A. (2019). *The social dynamics of discretionary green behavior at work in five countries* (pp. 1–57). Paper presented at the 79th Academy of Management Annual Meeting, Boston, MA, USA, August.

Jackson, S. E., Renwick, D. W. S., Jabbour, C. J. C., & Muller-Camen, M. (2011). State-of-the-art and future directions for green human resource management: Introduction to the special issue. *Zeitschrift fur Personalforschung, (German Journal of Research in Human Resource Management), 25*, 99–115.

Jiang, Y., Jackson, S. E., Shim, H., Budhwar, P., Renwick, D. W. S., Jabbour, C. J. C., Jabbour, A. B. L., Tang, G., Muller-Camen, M., Wagner, M., & Kim, A. (2022). Culture as context: A five-country study of discretionary green behavior in organizations. *Organization & Environment, 35*(4), 499–522.

Kagan, R. A., Gunningham, N., & Thornton, D. (2003). Explaining corporate environmental performance: How does regulation matter? *Law Society Review, 7*(1), 51–90.

Kim, A., Kim, Y., Han, K., Jackson, S. E., & Ployhart, R. E. (2017). Multilevel influences on voluntary workplace green behavior: Individual differences, leader behavior, and coworker advocacy. *Journal of Management, 43*(5), 1335–1358. https://doi.org/10.1177/0149206314547386

Kitazawa, S., & Sarkis, J. (2000). The relationship between ISO 14001 and continuous source reduction programs. *International Journal of Operations and Production Management, 20*, 225–248.

Li, W., Abdalla, A. A., Mohammad, T., Khassawneh, O., & Parveen, M. (2023). Towards examining the link between green HRM practices and employee green in-role behavior: Spiritual leadership as a moderator. *Psychology Research and Behavior Management, 16*, 383–396.

Llewellyn, B., & Golden, K. (2008). *Green jobs: A guide to eco-friendly employment*. Adams.

Madsen, H., & Ulhoi, J. P. (2001). Greening of human resources: Environmental awareness and training interests within the workforce. *Industrial Management and Data Systems, 101*, 57–63.

Manjula, T. V., & Parvathy, C. N. (2019). Go green through ethical HRM. *International Journal of Scientific & Engineering Research, 10*(3), 178–180.

May, D. R., & Flannery, B. L. (1995). Cutting waste with employee involvement teams. *Business Horizons, 38*, 28–38.

Mellahi, K., Morrell, K., & Wood, G. (2010). *The ethical business: Challenges and controversies* (pp. 1–553). Bloomsbury Publishing.

Miceli, M. P., & Near, J. P. (2001). Ethical issues in the management of human resources. *Human Resource Management Review, 11*, 1–9.

Milliman, J., & Clair, J. (1996). Best environmental HRM practices in the U.S. In W. Wehrmeyer (Ed.), *Greening people: Human resources and environmental management* (pp. 49–73). Greenleaf Publishing.

Ministry of Justice. (2007). *A guide to the Corporate Manslaughter and Corporate Homicide Act 2007*. Retrieved 2:56pm on June 21, 2012, from http://www.nio.gov.uk/guide_to_the_cmch_act_2007_web.pdf_oct_07-3.pdf

O'Higgins, E., & Kelleher, B. (2005). Comparative perspectives on the ethical orientations of human resources, marketing and finance functional managers. *Journal of Business Ethics, 26*, 275–288.

Oates, A. (1996). Industrial relations and the environment in the U.K. In W. Wehrmeyer (Ed.), *Greening people: Human resources and environmental management* (pp. 117–140). Greenleaf Publishing.

Parks, B., & Helmer, J. (2009). *Green careers*. Alpha Publishing.

Petrick, J. A., & Scherer, R. F. (2005). Management educators' expectations for professional ethics development. *Journal of Business Ethics, 61*, 301–314.

Provis, C. (2009). Ethics and HRM. In A. Wilkinson, N. Bacon, T. Redman, & S. Snell (Eds.), *The Sage handbook of human resource management* (pp. 475–489). Sage.

Rees, S. (1996). Action through ownership: Learning the way at Kent County Council. In W. Wehrmeyer (Ed.), *Greening people: Human resources and environmental management* (pp. 357–375). Greenleaf Publishing.

Remmen, A., & Lorentzen, B. (2000). Employee participation and cleaner technology: Learning processes in environmental teams. *Journal of Cleaner Production, 8*, 365–373.

Renwick, D., Redman, T., & Maguire, S. (2008). *Green HRM, a review, process model and research agenda* (pp. 1–46). Discussion paper series (Number 2008.01, April), Management School, University of Sheffield.

Renwick, D. W. S. (Ed.). (2018). *Green human resource management: Towards environmental sustainability in action? Routledge research in sustainability and business series* (pp. 1–199). Routledge.

Renwick, D. W. S. (Ed.). (2020). *Contemporary developments in green human resource management research: Towards sustainability in action? Routledge research in sustainability and business series* (1st ed., pp. 1–199). Routledge.

Renwick, D. W. S. (2024). *Work, politics and the green industrial revolution: A reflective analysis of the UK Green Jobs Taskforce* (1st ed., pp. 1–155). Palgrave MacMillan.

Renwick, D. W. S., Clark, I., & Crichton, R. (2024). The limitations of individual analysis and alternative framing of employee green behavior. In M. Y. Yusoff & D. W. S. Renwick (Eds.), *Green human resource management and sustainability: A view from the global south countries. Accounting, finance, sustainability, governance and fraud book series* (1st ed., pp. 3–19). Springer.

Renwick, D. W. S., Jabbour, C. J. C., Muller-Camen, M., Redman, T., & Wilkinson, A. (2016). Contemporary developments in Green (environmental) HRM scholarship. *The International Journal of Human Resource Management, 27*, 114–128.

Renwick, D. W. S., Redman, T., & Maguire, S. (2013). Green human resource management: A review and research agenda. *The International Journal of Management Reviews, 15*, 1–14.

Rothenberg, S. (2003). Knowledge content and worker participation in environmental management at NUMMI. *Journal of Management Studies, 40*, 1783–1802.

Ruiz-Palomino, P., Martinez-Ruiz, M. P., & Martinez-Canas, R. (2013). Assessing ethical behaviors in the Spanish banking and insurance industries: Evidence and challenges. *The International Journal of Human Resource Management, 24*(11), 2173–2196.

Russo, M., & Harrison, N. (2005). Organizational design and environmental performance: Clues from the electronics industry. *Academy of Management Journal, 48*, 583–593.

Segal, L., & Lehrer, M. (2013). The conflict of ethos and ethics: A sociological theory of business people's ethical values. *Journal of Business Ethics, 114*, 513–528.

Simpson, D., & Samson, D. (2008). Environmental strategy and low waste operations: Exploring complementarities. *Business Strategy and the Environment, 19*, 104–118.

Snape, E., Redman, T., & Bamber, G. (1994). *Managing managers: Strategies and techniques for human resource management*. Blackwell.

Starkey, K., & Crane, A. (2003). Toward green narrative: Management and the evolutionary epic. *Academy of Management Review, 28*, 220–237.

Stone, L. J. (2000). When case studies are not enough: The influence of corporate culture and employee attitudes on the success of cleaner production initiatives. *Journal of Cleaner Production, 8*, 353–359.

Thorpe, R., Gold, J., & Lawler, J. (2011). Locating distributed leadership. *The International Journal of Management Reviews, 13*, 239–250.

Tirno, R. R., Islam, N., & Happy, K. (2023). Green HRM and ecofriendly behavior of employees: Relevance of proecological climate and environmental knowledge. *Heliyon, 9*(4), e14632.

Torres, R. (2008). The social impact of policies to address climate change: A review of the issues. *International Labour Review, 147*, 275–279.

TUC. (2009). *Go green at work: A handbook for union green representatives*. TUC.

TUSDAC. (2014). *The union effect: Greening the workplace*. Discussion paper. Retrieved from https://www.tuc.org.uk/sites/default/files/The_Union_Effect_Greening_The_Workplace_Covers_2014_All.pdf

Ulrich, D. (1997). *Human resource champions: The next agenda for adding value and delivering results*. Harvard.

Ulrich, D., & Brockbank, W. (2005). *The HR value proposition*. Harvard Business School Press.

Valentine, S., & Hollingworth, D. (2014). Ethics-related selection and reduced ethical conflict as drivers of positive work attitudes: Delivering on employees' expectations for an ethical workplace. *Personnel Review, 43*(5), 692–716.

Van Riel, A. (2019, March 5). *The circular economy could create an enormous jobs boom* (pp. 1–2). Fast Company Compass. https://www.fastcompany.com/90343657/the-circular-economy-could-create-an-enormous-jobs-boom

Wang, Y.-D., & Hsieh, H.-H. (2013). Organizational ethical climate, perceived organizational support, and employee silence: A cross-level investigation. *Human Relations, 66*(6), 783–802.

Weaver, G. R., & Trevino, L. K. (2001). The role of human resources in ethics/compliance management: A fairness perspective. *Human Resource Management Review, 11*, 113–134.

Wehrmeyer, W. (1996). Introduction. In W. Wehrmeyer (Ed.), *Greening people: Human resources and environmental management* (pp. 11–32). Greenleaf Publishing.

Wehrmeyer, W., & McNeil, M. (2000). Activists, pragmatists, technophiles and tree-huggers? Gender differences in employees' environmental attitudes. *Journal of Business Ethics, 28*, 211–222.

Wehrmeyer, W., & Vickerstaff, S. (1996). Analysis for environmental training needs. In W. Wehrmeyer (Ed.), *Greening people: Human resources and environmental management* (pp. 339–356). Greenleaf Publishing.

Wells, D., & Schminke, M. (2001). Ethical development and human resources training: An integrative framework. *Human Resource Management Review, 11*, 135–158.

Wilton, N. (2013). *An introduction to human resource management* (2nd ed., pp. 1–455). Sage.

Winstanley, D., & Woodall, J. (2006). The ethical dimension of human resource management. *Human Resource Management Journal, 10*(2), 5–20.

Wooten, K. C. (2001). Ethical dilemmas in human resource management: An application of a multi-dimensional framework, unifying taxonomy, and applicable codes. *Human Resource Management Review, 11*, 159–175.

World Bank. (2009). *Public attitudes towards climate change: Findings from a multi-country poll. World development report 2010*. Development and Climate Change, Washington DC.

Zelezny, L. C., Chua, P., & Aldrich, C. (2000). Elaborating on gender differences in environmentalism. *Journal of Social Issues, 56*, 443–457.

The Role of Environmental Self-Identity, Supervisory, and Organizational Support in Predicting Employee Green Behavior

Anja Shadabi, Ina Aust, Douglas W. S. Renwick, and Carsten Herbes

Abstract This study investigates the lesser-known organizational and psychological mechanisms leading individuals in organizations to engage in employee green behavior. Results from a survey of 744 employees from different organizations based in Germany reveal that high levels of environmental self-identity weaken the influence of supervisory support on employee green behavior—when both supervisory support and environmental self-identity are low, no relation between perceived organizational support toward the environment and employee green behavior is seen, and the highest impact on employee green behavior occurs for environmental self-identity. Theoretical and empirical implications of the findings are discussed.

1 Introduction

Environmental degradation and the problems caused by a changing climate are putting pressure on societies globally. Organizations are considered to be a significant contributor to these negative developments (United Nations, 2021), yet can remedy

A. Shadabi
Fachbereich Wirtschaftsingenieurwesen, Technische Hochschule Mittelhessen University of Applied Sciences, Friedberg, Germany
e-mail: anja.shadabi@wi.thm.de

I. Aust
Louvain Research Institute in Management and Organizations (LouRIM), UCLouvain, Louvain-la-Neuve, Belgium
e-mail: ina.aust@uclouvain.be

D. W. S. Renwick (✉)
Department of HRM, Nottingham Business School, Nottingham Trent University, Nottingham, UK
e-mail: douglas.renwick@ntu.ac.uk

C. Herbes
Institute for Int. Research on Sustainable Management and Renewable Energy, Nuertingen-Geislingen University, Nuertingen, Germany
e-mail: carsten.herbes@hfwu.de

M. Y. Yusliza et al. (eds.), *Employee Green Behavior*,
https://doi.org/10.1007/978-981-95-1422-9_3

their impact through pro-environmental programs (Testa et al., 2014, 2017), such as ISO14001 (Boiral, 2007), corporate greening (Andersson & Bateman, 2000; Andersson et al., 2013; Temminck et al., 2015), and/or a variety of different environmental activities (Wagner, 2011). Employee green behavior (EGB) at the individual level has been identified as a means to establish environmental sustainability in organizations (Norton et al., 2015). EGB has been defined as "scalable actions and behaviors that employees engage in that are linked with and contribute to or detract from environmental sustainability" (Ones & Dilchert, 2012, p. 87). Prior research has grouped EGB into abstract and specific behaviors (Francoeur et al., 2021). This study refers to the abstract form of EGB and uses the subscales of eco-helping and eco-civic engagement of the organization citizenship behaviors for the environment (OCBE) scale developed by Paillé and Boiral (2013). While eco-helping encompasses behaviors such as encouraging colleagues to behave in an environmentally friendly manner and developing ideas directed at environmental issues, eco-civic engagement refers to participating in organizational environmental events and being informed about environmental initiatives at work (Paillé & Boiral, 2013).

Research in the field of corporate greening has advanced our knowledge of the influences of various personal and organizational factors on EGB. Among the organizational factors, perceived organizational support (Ramus, 2001), perceived supervisory support (Raineri & Paillé, 2016; Ramus & Steger, 2000), organizational green policy (Ramus & Steger, 2000), and leadership behavior (Cheng et al., 2021; Graves & Sarkis, 2018; Graves et al., 2013) have been found to play a particularly important role in encouraging EGB (Paillé & Raineri, 2015; Ramus & Steger, 2000; Robertson & Barling, 2013). Among the individual factors, attitudes (Baker et al., 2014), identification with nature (Schmitt et al., 2019), motivation to perform sustainable behaviors (Graves & Sarkis, 2018), environmental beliefs (Raineri & Paillé, 2016), and personal norms (Ciocirlan et al., 2020) have been identified as predictors of EGB. These individual factors are considered to play an equally important and influential role in EGB as organizational factors do.

Two aspects deserve further attention and have been so far neglected in studies on EGB. First, a recurring point of criticism refers to the lack of studies that combine individual and organizational factors in studies on EGB (Lo et al., 2012; Raineri & Paillé, 2016; Starik & Kanashiro, 2020). Most studies have isolated either individual factors or organizational factors, with the exception of a few studies that included both (Alt & Spitzeck, 2016; Paillé & Raineri, 2015). Another exception is the study on recycling behavior in a health organization, in which Tudor et al. (2007) included attitudes as individual determinants and organizational culture as an organizational determinant. Graves and Sarkis (2018) found that environmental leadership behavior, as well as internal motivation with strong environmental values facilitates EGB.

Second, many studies only explore the role of attitudes, beliefs, motivation, and norms in explaining EGB (Baker et al., 2014; Bissing-Olson et al., 2013; Ciocirlan et al., 2020; Graves & Sarkis, 2018). Prior research on green behaviors outside of the workplace has explored the role of environmental self-identity (EnvID) in explaining such behavior (Fielding et al., 2008; Whitmarsh & O'Neill, 2010), and research on identity theory found that self-identity is indirectly and directly linked to behavior (Biddle et al., 1987). EnvID is a psychological, intraindividual determinant that has been used as a predictor and moderator in studies on green behavior outside the workplace (Carfora et al., 2017; Fielding et al., 2008; Whitmarsh & O'Neill, 2010). The influence of social identity on sustainable workplace behaviors has been explored in a study by Carmeli et al. (2017). In cases where employee beliefs align with their organizations, it was found that employee identification with their workplace indirectly increases employee engagement in sustainability-related behaviors (Carmeli et al., 2017). With the exception of a few studies (e.g., Ruepert et al., 2016; van der Werff et al., 2021), researchers have neglected to study the relationship between EnvID and EGB (Ciocirlan, 2016). Unlike pro-environmental behavior in the private sphere, green workplace behavior is influenced by various organizational determinants, such as organizational support (Lamm et al., 2015), supervisory support (Ramus & Steger, 2000), and green policy (Raineri & Paillé, 2016). Therefore, the observation of organizational determinants demands the inclusion of individual variables to explain EGB and vice versa.

By exploring a model that investigates how EnvID and perceived supervisory as well as organizational support influence EGB, this study aims to address these gaps. By considering the conjoint influence of EnvID and supervisory support as a moderator on the relationship between organizational support and EGB, this study offers an extension to previous findings on the impact of EnvID on sustainable behavior in the private sphere (Carfora et al., 2017). Although similar antecedents, such as environmental beliefs (Raineri & Paillé, 2016), attitudes (Bissing-Olson et al., 2013), or values (Whitmarsh & O'Neill, 2010) have been investigated before, we offer a new explanation for the role of EnvID in the execution of EGB. This is because self-identity is seen as something that a person can grow into by repeated behavior; it persists and might even override attitudes: “Once such a role identity becomes central to the self, the importance of attitude and norms in determining behavioral intention diminishes” (Charng et al., 1988, p. 316).

By exploring both individual and organizational determinants, this study respects workplace-specific influences, such as supervisory and organizational behavior, on EGB. Therefore, a three-way interaction has been calculated: perceived supervisory support toward the environment (PSS-E), environmental self-identity (EnvID), and perceived organizational support toward the environment (POS-E) are studied as independent predictors of EGB. Our research is guided by the following research question: To what extent does the conjoint interaction between EnvID and PSS-E affect the relationship between POS-E and EGB?

We make the following contributions to research in the field of corporate greening. First, by studying the influence of EnvID on EGB under the influence of perceived supervisory and organizational support toward the environment, this study extends previous findings on EnvID to green behavior in the work context (Gatersleben et al., 2014; van der Werff et al., 2021; Whitmarsh & O’Neill, 2010). Second, in considering individual as well as organizational determinants, we address calls for more studies combining social–psychological and organizational factors to explain their influence on EGB (Lo et al., 2012).

The remainder of this chapter is organized as follows. In the next section, the theoretical background and the study's model are presented alongside the development of our research hypotheses. In the section that follows, our research methods and analytical procedure are laid out. This is followed by the result section. The findings and implications are discussed in the discussion section. The chapter concludes by offering study limitations and future research ideas.

2 Literature Review

2.1 Theoretical Framework

Research on corporate greening has made a convincing case that EGB leads to more effective greening processes in organizations and thus improves environmental performance and contributes to environmental sustainability in organizations (Andersson et al., 2013; Daily & Huang, 2001; Paillé, 2020). Under the framework of identity theory, this study explores the conjoint role of EnvID, POS-E, and PSS-E in predicting EGB, which we now detail.

Previous research has found that EnvID plays a decisive role in predicting green behavior (Ciocirlan, 2016; Whitmarsh & O'Neill, 2010), and self-identity is about the description of oneself, which might include values attached to that self, as by definition, "through the process of categorization or identification, an identity is formed" (Stets & Burke, 2000, p. 224). Whether an identity leads to a certain behavior depends on the salience of a particular identity, as research suggests that the more relevant an individual identity is, the more likely it is that the identity-congruent behavior is enacted (Laverie & Arnett, 2000). Self-identity fulfills two purposes: differentiation from others and confirmation of one's set of values, beliefs, and behaviors, which reinforce positive emotions (Christensen et al., 2004). Identity theory is thus not only about individual choices, but also includes the context or social structure in which a person is embedded (Stets & Biga, 2003). Stets and Burke (2000) stated that self-identity is only played out if it is activated, which depends on the individual's structural embeddedness or commitment: "The more persons one is tied to by holding an identity (i.e., the greater the embeddedness of the identity in the social structure), the more likely it is that the identity will be activated in a situation. In brief, the stronger the commitment, the greater the salience" (p. 230).

An EnvID implies that actions are in accordance with the self, that is, being a person who acts in environmentally friendly ways, and while doing so, establishes consistency between the individual parts of a person's self and their behavior (Festinger, 1957; Whitmarsh & O'Neill, 2010). Depending on the salience and strength of a person's self-identity, he or she is therefore more likely to act in a pro-environmental way (van der Werff et al., 2013). In previous studies, self-identity has been identified as a predictor and mediator that influences greening behaviors outside of work (Fielding et al., 2008; Gatersleben et al., 2014; Stets & Biga, 2003; Whitmarsh & O'Neill, 2010). For example, in a study on recycling behavior, Nigbur et al. (2010) concluded that the identity of being a "recycler" enhances recycling behavior, and a recent study by van der Werff et al. (2021) suggested that a stronger EnvID increases the likelihood of employees engaging in green behavior. Cheng et al. (2021) concluded that EnvID can function as a moderator between psychological ownership, i.e., if employees feel a connection to their workplace and feel responsible for the organization, and EGB. As such, we predict that environmental self-identity (EnvID) is positively related to EGB (H1).

In workplace settings, POS-E and PSS-E especially are found to encourage EGB (Cantor et al., 2012; Paillé et al., 2019; Raineri & Paillé, 2016; Ramus, 2001; Ramus & Killmer, 2007; Ramus & Steger, 2000). In general terms, perceived organizational support (POS) is formed by the organization's concern about its employees' well-being and acknowledgment of employee contributions (Eisenberger et al., 1986). In this study, its narrower form POS-E is used (Lamm et al., 2015), where POS-E is defined as "the specific belief held by employees concerning how much the organization values their contributions toward sustainability" (Lamm et al., 2015, p. 209). When employees feel supported by their organizations, they are more likely to engage in EGB (Manika et al., 2015; Ramus, 2001). Consequently, POS-E functions as a facilitator for EGB. Thus, based on findings in previous research (Lamm et al., 2015; Paillé & Raineri, 2015; Ramus, 2001), we predict that POS-E is positively related to EGB (H2).

Perceived supervisory support toward the environment (PSS-E) is another relevant predictor of EGB (Paillé et al., 2019, 2022; Raineri & Paillé, 2016; Ramus & Steger, 2000), as the probability of employees engaging in greening behaviors was found to rise with the regular support of their supervisors (Ramus & Steger, 2000). PSS-E refers here to "the employee's belief that the supervisor provides subordinates with the resources and feedback needed to participate in environmental initiatives" (Cantor et al., 2012, p. 37). Prior studies have shown that supervisory and organizational support are linked, as Eisenberger et al. (2002) found that they are connected by the supervisor's organizational status and through a temporal link that supervisory support precedes the perception of organizational support. Other studies have reported reliable relationships between these two concepts (Rhoades et al., 2001; Yoon & Thye, 2000), as in her study of eco-initiatives, Ramus (2001) found that supervisory behaviors are key to encouraging EGB.

Owing to their position in the organizational hierarchy and their role, individual supervisors play a decisive role in shaping and encouraging EGB (Cantor et al., 2012; Graves & Sarkis, 2018; Graves et al., 2013), and are also able to discourage any attempts to engage in EGB too (Cantor et al., 2012). A study by Cantor et al. (2012) confirmed that high levels of PSS-E make it more likely for employees to also perceive organizational support for EGB, making PSS-E crucial for employees who attempt to engage in greening behaviors.

In this study, it is thus assumed that PSS-E does not directly influence a person's behavior but affects a person's self-identity, which in turn leads to the desired behavioral outcome. This is because identity researchers refer to self-identity in explaining behaviors that are guided by how persons perceive themselves (Sparks & Shepherd, 1992; Whitmarsh & O'Neill, 2010), and previous research has pointed out that identity has a greater influence on behavior than attitudes when a specific role identity dictates a certain behavior (Charng et al., 1988). If a preferred self-identity cannot be lived out, it is possible that the affected identity is devalued (Pratt et al., 2006) and a person adapts to a social identity (Brown, 2015, 2017). Our theorizing thus suggests that perceived supervisory support affects the living out of one's preferred identity, meaning we expect that a three-way POS-E × EnvID × PSS-E interaction positively predicts EGB (H3).

However, a self-identity that is contrary to existing organizational identity can lead to identity tensions, as individuals who perceive a mismatch between their preferred identity and the organizational identity try to work out their identity tensions (Kira & Balkin, 2014). In the worst case, if personal identities do not match with organizational identities, employees may decide to leave their work for good. Only identities that match with organizational identities lead to positive outcomes: if a personal identity and an organizational identity match, employees are found to thrive (Kira & Balkin, 2014). In several studies, the identity tensions of sustainability managers with their work environment have been seen (Carollo & Guerci, 2018; Ghadiri et al., 2015; Wright et al., 2012). For example, Carollo and Guerci (2018) found in their study of 26 sustainability managers that the double demand of work-related business goals and their personal values regarding social and environmental goals caused tensions for sustainability managers. In conclusion, EGB is performed if employees' environmental beliefs and those upheld in the work environment are congruent (Muster & Schrader, 2011). To a certain extent, PSS-E can balance out a low EnvID. However, only if EnvID and PSS-E are strong do they have an impact on the relationship between POS-E and EGB. Thus, we predict: low levels of both EnvID and PSS-E will result in low EGB (H4).

High levels of PSS-E and high levels of EnvID increase the effect of POS-E on EGB (Raineri et al., 2016). Thus, we predict that low levels of either PSS-E or EnvID can be balanced out by high levels of EnvID or PSS-E and vice versa (H5).

In sum, EnvID as well as the PSS-E and POS-E are relevant for EGB. The influence of EnvID depends on the organizational context, which is studied by the concept of PSS-E and POS-E. Our five hypotheses are therefore as follows:

Hypothesis 1: EnvID has a positive effect on EGB.
Hypothesis 2: There is an overall positive effect of POS-E on EGB.
Hypothesis 3: The three-way POS-E × EnvID × PSS-E interaction positively predicts EGB.
Hypothesis 4: Low levels of both EnvID and PSS-E will result in low EGB.
Hypothesis 5: Low levels of either PSS-E or EnvID can be balanced out by high levels of EnvID or PSS-E and vice versa.

We propose the following model in Fig. 1.

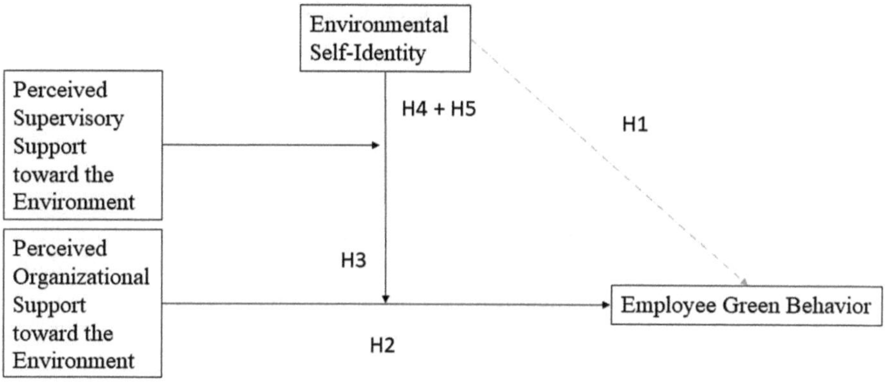

Fig. 1 Proposed conceptual model of hypothesized three-way interaction effect between perceived organizational support toward the environment (POS-E), environmental self-identity (EnvID), perceived supervisory support toward the environment (PSS-E), and employee green behavior (EGB). Source: Author's own work

3 Methodology

3.1 Sample and Procedure

The hypotheses were tested using a sample of adults working in a range of organizations and occupations, and data were collected as a part of a bigger research project in Germany in the form of an online survey, which was administered through an online panel (www.respondi.com) that focuses on online surveys and uses established procedures for ensuring data quality (e.g., recruiting suitable respondents, setting time stamps). For our study, the provider was instructed to apply a number of selection criteria during recruitment, and inclusion criteria were: white- and blue-collar workers, as well as civil servants aged 18–67 years working in an organization with at least 50 employees, resulting in a total sample size of 922 participants. Of the 922 questionnaires received, 178 were discarded because relevant data were missing, reducing the final sample to 744, of which 345 were women and 399 men. Additionally, gender and age quotas were applied, where age distribution of the sample was controlled to avoid selection effects, e.g., a bias of younger and more technologically adept Internet users (19% 18–25 years, 20% 26–34 years, 20% 35–44 years, 21% 45–54 years, 20% 55–64 years, and 0.5% 65 and older). Approximately 6% had worked for their organization for <1 year, 19% had between 1 and 3 years, 22% had between 3 and 6 years, 11% had between 6 and 9 years, and the largest group of participants (41%) had 9 years or more of organizational tenure. We also asked for their employment situation: 14% had limited and 86% had unlimited contracts. Finally, 22% worked for organizations with 50–99 employees, 16% for organizations with 100–199 employees, 10% for organizations with 200–299 employees, 11% for organizations with 300–499 employees, 15% for organizations with 500–999 employees, 10% for organizations with 1000–2499 employees, 5%

for organizations with 2500–4999 employees, and 11% for organizations with 5000 and more employees.

3.2 Measures

For all items, a 7-point Likert scale ranging from 1 = fully agree to 7 = fully disagree or 1 = always to 7 = never was applied (the scales were reversed before calculation for easier interpretation), and exceptions are indicated below for the individual measures. In all scales, a "don't know" or if necessary a "not applicable" option was added to increase reliability and to prevent participants from opting for the middle category, and "don't know" and "not applicable" categories were coded as missing data. For single missing values within a scale, the mean values were calculated. All items were translated from German into English, and the translation was controlled by a back translation by another team member as per Brislin (1990). Scale reliability was tested by coefficient alphas, with over 0.70 indicating acceptable reliability.

3.2.1 Environmental Self-Identity (EnvID)

An EnvID scale was developed by using items from previous research, where items 1, 2, 4, and 5 were derived from a scale by Whitmarsh and O'Neill (2010) and items 3 and 6 from one by Fielding et al. (2008). The items were: (1) "I consider myself an environmentally friendly person"; (2) "I think of myself as someone who is very concerned with environmental issues"; (3) "I would be embarrassed to be seen as having an environmentally friendly lifestyle"; (4) "I would not want my family or friends to think of me as someone who is concerned about environmental issues"; (5) "I am not the type of person who would be involved in environmental and climate protection"; and (6) "It is important to me to work for environmental and climate protection in my free time." The factor loadings of item 4 (0.04) and item 5 (−0.05) were below the standard cutoff (Hair et al., 2010) and thus dropped. The remaining factor loadings ranged from 0.50 to 0.94 and formed a reliable scale (coefficient alpha = 0.85, M = 4.57, SD = 1.42).

3.2.2 Perceived Supervisory Support Toward the Environment (PSS-E)

PSS-E was measured by five items developed by Ramus (2001):

My immediate superior: (1) "…encourages environmental initiatives"; (2) "…makes sure that employees develop environmental skills"; (3) "…listens carefully to and values inputs on environmental topics"; (4) "…gives complete and accurate information regarding environmental issues"; and (5) "…involves employees in environmental problem solving." The items' factor loadings ranged from 0.90 to 0.93, and the coefficient alpha was 0.96, M = 3.30, SD = 1.79.

3.2.3 Employee Green Behavior (EGB)

EGB was investigated through the construct of OCBE, using the scale developed by Boiral and Paillé (2012), which is among the five most common measurement scales in EGB research (Francoeur et al., 2021). OCBEs are described as "discretionary behaviors performed by employees whereby they demonstrate their willingness to cooperate with their organization and its members by performing behaviors in the workplace that benefit the natural environment" (Paillé & Boiral, 2013, p. 119). The OCBE concept can be split into three dimensions: eco-initiatives, eco-civic engagement, and eco-helping (Boiral & Paillé, 2012). Although eco-civic engagement was considered, it was not used in the final analysis because it had a very high nonresponse rate. Therefore, this study focused only on the subscale eco-helping to measure the encouragement of colleagues to consider environmental concerns (Paillé & Boiral, 2013). Three items (1–3) measured eco-helping and three items eco-civic engagement (4–6): (1) "I encourage my colleagues to behave in a climate and environmentally friendly way in the workplace"; (2) "I encourage my colleagues to express their ideas and opinions on environmental and climate protection issues"; (3) "I discuss environmental and climate protection issues with colleagues, even if they have a negative or skeptical attitude towards the issues"; (4) "I actively participate in environmental or climate protection events organized by my employer"; (5) "I stay informed about the latest developments on environmental and climate protection issues from my employer"; and (6) "I volunteer for environmental and climate protection projects in my company." Participants were given a "not applicable" option to indicate behaviors that they had no opportunity to perform in their workplace (e.g., in case there were no events). The "not applicable" option was coded as missing data. The coefficient alpha for all items was measured as 0.94 (eco-civic engagement only: coefficient alpha = 0.92; eco-helping only: coefficient alpha = 0.91). The items were measured on a 7-point scale ranging from 1 = never to 7 = always, which were reversed for the analysis. Items 4, 5, and 6 were dropped for analysis, because of the high numbers of respondents who indicated that there was no option to participate (N = 245), inform about the latest developments (N = 245), or volunteer (N = 281) in environmental events at their workplace. The items' factor loadings ranged from 0.84 to 0.93 (coefficient alpha: 0.91, M = 4.05, SD = 1.79).

3.2.4 Perceived Organizational Support Toward the Environment (POS-E)

Four items measuring organizational support for environmental behavior and the possibility of suggesting and implementing employees' own ideas for environmental and climate protection were taken and slightly adapted from Lamm et al. (2015). The items assessed the opportunities and freedom to engage in environmental activities in the workplace and how they are received in the organization. A 7-item scale was used to measure POS-E ranging from 1 = strongly agree to 7 = strongly disagree (the scale was reversed for the analysis). The items were: (1) "At my workplace I can behave as environmentally and climate-friendly as I want without restrictions"; (2) "Anyone who brings ideas about environmental or climate protection to my employer will be rewarded (e.g., through vouchers, bonuses)"; (3) "My organization does not care about whether I behave in a sustainable manner or not" (reverse-scored); and (4) "At my workplace I have the opportunity to share ideas and initiatives on environmental and climate protection." Since the scale was slightly adapted, we conducted an exploratory factor analysis using principal component extraction with varimax rotation to assess the measure's reliability. This resulted in the expected one-factor solution (Kaiser-Meyer-Olkin: 0.70) accounting for 53% of common variance with factor loadings between 0.40 and 0.85. Item 3 "My organization does not care about whether I behave in a sustainable manner or not" (reverse-scored) was dropped because it was below the cutoff (0.40). After dropping item 3, the coefficient alpha increased to an acceptable 0.73, M = 3.69, SD = 1.58. Individual factor loadings ranged from 0.62 to 0.77.

3.2.5 Control Variables

Since demographic characteristics have been suggested as potential reasons for green behaviors (Lamm et al., 2013), in this study the following control variables were used: gender (1 = male, 2 = female), age (1 = 18–25, 2 = 26–34, 3 = 35–44, 4 = 45–54, 5 = 55–64, and 6 = 65 and older), organizational size (1–4 were excluded; 5 = 50–99, 6 = 100–199, 7 = 200–299, 8 = 300–499, 9 = 500–999, 10 = 1000–2499, 11 = 2500–4999, and 12 = 5000 or more), and tenure (1 = <1 year, 2 = 1 to <3 years, 3 = 3 to <6 years, 4 = 6 to <9 years, 5 = 9 or more years). Correlations between control variables and model variables are reported in Table 1. It can be noted that age correlates with EnvID and EGB, firm size with PSS-E, and tenure with EnvID.

3.2.6 Common Method Variance

Since the study was cross-sectional and used self-reports, we controlled for common method variance (CMV), defined as "variance that is attributable to the measurement method rather than to the constructs the measures represent" (Podsakoff et al., 2003, p. 879). To assess if CMV inflated the data, a confirmatory factor analysis (CFA) was conducted. The first CFA included all items mentioned above, and the second CFA was run without the items with low factor loadings. In the original model, the hypothesized factor loadings were not all statistically significant ($p < 0.05$) and ranged from −0.09 to 0.94. Therefore, items below the cutoff value of 0.5 (Hair et al., 2010) were removed: item 3, "My organization does not care about whether I behave in a sustainable manner or not" (reverse-scored) of the latent factor POS-E, and items 4, "I would not want my family or friends to think of me as someone who is concerned about environmental issues," and 5, "I am not the type of person who would be involved in environmental and climate protection" of the

Table 1 Descriptive statistics, correlations, and coefficient alphas for study variables

No.	Variable	Min–Max	M	SD	1	2	3	4	5	6	7	
1	Gender	1–2	1.46	0.5								
2	Age	1–6	3.04	1.41	−0.165**							
3	Firm size	5–12	3.62	1.35	−0.050	−0.024						
4	Tenure	1–5	7.84	2.33	−0.167**	0.640**	0.093*					
5	PSS-E	1–7	3.30	1.79	−0.167**	−0.016	0.115**	−0.037	(0.96)			
6	EnvID	1–7	4.57	1.42	−0.044	0.146**	0.009	0.076*	0.308**	(0.85)		
7	POS-E	1–7	3.69	1.58	−0.171**	−0.041	0.157**	−0.017	0.689**	0.284**	(0.73)	
8	EGB	1–7	4.05	1.71	−0.101**	0.109**	0.019	0.056	0.539**	0.619**	0.513**	(0.91)

**$p < 0.01$, *$p < 0.05$

Gender (1 = male, 2 = female), Age (1 = 18–25, 2 = 26–34, 3 = 35–44, 4 = 45–54, 5 = 55–64, and 6 = 65 or older), firm size (not included in sample: 1 = 1–5, 2 = 6–9, 3 = 10–19, 4 = 20–49; included in sample: 5 = 50–99, 6 = 100–199, 7 = 200–299, 8 = 300–499, 9 = 500–999, 10 = 1000–2499, 11 = 2500–4999, and 12 = 5000 or more employees), tenure (1 = <1 year, 2 = 1 to <3 years, 3 = 3 to <6 years, 4 = 6 to <9 years, 5 = 9 or more years). Source: Author's own work

Source: Author's own work

latent EnvID factor we recalculated the model without these items. As a result, the hypothesized four-factor model indicated an adequate fit to the data (Marsh et al., 2004), a chi-square (χ^2) of 435.384 (df = 84, N = 744), an acceptable comparative fit index (CFI) of 0.96, an acceptable root mean square error of approximation (RMSEA) of 0.08, and a RMSEA confidence interval of [0.07, 0.08] as well as a root mean square residual (SRMR) of 0.06 (Bentler, 1990). Previous tests to see if the data met the assumption of collinearity indicated that multicollinearity was not a concern [POS-E: tolerance 0.52; variance inflation factor (VIF) 1.93; EnvID: tolerance 0.89; VIF 1.135; PSS-E: tolerance 0.51; VIF 1.97] (Field, 2018). Inspection of the correlation table revealed some high correlations between variables.

As such, different factor models were computed, as per Table 2. The four-factor model had a better fit than other computed models. In the case of POS-E and EGB, the two-factor solution showed a better fit [χ^2 of 13.77 (df = 8), CFI: 0.997, RMSEA: 0.03, SRMR: 0.02] than the calculated one-factor solution [χ^2 of 297.08 (df = 9), CFI: 0.872, RMSEA: 0.21, SRMR: 0.09]. For EnvID and EGB, the two-factor solution also had a better fit [χ^2 of 134.82 (df = 13), CFI: 0.97, RMSEA: 0.11, SRMR: 0.04] than the calculated one-factor solution [χ^2 of 848.72 (df = 14), CFI: 0.78, RMSEA: 0.28, SRMR: 0.08]. The calculated two-factor solution for PSS-E and POS-E showed a better fit [χ^2 of 95.25 (df = 19), CFI: 0.98, RMSEA: 0.09, SRMR: 0.03] than if both were loaded on one factor [χ^2 of 250.02 (df = 20), CFI: 0.96, RMSEA: 0.12, SRMR: 0.05]. Thus, although POS-E and PSS-E are highly correlated, for theoretical and empirical reasons, treating them as two independent latent factors was found justifiable.

Table 2 Fit indices for alternative measurement models

Model	χ2	df	χ2/df	CFI	TLI	RSMEA	SRMR
4-factor model	435.38**	84	5.18	0.96	0.95	0.08 [0.07; 0.08]	0.06
3-factor model: PSS-E+POS-E, EnvID, EGB	562.70**	87	6.05	0.95	0.94	0.08 [0.08; 0.09]	0.07
3-factor model: EnvID+POS-E, PSS-E, EGB	1200.71**	87	13.8	0.88	0.86	0.13 [0.13; 0.14]	0.16
3-factor model: PSS-E+EnvID, POS-E, EGB	2188.94**	87	25.16	0.78	0.74	0.18 [0.17; 0.19]	0.16
2-factor model: PSS-E+POS-E+EnvID, EGB	2305.91**	89	25.91	0.77	0.73	0.18 [0.18; 0.19]	0.16
1-factor model: PSS-E+POS-E+EnvID+EGB	3348.17**	90	37.2	0.66	0.60	0.22 [0.21; 0.23]	0.17

**$p < 0.001$
Source: Author's own work

4 Findings

PSS-E as an independent variable, EnvID as a moderator Z, and POS-E as a moderator W were standardized and together with EGB as a dependent variable entered in model 3 in the PROCESS macro (Hayes, 2013) in SPSS. The overall model measures were: $F(7, 736) = 129.57$, $p < 0.001$; $R^2 = 0.55$. The model revealed a significant effect of EnvID [$\beta = 0.52$, $t\ (736) = 15.96$, $p < 0.001$, LLCI = 0.46, ULCI = 0.58] and of POS-E [$\beta = 0.21$, $t\ (736) = 5.86$, $p < 0.001$, LLCI = 0.14, ULCI = 0.28]. Hypotheses 1 and 2 are thus confirmed. A significant effect was also reported for PSS-E [$\beta = 0.27$, $t\ (736) = 7.59$, $p < 0.001$].

We also tested for the individual interactions. The three-way interaction between POS-E × PSS-E × EnvID was significant with $\beta = -0.02$, $t\ (736) = -2.0$, $p = 0.04$, LLCI = −0.04, ULCI = −0.001. Therefore, H3 can be confirmed. The two-way interaction between POS-E and PSS-E was not significant with $\beta = 0.03$, $t\ (736) = 1.01$, $p = 0.31$, LLCI = −0.03, ULCI = 0.08, and so was the interaction between POS-E and EnvID ($\beta = 0.05$, $t\ (736) = 1.31$, $p = 0.19$, LLCI = −0.02, ULCI = 0.11). The two-way interaction between PSS-E and EnvID was marginally significant with $\beta = -0.06$, $t\ (736) = -1.88$, $p = 0.06$, LLCI = −0.13, ULCI = 0.003. At different levels of PSS-E and EnvID, POS-E was found to predict EGB—see Table 3.

However, when PSS-E and EnvID were low, no significant relation between POS-E and EGB could be identified. Ergo, H4 cannot be confirmed. In all other cases, with either PSS-E or EnvID showing a high level, POS-E predicted EGB. However, the effect of EnvID decreased when PSS-E levels were high. The largest observed effect can be reported for low levels of PSS-E and high levels of EnvID (effect: 0.27, $p < 0.001$). At high levels of PSS-E, the effect slightly decreased with increasing levels of EnvID, ranging from 0.24 to 0.23 ($p < 0.001$). In all other cases, the effect increased with increasing levels of EnvID (effects ranging from 0.16 to 0.25, $p < 0.001$). Overall, it can be said that if EnvID increases, the effect of PSS-E on EGB decreases, and vice versa. H5 can therefore be confirmed.

Table 3 Interaction effects at different levels of PSS-E and EnvID on POS-E

PSS-E	EnvID	β	se	T	P	LLCI	ULCI
Low	Low	0.08	0.05	1.58	0.11	−0.02	0.18
Low	High	0.28	0.06	4.53	0.00	0.16	0.40
High	Low	0.24	0.08	3.21	0.00	0.10	0.39
High	High	0.23	0.06	3.76	0.00	0.11	0.34

High: +1SD, Low: −1SD
Source: Author's own work

5 Discussion of Implications

The purpose of this study was to investigate the conjoint influence of environmental self-identity (EnvID) and perceived supervisory support toward the environment (PSS-E) on the link between perceived organizational support toward the environment (POS-E) and EGB, and the study results provide evidence that PSS-E and EnvID conjointly moderate the link between POS-E and EGB. However, it was observed that the link is rather weak and EnvID individually is a strong predictor of EGB. In line with research on self-identity in predicting private green behavior (Lalot et al., 2019; Sparks & Shepherd, 1992; Whitmarsh & O'Neill, 2010) and EGB (Cheng et al., 2021; van der Werff et al., 2021), EnvID was found to be positively connected with EGB (H1). As such, employees with higher levels of EnvID are more likely to engage in EGB, which also confirms expectations that intrinsic motivation guides employees' green behaviors (Ciocirlan, 2016).

Analyzing the influence of organizational determinants on EGB, as expected, POS-E positively predicted EGB overall (H2), consistent with previous findings on the influence of POS-E on EGB (Paillé & Raineri, 2015; Ramus, 2001). Under the conjoint influence of PSS-E and EnvID, POS-E predicts EGB. However, the link between POS-E and EGB disappeared when a person reported low levels of PSS-E and EnvID (H4). This is unexpected, since it was assumed that it would only result in weak EGB, and differs from previous findings, which found that POS-E is directly related to EGB, even when moderated by psychological empowerment of employees (Lamm et al., 2015). However, this study adds to previous research finding that a relationship between POS-E and EGB only holds for employees who are committed to the organization (Temminck et al., 2015) or who show environmental commitment and beliefs (Raineri & Paillé, 2016). If either PSS-E or the reported EnvID was high (i.e., the weaker one of the variables, the more positive the effect of the other variable), we expected the link between POS-E and EGB to be reinforced, which was confirmed (H5). The strongest influence on EGB was found if employees reported high levels of all three independent variables (H3)—see Fig. 2. Consequently, this finding confirms the combined effect of individual and organizational determinants influence EGB (Lo et al., 2012; Raineri & Paillé, 2016).

This study's findings advance theory on EGB in several ways. First, it contributes to recent but scarce research on the role of EnvID in predicting EGB (Ruepert et al., 2016; van der Werff et al., 2021), as EnvID can be regarded as a new influential factor in explaining EGB. This study thus extends previous findings on social-psychological mechanisms (Paillé & Mejía-Morelos, 2014; Paillé & Raineri, 2015; Raineri et al., 2016), and extends previous research on EnvID in the private sphere (Fielding et al., 2008; van der Werff et al., 2013; Whitmarsh & O'Neill, 2010) and studies using self-identity as an extension to the theory of planned behavior to explain pro-environmental behavior in general (van der Werff et al., 2021), by combining elements from both streams of research.

Second, combining EnvID and supervisory as well as organizational support in explaining EGB responds to calls to combine individual and organizational factors

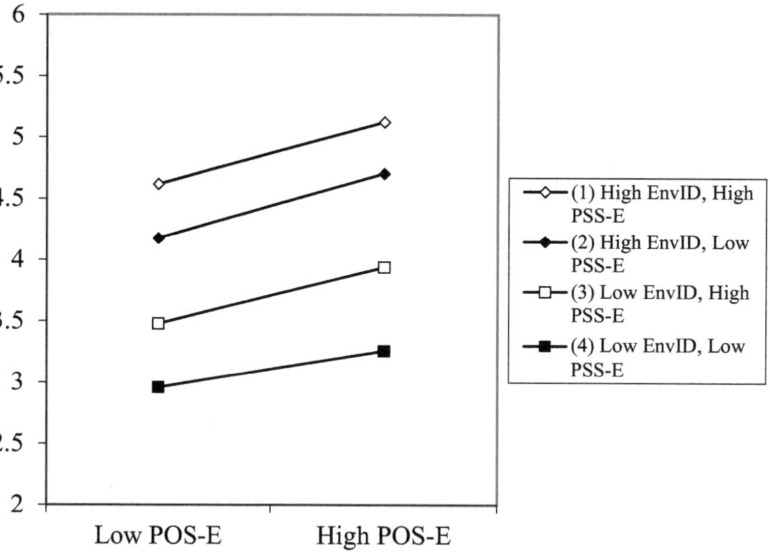

Fig. 2 Strength of employee green behavior (EGB) as a function of organizational support (POS-E), environmental self-identity (EnvID), and supervisory support (PSS-E)*. *All independent variables were mean centered. Source: Author's own work

(Lo et al., 2012), and makes a contribution to studies that have started to combine supervisory and organizational support with psychological-value-related attributes in EGB studies (Paillé et al., 2019; Paillé & Raineri, 2015; Raineri & Paillé, 2016).

This study adds to the scarce studies addressing the role of EnvID in workplace settings. In particular, green employees with EnvID are found to be intrinsically motivated to engage in EGB (Ciocirlan, 2016), as employees connect private environmental interests with their engagement in environmental goals at work (Clayton, 2003; Muster & Schrader, 2011), and also underpins findings in green HRM acknowledging the role of empowerment through supervisors and attracting "green job candidates" (Renwick et al., 2013, p. 23).

However, whether an identity is enacted depends on the identity's salience, the relative importance that a person assigns to an individual identity, and the context (Charng et al., 1988; Stryker & Serpe, 1994). In this study, the role of EnvID in EGB has been studied under the influence of supervisory and organizational support toward the environment. Although low levels of EnvID can be balanced with PSS-E, thus confirming previous findings on the influential role of supervisors on EGB (Paillé et al., 2022; Raineri & Paillé, 2016; Ramus & Steger, 2000), this study shows that high levels of EnvID decrease the influence of PSS-E on EGB. It is possible that PSS-E might even have countereffects for employees with high EnvID (Davis et al., 2020), discouraging their engagement in EGB. Essentially, there is a need to leverage the salience of EnvID if employees are to stay connected and committed to organizations, as it is found that the activation of a preferred identity is connected with positive emotions (Stets, 2005). Therefore, employees with high levels of EnvID play an important role in fostering EGB.

A key question for HR professionals and practitioners arises on assessing levels of employee environmental self-identity, and the role of supervisory support on EGB, and the role of both in supervisor-employee relationships, and organizational support for environmental self-identity in them too.

6 Future Directions

6.1 Research Limitations

Some limitations are present in this study. First, the study's item selection did not fulfill common standards. Although, coefficient alphas and other reliability estimates indicate no concern, the implemented item battery of POS-E cannot be fully compared to the variables used in the original scale to measure POS-E (Lamm et al., 2015) since items were changed and adapted to fit the German context, we also need to report the high correlation between POS-E and PSS-E which could raise some concern regarding the independence of these two variables. It is imaginable that the respondents perceived no significant difference between organizational support and the supervisory support toward the environment in their work environments, thus leading to a high overlap between the two variables.

Second, our study findings are based on self-reported data, which resulted in CMV, and raise calls for a study with more objective measures to capture EGB. Therefore, a future study should try to measure actual behavior and/or collect data in waves or measure the dependent and independent variables from different respondents as recommended by Podsakoff et al. (2003).

Third, the data stem from a convenience sample of an online panel, and it is a concern that the respondents of such panels are "trained" in filling out questionnaires (although, according to the provider, protections are in place to avoid such a scenario) and, therefore, lead to biased data. In particular, reverse coded, negatively formulated items in general loaded low on the respective factors. The sample was further composed of participants working in a range of different companies and industries that might have reduced the possibility of engaging in eco-civic behaviors (Boiral & Paillé, 2012). It is imaginable that certain types of organizations do not offer any environmental activities in which participants can actually take part (which was also shown in the high number of participants indicating that there were no such options at their workplace). This problem should be addressed in future research studies measuring eco-civic engagement.

6.2 *Future Research*

More research is needed on the influence of EnvID in predicting EGB, as earlier research has pointed at a possible influence of past behaviors on identity (Smith et al., 2007), yet in a recent study, while controlling for past behavior, interaction effects rose with rising levels of identity (Carfora et al., 2017). However, in the light of the present study, this could be interpreted as behavioral spillover from private or public green activism to the workplace (Blazejewski et al., 2018), i.e., past green behavior outside the workplace strengthens workplace EnvID.

Future research could also deepen the study of organizational context on the unfolding and exertion of EnvID. In particular, in light of that EnvID is positively related to EGB and beyond (Whitmarsh & O'Neill, 2010), it is of great theoretical and practical interest to understand the mechanisms beyond the influence of EnvID in the work context. Although research on self-identity asserts that identity activation/salience is connected with positive emotions (Stryker, 1987), the contextual factors that lead to such an activation in the context of EGB are still to be explored.

7 Conclusions

Organizations are tasked with demands to achieve environmental sustainability, and this study provides an understanding of the role of EnvID in predicting EGB, and our results show that supervisory and organizational support plays a decisive role in EGB. A strong EnvID together with a high level of supervisory and organizational support also resulted in the strongest levels of EGB, while low levels of either PSS-E or EnvID can be balanced out by high levels of EnvID or PSS-E in moderating the relationship between POS-E and EGB.

Academic Contribution This chapter contributes to the literature on EGB and green HRM by addressing the call for more research on it from Renwick (2018), and Renwick et al. (2024), and extends it by branching out to explore employee green behavior in a relatively large sample of employees in Germany, which is less seen in the relevant literature.

References

Alt, E., & Spitzeck, H. (2016). Improving environmental performance through unit-level organizational citizenship behaviors for the environment: A capability perspective. *Journal of Environmental Management, 182*, 48–58. https://doi.org/10.1016/j.jenvman.2016.07.034

Andersson, L. M., & Bateman, T. S. (2000). Individual environmental initiative: Championing natural environmental issues in U.S. business organizations. *Academy of Management Journal, 43*(4), 548–570. https://doi.org/10.5465/1556355

Andersson, L. M., Jackson, S. E., & Russell, S. V. (2013). Greening organizational behavior: An introduction to the special issue. *Journal of Organizational Behavior, 34*(2), 151–155. https://doi.org/10.1002/job.1854

Baker, M. A., Davis, E. A., & Weaver, P. A. (2014). Eco-friendly attitudes, barriers to participation, and differences in behavior at green hotels. *Cornell Hospitality Quarterly, 55*(1), 89–99. https://doi.org/10.1177/1938965513504483

Bentler, P. M. (1990). Comparative fit indexes in structural models. *Psychological Bulletin, 107*(2), 238–246. https://doi.org/10.1037/0033-2909.107.2.238

Biddle, B. J., Bank, B. J., & Slavings, R. L. (1987). Norms, preferences, identities and retention decisions. *Social Psychology Quarterly, 50*(4), 322. https://doi.org/10.2307/2786817

Bissing-Olson, M. J., Iyer, A., Fielding, K. S., & Zacher, H. (2013). Relationships between daily affect and pro-environmental behavior at work: The moderating role of pro-environmental attitude. *Journal of Organizational Behavior, 34*(2), 156–175. https://doi.org/10.1002/job.1788

Blazejewski, S., Gräf, A., Buhl, A., & Dittmer, F. (2018). Enabling green spillover: How firms can benefit from employees' private green activism. In D. W. Renwick (Ed.), *Routledge research in sustainability and business. Contemporary developments in green human resource management research: Towards sustainability in action?* (pp. 73–94). Routledge.

Boiral, O. (2007). Corporate greening through ISO 14001: A rational myth? *Organization Science, 18*(1), 127–146. https://doi.org/10.1287/orsc.1060.0224

Boiral, O., & Paillé, P. (2012). Organizational citizenship behaviour for the environment: Measurement and validation. *Journal of Business Ethics, 109*(4), 431–445. https://doi.org/10.1007/s10551-011-1138-9

Brislin, R. (1990). Applied cross-cultural psychology: An introduction. In R. Brislin (Ed.), *Applied cross-cultural psychology* (pp. 9–33). SAGE Publications. https://doi.org/10.4135/9781483325392.n1

Brown, A. D. (2015). Identities and identity work in organizations. *International Journal of Management Reviews, 17*(1), 20–40. https://doi.org/10.1111/ijmr.12035

Brown, A. D. (2017). Identity work and organizational identification. *International Journal of Management Reviews, 19*(3), 296–317. https://doi.org/10.1111/ijmr.12152

Cantor, D. E., Morrow, P. C., & Montabon, F. (2012). Engagement in environmental behaviors among supply chain management employees: An organizational support theoretical perspective. *Journal of Supply Chain Management, 48*(3), 33–51. https://doi.org/10.1111/j.1745-493X.2011.03257.x

Carfora, V., Caso, D., Sparks, P., & Conner, M. (2017). Moderating effects of pro-environmental self-identity on pro-environmental intentions and behaviour: A multi-behaviour study. *Journal of Environmental Psychology, 53*, 92–99. https://doi.org/10.1016/j.jenvp.2017.07.001

Carmeli, A., Brammer, S., Gomes, E., & Tarba, S. Y. (2017). An organizational ethic of care and employee involvement in sustainability-related behaviors: A social identity perspective. *Journal of Organizational Behavior, 38*(9), 1380–1395. https://doi.org/10.1002/job.2185

Carollo, L., & Guerci, M. (2018). 'Activists in a suit': Paradoxes and metaphors in sustainability managers' identity work. *Journal of Business Ethics, 148*(2), 249–268. https://doi.org/10.1007/s10551-017-3582-7

Charng, H.-W., Piliavin, J. A., & Callero, P. L. (1988). Role identity and reasoned action in the prediction of repeated behavior. *Social Psychology Quarterly, 51*(4), 303. https://doi.org/10.2307/2786758

Cheng, Z., Liu, W., Zhou, K., Che, Y., & Han, Y. (2021). Promoting employees' pro-environmental behaviour through empowering leadership: The roles of psychological ownership, empowerment role identity, and environmental self-identity. *Business Ethics, the Environment & Responsibility, 30*(4), 604–618. https://doi.org/10.1111/beer.12366

Christensen, P. N., Rothgerber, H., Wood, W., & Matz, D. C. (2004). Social norms and identity relevance: A motivational approach to normative behavior. *Personality & Social Psychology Bulletin, 30*(10), 1295–1309. https://doi.org/10.1177/0146167204264480

Ciocirlan, C. E. (2016). Environmental workplace behaviors. *Organization & Environment, 30*(1), 1–20. https://doi.org/10.1177/1086026615628036

Ciocirlan, C. E., Gregory-Smith, D., Manika, D., & Wells, V. (2020). Using values, beliefs, and norms to predict conserving behaviors in organizations. *European Management Review, 17*(2), 543–558. https://doi.org/10.1111/emre.12388

Clayton, S. (2003). Environmental identity: A conceptual and an operational definition. In S. Clayton & S. Opotow (Eds.), *Identity and the natural environment: The psychological significance of nature* (pp. 45–65). MIT Press.

Daily, B. F., & Huang, S. (2001). Achieving sustainability through attention to human resource factors in environmental management. *International Journal of Operations & Production Management, 21*(12), 1539–1552. https://doi.org/10.1108/01443570110410892

Davis, M. C., Unsworth, K. L., Russell, S. V., & Galvan, J. J. (2020). Can green behaviors really be increased for all employees? Trade-offs for "deep greens" in a goal-oriented green human resource management intervention. *Business Strategy and the Environment, 29*(2), 335–346. https://doi.org/10.1002/bse.2367

Eisenberger, R., Huntington, R., Hutchison, S., & Sowa, D. (1986). Perceived organizational support. *The Journal of Applied Psychology, 71*(3), 500–507. https://doi.org/10.1037/0021-9010.71.3.500

Eisenberger, R., Stinglhamber, F., Vandenberghe, C., Sucharski, I. L., & Rhoades, L. (2002). Perceived supervisor support: Contributions to perceived organizational support and employee retention. *Journal of Applied Psychology, 87*(3), 565–573. https://doi.org/10.1037/0021-9010.87.3.565

Festinger, L. (1957). *A theory of cognitive dissonance*. Stanford Univ. Press.

Field, A. (2018). *Discovering statistics using IBM SPSS statistics* (5th ed.). SAGE edge. Sage.

Fielding, K. S., McDonald, R., & Louis, W. R. (2008). Theory of planned behaviour, identity and intentions to engage in environmental activism. *Journal of Environmental Psychology, 28*(4), 318–326. https://doi.org/10.1016/j.jenvp.2008.03.003

Francoeur, V., Paillé, P., Yuriev, A., & Boiral, O. (2021). The measurement of green workplace behaviors: A systematic review. *Organization & Environment, 34*(1), 18–42. https://doi.org/10.1177/1086026619837125

Gatersleben, B., Murtagh, N., & Abrahamse, W. (2014). Values, identity and pro-environmental behaviour. *Contemporary Social Science, 9*(4), 374–392. https://doi.org/10.1080/21582041.2012.682086

Ghadiri, D. P., Gond, J.-P., & Brès, L. (2015). Identity work of corporate social responsibility consultants: Managing discursively the tensions between profit and social responsibility. *Discourse & Communication, 9*(6), 593–624. https://doi.org/10.1177/1750481315600308

Graves, L. M., & Sarkis, J. (2018). The role of employees' leadership perceptions, values, and motivation in employees' provenvironmental behaviors. *Journal of Cleaner Production, 196*, 576–587. https://doi.org/10.1016/j.jclepro.2018.06.013

Graves, L. M., Sarkis, J., & Zhu, Q. (2013). How transformational leadership and employee motivation combine to predict employee proenvironmental behaviors in China. *Journal of Environmental Psychology, 35*, 81–91. https://doi.org/10.1016/j.jenvp.2013.05.002

Hair, J. F., Black, W. C., Babin, B. J., & Anderson, R. E. (2010). *Multivariate data analysis: A global perspective* (7th ed.). Pearson.

Hayes, A. F. (2013). *Introduction to mediation, moderation, and conditional process analysis: A regression-based approach. Methodology in the social sciences*. Guilford Press.

Kira, M., & Balkin, D. B. (2014). Interactions between work and identities: Thriving, withering, or redefining the self? *Human Resource Management Review, 24*(2), 131–143. https://doi.org/10.1016/j.hrmr.2013.10.001

Lalot, F., Quiamzade, A., Falomir-Pichastor, J. M., & Gollwitzer, P. M. (2019). When does self-identity predict intention to act green? A self-completion account relying on past behaviour and majority-minority support for pro-environmental values. *Journal of Environmental Psychology, 61*, 79–92. https://doi.org/10.1016/j.jenvp.2019.01.002

Lamm, E., Tosti-Kharas, J., & Williams, E. G. (2013). Read This Article, but Don't Print It. *Group & Organization Management, 38*(2), 163–197. https://doi.org/10.1177/1059601112475210

Lamm, E., Tosti-Kharas, J., & King, C. E. (2015). Empowering employee sustainability: Perceived organizational support toward the environment. *Journal of Business Ethics, 128*(1), 207–220. https://doi.org/10.1007/s10551-014-2093-z

Laverie, D. A., & Arnett, D. B. (2000). Factors affecting fan attendance: The influence of identity salience and satisfaction. *Journal of Leisure Research, 32*(2), 225–246. https://doi.org/10.1080/00222216.2000.11949915

Lo, S. H., Peters, G.-J. Y., & Kok, G. (2012). A review of determinants of and interventions for proenvironmental behaviors in organizations. *Journal of Applied Social Psychology, 42*(12), 2933–2967. https://doi.org/10.1111/j.1559-1816.2012.00969.x

Manika, D., Wells, V. K., Gregory-Smith, D., & Gentry, M. (2015). The impact of individual attitudinal and organisational variables on workplace environmentally friendly behaviours. *Journal of Business Ethics, 126*(4), 663–684. https://doi.org/10.1007/s10551-013-1978-6

Marsh, H. W., Hau, K.-T., & Wen, Z. (2004). In search of golden rules: Comment on hypothesis-testing approaches to setting cutoff values for fit indexes and dangers in overgeneralizing Hu and Bentler's (1999) findings. *Structural Equation Modeling: A Multidisciplinary Journal, 11*(3), 320–341. https://doi.org/10.1207/s15328007sem1103_2

Muster, V., & Schrader, U. (2011). Green work-life balance: A new perspective for green HRM. *German Journal of Human Resource Management, 25*(2), 140–156. https://doi.org/10.1177/239700221102500205

Nigbur, D., Lyons, E., & Uzzell, D. (2010). Attitudes, norms, identity and environmental behaviour: Using an expanded theory of planned behaviour to predict participation in a kerbside recycling programme. *The British Journal of Social Psychology, 49*(Pt 2), 259–284. https://doi.org/10.1348/014466609X449395

Norton, T. A., Parker, S. L., Zacher, H., & Ashkanasy, N. M. (2015). Employee green behavior: A theoretical framework, multilevel review, and future research agenda. *Organization & Environment, 28*(1), 103–125. https://doi.org/10.1177/1086026615575773

Ones, D. S., & Dilchert, S. (2012). Employee green behaviors. In S. E. Jackson, D. S. Ones, & S. Dilchert (Eds.), *Managing human resources for environmental sustainability* (pp. 85–116). John Wiley & Sons.

Paillé, P. (2020). Greening the workplace through employees: An integrative model. In *Greening the workplace* (Vol. 1). Palgrave Pivot. https://doi.org/10.1007/978-3-030-58388-0_8

Paillé, P., & Boiral, O. (2013). Pro-environmental behavior at work: Construct validity and determinants. *Journal of Environmental Psychology, 36*, 118–128. https://doi.org/10.1016/j.jenvp.2013.07.014

Paillé, P., Mejía Morelos, J. H., Raineri, N., & Stinglhamber, F. (2019). The influence of the immediate manager on the avoidance of non-green behaviors in the workplace: A three-wave moderated-mediation model. *Journal of Business Ethics, 155*(3), 723–740. https://doi.org/10.1007/s10551-017-3519-1

Paillé, P., & Mejía-Morelos, J. H. (2014). Antecedents of pro-environmental behaviours at work: The moderating influence of psychological contract breach. *Journal of Environmental Psychology, 38*, 124–131. https://doi.org/10.1016/j.jenvp.2014.01.004

Paillé, P., Mejía-Morelos, J. H., Amara, N., & Norrin, H. (2022). Greening the workplace through supervisory behaviors: Assessing what really matters to employees. *The International Journal of Human Resource Management, 33*, 1754–1781. https://doi.org/10.1080/09585192.2020.1819857

Paillé, P., & Raineri, N. (2015). Linking perceived corporate environmental policies and employees eco-initiatives: The influence of perceived organizational support and psychological contract breach. *Journal of Business Research, 68*(11), 2404–2411.

Podsakoff, P. M., MacKenzie, S. B., Lee, J.-Y., & Podsakoff, N. P. (2003). Common method biases in behavioral research: A critical review of the literature and recommended remedies. *The Journal of Applied Psychology, 88*(5), 879–903. https://doi.org/10.1037/0021-9010.88.5.879

Pratt, M. G., Rockmann, K. W., & Kaufmann, J. B. (2006). Constructing professional identity: The role of work and identity learning cycles in the customization of identity among medical residents. *Academy of Management Journal, 49*(2), 235–262. https://doi.org/10.5465/amj.2006.20786060

Raineri, N., Mejía-Morelos, J. H., Francoeur, V., & Paillé, P. (2016). Employee eco-initiatives and the workplace social exchange network. *European Management Journal, 34*(1), 47–58. https://doi.org/10.1016/j.emj.2015.10.006

Raineri, N., & Paillé, P. (2016). Linking corporate policy and supervisory support with environmental citizenship behaviors: The role of employee environmental beliefs and commitment. *Journal of Business Ethics, 137*(1), 129–148. https://doi.org/10.1007/s10551-015-2548-x

Ramus, C. A. (2001). Organizational support for employees: Encouraging creative ideas for environmental sustainability. *California Management Review, 43*(3), 85–105. https://doi.org/10.2307/41166090

Ramus, C. A., & Killmer, A. B. C. (2007). Corporate greening through prosocial extrarole behaviours – A conceptual framework for employee motivation. *Business Strategy and the Environment, 16*(8), 554–570. https://doi.org/10.1002/bse.504

Ramus, C. A., & Steger, U. (2000). The roles of supervisory support behaviors and environmental policy in employee "ecoinitiatives" at leading-edge European companies. *Academy of Management Journal, 43*(4), 605–626. https://doi.org/10.5465/1556357

Renwick, D. W. (2018). From green HRM towards workforce sustainability? In D. W. Renwick (Ed.), *Routledge research in sustainability and business. Contemporary developments in green human resource management research: Towards sustainability in action?* (pp. 173–190). Routledge.

Renwick, D. W., Clark, I., & Crichton, R. (2024). The limitations of individual analysis and alternative framing of employee green behaviour. In M. Y. Yusliza & D. Renwick (Eds.), *Green human resource management* (pp. 3–19). Springer. https://doi.org/10.1007/978-981-99-7104-6_1

Renwick, D. W., Redman, T., & Maguire, S. (2013). Green human resource management: A review and research agenda. *International Journal of Management Reviews, 15*(1), 1–14. https://doi.org/10.1111/j.1468-2370.2011.00328.x

Rhoades, L., Eisenberger, R., & Armeli, S. (2001). Affective commitment to the organization: The contribution of perceived organizational support. *The Journal of Applied Psychology, 86*(5), 825–836. https://doi.org/10.1037/0021-9010.86.5.825

Robertson, J. L., & Barling, J. (2013). Greening organizations through leaders' influence on employees' pro-environmental behaviors. *Journal of Organizational Behavior, 34*(2), 176–194. https://doi.org/10.1002/job.1820

Ruepert, A., Keizer, K., Steg, L., Maricchiolo, F., Carrus, G., Dumitru, A., García Mira, R., Stancu, A., & Moza, D. (2016). Environmental considerations in the organizational context: A pathway to pro-environmental behaviour at work. *Energy Research & Social Science, 17*, 59–70. https://doi.org/10.1016/j.erss.2016.04.004

Schmitt, M. T., Mackay, C. M., Droogendyk, L. M., & Payne, D. (2019). What predicts environmental activism? The roles of identification with nature and politicized environmental identity. *Journal of Environmental Psychology, 61*, 20–29. https://doi.org/10.1016/j.jenvp.2018.11.003

Smith, J. R., Terry, D. J., Manstead, A. S. R., Louis, W. R., Kotterman, D., & Wolfs, J. (2007). Interaction effects in the theory of planned behavior: The interplay of self-identity and past behavior. *Journal of Applied Social Psychology, 37*(11), 2726–2750. https://doi.org/10.1111/j.1559-1816.2007.00278.x

Sparks, P., & Shepherd, R. (1992). Self-identity and the theory of planned behavior: Assesing the role of identification with "green consumerism". *Social Psychology Quarterly, 55*(4), 388. https://doi.org/10.2307/2786955

Starik, M., & Kanashiro, P. (2020). Advancing a multi-level sustainability management theory. In D. M. Wasieleski & J. Weber (Eds.), *Business and society 360: Volume 4. Sustainability* (Vol. 34, 1st ed., pp. 17–42). Emerald Publishing. https://doi.org/10.1108/S2514-175920200000004003

Stets, J. E. (2005). Examining emotions in identity theory. *Social Psychology Quarterly, 68*(1), 39–56. https://doi.org/10.1177/019027250506800104

Stets, J. E., & Biga, C. (2003). Bringing identity theory into environmental sociology. *Sociological Theory, 21*(4), 398–423. https://doi.org/10.1046/j.1467-9558.2003.00196.x

Stets, J. E., & Burke, P. J. (2000). Identity theory and social identity theory. *Social Psychology Quarterly, 63*(3), 224. https://doi.org/10.2307/2695870

Stryker, S. (1987). Identity theory: Developments and extensions. In K. Yardley & T. Honess (Eds.), *Self and identity: Psychosocial perspectives* (pp. 89–103). Wiley.

Stryker, S., & Serpe, R. T. (1994). Identity salience and psychological centrality: Equivalent, overlapping, or complementary concepts? *Social Psychology Quarterly, 57*(1), 16. https://doi.org/10.2307/2786972

Temminck, E., Mearns, K., & Fruhen, L. (2015). Motivating employees towards sustainable behaviour. *Business Strategy and the Environment, 24*(6), 402–412. https://doi.org/10.1002/bse.1827

Testa, F., Iraldo, F., & Daddi, T. (2017). The effectiveness of EMAS as a management tool: A key role for the internalization of environmental practices. *Organization & Environment, 31*(1), 48–69. https://doi.org/10.1177/1086026616687609

Testa, F., Rizzi, F., Daddi, T., Gusmerotti, N. M., Frey, M., & Iraldo, F. (2014). EMAS and ISO 14001: The differences in effectively improving environmental performance. *Journal of Cleaner Production, 68*, 165–173. https://doi.org/10.1016/j.jclepro.2013.12.061

Tudor, T. L., Barr, S. W., & Gilg, A. W. (2007). Strategies for improving recycling behaviour within the Cornwall National Health Service (NHS) in the UK. *Waste Management & Research: The Journal of the International Solid Wastes and Public Cleansing Association, ISWA, 25*(6), 510–516. https://doi.org/10.1177/0734242X07082030

United Nations. (2021). *The heat is on: A world of climate promises not yet delivered* [Emissions gap report 2021]. Retrieved from https://www.unep.org/resources/emissions-gap-report-2021

van der Werff, E., Steg, L., & Keizer, K. (2013). The value of environmental self-identity: The relationship between biospheric values, environmental self-identity and environmental preferences, intentions and behaviour. *Journal of Environmental Psychology, 34*, 55–63. https://doi.org/10.1016/j.jenvp.2012.12.006

van der Werff, E., Steg, L., & Ruepert, A. (2021). My company is green, so am I: The relationship between perceived environmental responsibility of organisations and government, environmental self-identity, and pro-environmental behaviours. *Energy Efficiency, 14*(5). https://doi.org/10.1007/s12053-021-09958-9

Wagner, M. (2011). Environmental management activities and sustainable HRM in German manufacturing firms – Incidence, determinants, and outcomes. *German Journal of Human Resource Management: Zeitschrift Für Personalforschung, 25*(2), 157–177. https://doi.org/10.1177/239700221102500206

Whitmarsh, L. E., & O'Neill, S. (2010). Green identity, green living? The role of pro-environmental self-identity in determining consistency across diverse pro-environmental behaviours. *Journal of Environmental Psychology, 30*(3), 305–314. https://doi.org/10.1016/j.jenvp.2010.01.003

Wright, C., Nyberg, D., & Grant, D. (2012). "Hippies on the third floor": Climate change, narrative identity and the micro-politics of corporate environmentalism. *Organization Studies, 33*(11), 1451–1475. https://doi.org/10.1177/0170840612463316

Yoon, J., & Thye, S. (2000). Supervisor support in the work place: Legitimacy and positive affectivity. *The Journal of Social Psychology, 140*(3), 295–316. https://doi.org/10.1080/00224540009600472

Cultivating Environmental Champions: Exploring the Influence of Environmental, Social, and Governance (ESG) Principles on Employee Green Behavior

Jing Yi Yong, Woon Leong Lin, Nelvin XeChung Leow, Chuen Khee Pek, and Kelvin Yong Ming Lee

Abstract In the quest for sustainable organizational practices, the cultivation of environmental champions among employees emerges as a pivotal strategy. This conceptual chapter endeavors to elucidate the intricate interplay between environmental, social, and governance (ESG) principles and the development of employee green behavior within organizational contexts. Drawing on Social Cognitive Theory, it identifies how ESG principles shape employees' actions, with leadership and organizational culture playing central roles in fostering environmental stewardship. The chapter offers a theoretical framework illustrating how ESG principles can drive actionable employee behaviors and provides practical insights for enhancing engagement in sustainability initiatives, fostering a workforce committed to environmental stewardship. It argues that organizations with strong governance structures and clear sustainability goals are better equipped to create environments conducive to green behavior. Additionally, it emphasizes the mediating effects of leadership and organizational culture, positioning leaders as key drivers in modeling sustainable behaviors and embedding a culture of sustainability. The discussion provides actionable insights for practitioners, focusing on the operationalization of ESG principles to enhance employee engagement in sustainability initiatives, thereby fostering a workforce committed to environmental stewardship.

J. Y. Yong (✉) · W. L. Lin · N. X. Leow · C. K. Pek
School of Management and Marketing, Faculty of Business and Law, Taylor's University, Subang Jaya, Selangor, Malaysia
e-mail: serene.yong@taylors.edu.my; woonleong.lin@taylors.edu.my; nelvinxechung.leow@taylors.edu.my; eugene.pek@taylors.edu.my

K. Y. M. Lee
School of Accounting and Finance, Faculty of Business and Law, Taylor's University, Subang Jaya, Selangor, Malaysia
e-mail: kelvinyongming.lee@taylors.edu.my

M. Y. Yusliza et al. (eds.), *Employee Green Behavior*,
https://doi.org/10.1007/978-981-95-1422-9_4

1 Introduction

In recent years, sustainability has evolved from a peripheral concern to a central priority for businesses and global economies alike. Growing concerns about climate change, resource depletion, and social inequality are reshaping how corporate success is defined. Today's businesses are no longer focused solely on maximizing profits but are also expected to contribute to environmental conservation, social progress, and ethical governance (Aguilera et al., 2021; Rajesh & Rajendran, 2020). This shift is partly driven by increasing consumer awareness (Boufounou et al., 2023), investor expectations (Lei & Yu, 2024), and regulatory demands (Deng et al., 2024), all of which highlight the need for long-term resilience through sustainable practices.

As pressures to address global challenges increase, companies are reevaluating their strategies to integrate sustainability into their core business models. This move aligns with global initiatives such as the Paris Agreement and the United Nations' Sustainable Development Goals (SDGs). Sustainability is no longer viewed as a mere compliance exercise but as a strategic imperative to drive innovation (Santa-Maria et al., 2022) and operational efficiency (Kumar et al., 2021). For sustainability efforts to succeed, they must be deeply embedded throughout the organization, including at the employee level.

In response to the evolving sustainability landscape, environmental, social, and governance (ESG) principles have become essential for addressing these challenges. Businesses are increasingly aware that long-term success depends on embedding environmental protection, social welfare, and ethical governance into their core strategies. This approach not only meets stakeholder expectations but also contributes to broader societal and environmental goals (MacNeil & Esser, 2022). Research shows that implementing ESG practices can enhance corporate financial performance (Chen et al., 2023) and improve stock liquidity (Meng-tao et al., 2023), highlighting the increasing stakeholder interest in how businesses manage these elements.

While much of the attention has historically been on the external impact of organizations, there is an increasing recognition of the internal dynamics that contribute to sustainability, particularly the role of employees. In recent years, the concept of the environmental champion has gained significant attention as organizations strive to encourage more employee involvement in sustainability efforts (Lasrado & Zakaria, 2020; Murphy et al., 2020). According to Andersson and Bateman (2000), environmental champions are characterized by their voluntary, discretionary actions aimed at improving environmental outcomes. They often act as change agents who facilitate the integration of environmental practices into organizational strategies by mobilizing resources, raising awareness, and advocating for greener business practices. These individuals do more than comply with sustainability policies; they actively advocate for, implement, and lead initiatives that align with the organization's ESG goals. However, cultivating such champions is not merely about hiring environmentally conscious individuals; it requires creating a supportive organizational environment that motivates all employees to adopt green behaviors.

The challenge arises when, despite organizations adopting ESG frameworks, many struggle to translate these principles into active, sustained employee engagement (Huang, 2022). While some employees enthusiastically advocate for green practices, others tend to be more passive (Pham et al., 2019), which can prevent ESG initiatives from reaching their full potential. In addition, although business models are often viewed as essential for sustainable entrepreneurship and innovations, there is a lack of understanding regarding the mechanisms that integrate these elements (Lüdeke-Freund, 2020). This gap is evident in the integration of ESG principles into organizational practices, where the lack of clear mechanisms hinders effective implementation and limits employee engagement in sustainability initiatives.

This chapter aims to explore the interplay between ESG principles and employee green behavior (EGB) within organizational contexts. Recognizing that sustainability begins from within, it seeks to identify the mechanisms and dynamics that enable the effective integration of ESG principles into organizational practices, thereby enhancing employee engagement and fostering environmental champions. Employees often serve as the bridge between corporate ESG strategies and their execution, making the roles of leadership and organizational culture pivotal in shaping how they engage with sustainability efforts. This process involves change management, which actively engages employees in the transition, building their commitment, reducing resistance, and increasing the likelihood of successful implementation (Dathe et al., 2024).

Through the lens of social cognitive theory (SCT), this chapter examines how employees internalize ESG principles and the roles of leadership and organizational culture as mediators in shaping EGB. It highlights how organizations can cultivate a culture of sustainability that empowers all employees to actively participate in and advocate for environmental initiatives. Ultimately, this chapter proposes a conceptual framework that offers actionable insights for organizations aiming to enhance EGB behavior through the effective integration of ESG principles.

2 Theoretical Foundation – Social Cognitive Theory

Social cognitive theory (SCT), developed by Albert Bandura and his colleagues (Bandura, 1986, 2001; Bandura et al., 2003; Bussey & Bandura, 1999), offers an insightful approach to understanding how people learn and behave within social environments, particularly within organizations. According to SCT, behavior is shaped by an ongoing interaction among personal characteristics, the surrounding environment, and behavioral responses. This relationship is known as reciprocal determinism (Bandura, 1986). In this view, individuals actively process information from their surroundings, learning through observation and experience. The principles of SCT are especially relevant when examining how employees' behavior may be influenced by the implementation of ESG principles within an organization.

One key idea in SCT is observational learning, where individuals imitate the behaviors of others. According to Bandura (1974), this learning occurs naturally as people witness how others' actions result in various outcomes in social environments. Through observation, people acquire attitudes, values, and behavior patterns (Bandura, 2008). In the ESG context, leaders act as role models for sustainable behaviors. For example, when employees observe leaders engaging in environmentally friendly practices, such as reducing energy consumption or promoting recycling, they are more likely to adopt similar actions. This effect is amplified when employees perceive that such behaviors are rewarded or recognized (Bandura, 1986), thus reinforcing their commitment to ESG values.

SCT also highlights the role of personal factors such as self-efficacy, or an individual's belief in their ability to complete tasks and reach specific goals (Bandura, 1986, 1997a, b). Employees who feel capable of making meaningful contributions to sustainability are more likely to engage in behaviors aligned with these goals. This motivation increases when they foresee positive outcomes from their efforts, like improved organizational reputation or personal satisfaction. Schutte and Bhullar (2017) found that individuals with higher self-efficacy and belief in their capacity to change their sustainability behaviors demonstrated greater motivation and more frequent engagement in such actions. Hence, the present study suggests that in organizations prioritizing ESG principles, employees with strong self-efficacy are particularly driven to adopt environmentally friendly behaviors, as they trust that their actions will significantly advance sustainability objectives.

Lastly, environmental factors in SCT highlight the role of organizational settings in shaping behavior through reinforcement. Individuals are more likely to repeat actions that are reinforced, whether through direct rewards, observing others being rewarded (vicarious reinforcement), or by self-rewarding for achieving personal goals. Positive reinforcement, which offers rewards, and negative reinforcement, which eliminates undesirable outcomes, both effectively motivate behavior (Bandura, 1986; Schunk, 2019). Leeuwis et al. (2022) further illustrate that using rewards or punishments activates brain structures linked to the reward system, thereby encouraging engagement in climate change action. This study posits that in organizational settings, reinforcement mechanisms are pivotal in fostering sustainable behaviors. By aligning individual actions with broader ESG objectives, these reinforcement strategies can drive the widespread adoption of environmentally friendly practices within the workplace.

3 Literature Review

3.1 Employee Green Behavior

Employee green behavior (EGB) has gained significant attention as organizations strive to enhance sustainability efforts and reduce their environmental impact. Ones and Dilchert (2012) were the first to introduce the concept, defining EGB as environmentally beneficial actions within organizations. They identified five types of EGB: working sustainably, conserving resources, influencing others, taking initiative, and avoiding harm. However, these categories can sometimes overlap, as individual behaviors may fall into more than one category due to vague or broad definitions (Norton et al., 2015). To address this, Norton et al. (2015) simplified the concept into two dimensions: required EGB, which includes task-related behaviors aligned with job duties, and voluntary EGB, which refers to discretionary actions that go beyond expectations, such as double-sided printing and encouraging colleagues to adopt green practices.

Moreover, Tian et al. (2020) emphasize that EGB plays a vital role in translating an organization's broader sustainability strategies into practical actions within the workplace. It encompasses both in-role behaviors, such as adhering to environmental policies, and extra-role behaviors, which involve proactive, voluntary efforts. Employees not only follow established guidelines but also take the initiative in environmental stewardship, such as reducing energy consumption and identifying opportunities to conserve resources (Ababneh, 2021; Dumont et al., 2017). This blend of required and voluntary actions makes EGB an essential component in fostering a sustainable workplace culture.

EGB has been shaped by various determinants identified in the literature. Numerous studies have established key factors that influence EGB, including organizational support, individual attitudes, and external motivations. For instance, Fawehinmi et al. (2020) highlighted the impact of green human resource management (HRM) on EGB, underscoring how environmental information can mediate this relationship. Similarly, Salvador and Burciaga (2020) found a positive association between an organization's internal environmental orientation and employee recycling behaviors, emphasizing the role of management in fostering these practices. Furthermore, elements such as green self-identity, personal norms, and environmental attitudes have been shown to significantly influence individuals' ecological actions (Arshad et al., 2021; Banwo & Du, 2019; Hansmann et al., 2020).

Despite the robust body of research identifying these factors, the introduction of ESG principles represents a crucial new dimension influencing EGB. ESG frameworks encourage organizations to align their operations with sustainability goals, thereby creating a conducive environment for employees to engage in green practices. Research indicates that CSR activities not only catalyze organizational commitment but also positively affect employees' environmental behaviors (AlSuwaidi et al., 2021; Shah et al., 2021). By integrating ESG principles, organizations can enhance the motivational landscape for EGB, promoting both required and voluntary green behaviors. This alignment between personal values and organizational missions fosters a culture where employees are more likely to embrace environmentally friendly practices, ultimately bridging the gap between individual actions and organizational sustainability efforts.

3.2 ESG Principles and Employee Green Behavior

Organizations are increasingly incorporating ESG principles into their operational frameworks. The United Nations Principles for Responsible Investment highlight ESG through three key dimensions: (1) the environment dimension, which includes climate change, resource depletion, waste management, pollution and deforestation; (2) the social dimension, which addresses working conditions, health and safety, employee relations, and human rights; and (3) the governance dimension, which encompass issues like bribery and corruption, board diversity and structure, executive compensation, and tax strategies (United Nations Global Compact, 2021). While ESG principles are often viewed as external strategies reflecting a company's engagement with stakeholders, their internal application can effectively influence employee behavior.

3.3 Environmental Principles and Employee Green Behavior

Environmental principles within the ESG framework directly address the ecological impact of an organization's operations. These principles are designed to minimize environmental harm through measures like resource efficiency, pollution reduction, and the adoption of renewable energy sources (Zhou & Liu, 2023). Employees are more likely to engage in these efforts when they perceive that the organization is genuinely committed to sustainability, rather than merely pursuing it as a form of greenwashing (Tahir et al., 2020). When employees are aware of and engaged in these environmental goals, they become active participants in reducing the organization's ecological footprint.

Employees participating in initiatives like recycling programs or carpooling often develop a stronger sense of ownership over environmental outcomes. This effect is heightened when they can observe tangible results, such as lower energy consumption or reductions in the company's carbon footprint (Yuriev & Sierra-Barón, 2020). These experiences can positively reinforce their pro-environmental behavior and foster the emergence of environmental champions (Ali et al., 2023). Furthermore, environmental education and training are essential in fostering employee engagement. Organizations that offer training programs to enhance employees' understanding of environmental issues and demonstrate how their actions contribute to sustainability significantly boost motivation for green behavior

(Ababneh, 2021). Drawing from the insights gained through the literature review, this study puts forth the following proposition:

Proposition 1 The adoption of environmental principles within the ESG framework has a positive effect on EGB.

3.4 Social Principles and Employee Green Behavior

The social dimension of ESG focuses on the well-being of employees, customers, and communities. For employees, this includes creating a supportive work environment that promotes diversity, equity, inclusion, and fair treatment (Torres et al., 2023). While these issues may seem distinct from environmental concerns, Ahmed et al. (2020) demonstrate a link between employee welfare and their willingness to engage in environmental initiatives. The authors argued that organizations should prioritize meeting employees' psychological needs by cultivating a positive work environment and providing opportunities for skill development. This approach helps employees feel a deeper connection to the organization, motivating them to actively engage in sustainable green behaviors.

A culture of social responsibility can enhance employee engagement in sustainability efforts by creating an environment where employees feel valued and empowered. For instance, companies that provide flexible work options and promote a healthy work–life balance tend to foster higher levels of job satisfaction, which in turn encourages employees to go above and beyond in their work, including participating in environmental initiatives (Paillé et al., 2013). Building upon the findings of the literature review, this study advances the following proposition:

Proposition 2 The adoption of social principles within the ESG framework has a positive effect on EGB.

3.5 Governance Principles and Employee Green Behavior

Finally, the governance dimension, which refers to corporate transparency, ethical leadership, and accountability (Khaireddine et al., 2020), plays a pivotal role in building trust between employees and management. When employees trust their leaders and believe in the governance structures in place, they are more inclined to engage in behaviors that support organizational goals, including environmental sustainability (Dey et al., 2022). This trust motivates employees to adopt green practices, knowing that their leaders are dedicated to upholding environmental commitments.

In addition, organizations with robust governance structures are more likely to set clear sustainability goals and metrics, which can be integrated into employee performance evaluations. This transparency ensures that employees understand the environmental objectives the organization is aiming to achieve and how their individual roles contribute to these outcomes (Eccles et al., 2014). When sustainability metrics are tied to performance reviews, and rewards like bonuses or other performance-related incentives are linked to sustainability achievements, employees are often more motivated to engage actively in these initiatives (Marrucci et al., 2024). In light of the reviewed literature, this study proposes the following:

Proposition 3 The adoption of governance principles within the ESG framework has a positive effect on EGB.

3.6 The Mediating Role of Leadership

This study asserts that leadership plays a crucial mediating role in translating ESG principles into EGB by influencing the organizational climate and embedding sustainability as a fundamental value. While leadership is frequently regarded as a direct factor influencing EGB (Ashraf et al., 2024; Nawaz Khan, 2023), its primary impact stems from its capacity to facilitate the incorporation of ESG principles across various organizational levels. Leaders serve as vital conduits for these principles, effectively modeling and integrating them within the organization's culture. Research by Aung and Hallinger (2023) indicates that sustainability leadership in higher education can significantly alter institutional practices, strengthen stakeholder commitment, and promote long-term resilience in society. These findings underline the notion that leadership mediates the link between ESG principles and EGB, translating sustainability objectives into practical actions.

Particularly notable is the role of transformational leadership in motivating employees to adopt sustainable practices. By clearly expressing the organization's green vision and fostering an uplifting environment, transformational leaders encourage employees to align their efforts with the larger goals of the organization and society (Çop et al., 2021). When leaders actively support green initiatives, they set a powerful example that inspires others to synchronize their actions with the organization's sustainability objectives (Al-Swidi et al., 2021). Furthermore, leaders bolster pro-environmental behaviors by serving as role models; employees are more inclined to engage in sustainable actions when they see their leaders practicing these behaviors. Additionally, incorporating sustainability targets into performance appraisal enhances motivation by helping employees understand the connection between their efforts and the organization's success (Amjad et al., 2021).

In view of the literature support given above, effective leadership not only directs employees toward sustainable practices but also acts as a mediator in the relationship between ESG principles and EGB. Through the demonstration of sustainable behaviors, open communication, and the alignment of rewards with environmental objectives, leaders foster an organizational culture that not only promotes but also anticipates green behavior, thereby serving as a key component in the successful realization of EGB within the organization. Therefore, this study proposes the possible mediating effect of leadership on the relationship between ESG principles and EGB:

Proposition 4 Leadership positively mediates the relationship between environmental principles within the ESG framework and EGB.

Proposition 5 Leadership positively mediates the relationship between social principles within the ESG framework and EGB.

Proposition 6 Leadership positively mediates the relationship between governance principles within the ESG framework and EGB.

3.7 The Mediating Role of Organizational Culture

Organizational culture is suggested as another key mediator in the relationship between ESG principles and EGB. A culture focused on sustainability fosters an environment that supports eco-friendly practices within an organization. According to Muisyo et al. (2021), individuals are influenced by the norms and values prevalent in their environment, and an organizational culture that prioritizes environmental responsibility can encourage employees to adopt and maintain green behaviors. By embedding environmental values into the organizational culture, companies can inspire employees to innovate green solutions, reduce carbon emissions, minimize waste, and promote eco-friendly practices (Xiaoyi et al., 2023).

A strong organizational culture that integrates ESG principles can guide employees' perceptions of which behaviors are expected and rewarded. When sustainability becomes a core cultural value, employees are more inclined to adopt and incorporate these behaviors into their everyday work (Renwick et al., 2016). Additionally, by fostering transparent communication around sustainability and creating avenues for employee participation in environmental initiatives, organizations establish a culture that emphasizes relevant training and effective appraisal systems. These approaches nurture a collaborative environment that promotes continuous learning and strengthens the reinforcement of sustainable behavior (Dumont et al., 2017).

Building on the aforementioned literature, organizational culture emerges as a crucial mediator between ESG principles and EGB. Organizations that cultivate environmental champions often establish a culture of stewardship where sustainability becomes integral to their identity. This is achieved through consistent communication of sustainability goals, celebrating green achievements, and fostering a sense of collective responsibility for environmental impact. Consequently, organizational culture not only reinforces individual commitment to sustainability but also translates ESG principles into actionable employee behaviors. Thus, it serves as a conduit, mediating the relationship between ESG principles and EGB. Therefore, this study posits that organizational culture mediates the link between ESG principles and EGB.

Proposition 7 Organizational culture positively mediates the relationship between environmental principles within the ESG framework and EGB.

Proposition 8 Organizational culture positively mediates the relationship between social principles within the ESG framework and EGB.

Proposition 9 Organizational culture positively mediates the relationship between governance principles within the ESG framework and EGB.

4 Proposed Theoretical Framework

This study introduces a theoretical framework that illustrates the relationship between ESG principles as the independent variable and EGB as the dependent variable, with leadership and organizational culture acting as mediators, as depicted in Fig. 1. Drawing on Social Cognitive Theory, which emphasizes the interplay between individual behaviors and the social environment, this framework suggests that effective leadership can influence employees' perceptions of sustainability, guiding their behaviors toward eco-friendly practices. Additionally, a strong organizational culture that embeds ESG principles can create a supportive environment for these behaviors to flourish.

Organizations operationalize ESG principles to shape EGB by embedding these principles into their core practices. Environmental principles encourage resource

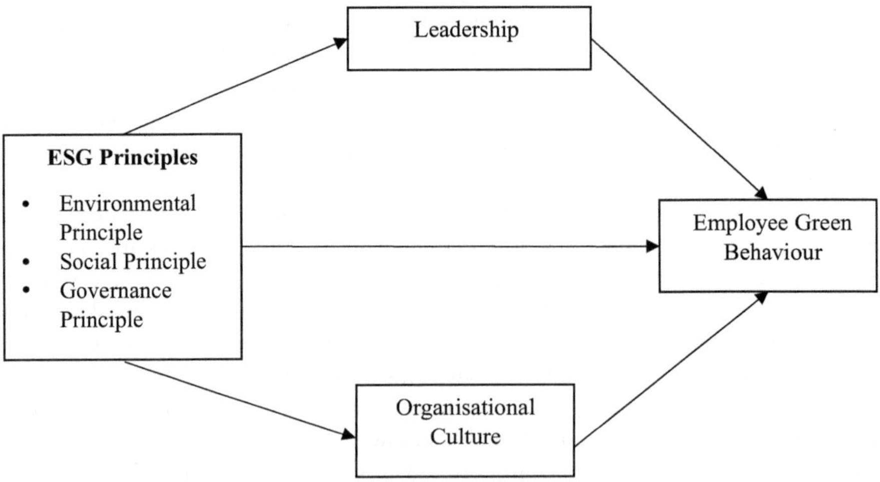

Fig. 1 Proposed theoretical framework. Source: Author's own work

conservation and renewable energy use, leading to specific behaviors like waste reduction and participation in recycling programs. Social principles, focusing on diversity and inclusion, create positive workplace dynamics that motivate employees to engage in voluntary pro-environmental actions. Governance principles, tied to transparent sustainability metrics and incentives, drive accountability and reinforce desired behaviors.

Leadership and organizational culture amplify these effects. Leaders act as role models, demonstrating and rewarding green behaviors that align with ESG principles. Organizational culture fosters collective responsibility by embedding sustainability values into shared norms and practices. For instance, green committees can champion ESG initiatives, and celebrating milestones like energy-saving achievements helps connect ESG goals to tangible outcomes. By integrating these pathways, the framework underscores the importance of a holistic approach to fostering EGB, ultimately contributing to a sustainable future.

5 Theoretical Implications

This chapter extends Social Cognitive Theory (SCT) by illustrating how ESG principles shape EGB through observational learning, reinforcement, and self-efficacy. Unlike prior ESG research that focuses on external pressures, this study highlights the internal behavioral mechanisms that drive sustainability engagement.

A key contribution is the mediating role of leadership and organizational culture, demonstrating how leaders act as role models and how a sustainability oriented culture reinforces ESG-driven behaviors. This expands SCT by showing how macro-level governance structures influence micro-level employee actions. Additionally, this study repositions ESG principles as behavioral enablers, rather than just corporate reporting tools, offering a fresh perspective on their impact within organizations.

By bridging corporate sustainability and behavioral psychology, this chapter provides a conceptual foundation for future empirical studies examining how ESG integration fosters long-term pro-environmental behaviors in the workplace.

6 Practical Implications

Building on these theoretical insights, organizations can operationalize ESG principles by embedding sustainability into daily operations. Training and awareness programs are essential, offering ESG-centric workshops that emphasize how individual actions contribute to overarching sustainability goals. Incentive systems should also be developed, rewarding employees who consistently demonstrate green behaviors, such as offering bonuses for waste reduction or resource conservation. Leadership involvement is crucial; equipping leaders with tools to model sustainable behaviors ensures they act as role models for employees. Collaborative platforms, such as cross-functional teams or green committees, can be established to foster idea-sharing and innovation in sustainability practices. Additionally, performance evaluations should be tied to sustainability metrics, ensuring that employee performance is aligned with organizational ESG objectives. By embedding these strategies into organizational structures, practitioners can ensure that ESG principles transition from conceptual frameworks into actionable initiatives, driving meaningful environmental outcomes.

7 Future Directions

Future investigations should aim to empirically test the proposed theoretical framework, with a particular focus on how leadership and organizational culture influence the translation of ESG principles into EGB. It would be beneficial to examine the dynamics between different organizational cultures, such as innovative versus traditional settings, to better understand their effects on engaging employees in sustainability initiatives. Longitudinal studies would also be valuable in revealing how changes in leadership and organizational culture over time impact green employee behaviors, offering essential insights for practitioners.

Moreover, expanding research to include diverse industries and geographic regions would enhance our comprehension of the factors that shape EGB across different contexts. By addressing these areas, future research can contribute to the development of more effective strategies for fostering sustainable practices within organizations, ultimately improving both organizational performance and environmental outcomes.

8 Conclusion

This chapter identifies the critical role of ESG principles in fostering EGB. It posits that through effective leadership and a sustainability focused organizational culture, companies can empower employees to act as environmental champions. The discussion highlights practical strategies, such as integrating ESG into training, performance reviews, and incentives, to ensure these principles are not just theoretical but actionable. Ultimately, embedding ESG into organizational identity drives sustainability, enhances employee engagement, and contributes to a resilient, eco-conscious future.

Acknowledgments This research is supported by Fundamental Research Grant Scheme (FRGS) from the Ministry of Higher Education Malaysia [FRGS/1/2022/SS01/TAYLOR/02/4].

References

Ababneh, O. M. A. (2021). How do green HRM practices affect employees' green behaviors? The role of employee engagement and personality attributes. *Journal of Environmental Planning and Management, 64*(7), 1204–1226.

Aguilera, R. V., Aragón-Correa, J. A., Marano, V., & Tashman, P. A. (2021). The corporate governance of environmental sustainability: A review and proposal for more integrated research. *Journal of Management, 47*(6), 1468–1497.

Ahmed, M., Zehou, S., Raza, S. A., Qureshi, M. A., & Yousufi, S. Q. (2020). Impact of CSR and environmental triggers on employee green behavior: The mediating effect of employee well-being. *Corporate Social Responsibility and Environmental Management, 27*(5), 2225–2239.

Ali, M. A., Khan, A. Z., Azeem, M. U., & Inam, U. H. (2023). How does environmental corporate social responsibility contribute to the development of a green corporate image? The sequential mediating roles of employees' environmental passion and pro-environmental behavior. *Business Ethics, the Environment & Responsibility, 32*(3), 896–909.

AlSuwaidi, M., Eid, R., & Agag, G. (2021). Understanding the link between CSR and employee green behaviour. *Journal of Hospitality and Tourism Management, 46*, 50–61.

Al-Swidi, A. K., Gelaidan, H. M., & Saleh, R. M. (2021). The joint impact of green human resource management, leadership and organizational culture on employees' green behaviour and organisational environmental performance. *Journal of Cleaner Production, 316*, 128112.

Amjad, F., Abbas, W., Zia-Ur-Rehman, M., Baig, S. A., Hashim, M., Khan, A., & Rehman, H. U. (2021). Effect of green human resource management practices on organizational sustainability: The mediating role of environmental and employee performance. *Environmental Science and Pollution Research, 28*, 28191–28206.

Andersson, L. M., & Bateman, T. S. (2000). Individual environmental initiative: Championing natural environmental issues in U.S. business organizations. *Academy of Management Journal, 43*(4), 548–570.

Arshad, M., Abid, G., Ahmad, J., Anum, L., & Khan, M. M. (2021). Impact of employee job attitudes on ecological green behavior in hospitality sector. *Journal of Open Innovation: Technology, Market, and Complexity, 7*(1), 31.

Ashraf, S., Ali, S. Z., Khan, T. I., Azam, K., & Afridi, S. A. (2024). Fostering sustainable tourism in Pakistan: Exploring the influence of environmental leadership on employees' green behavior. *Business Strategy & Development, 7*(1), e328.

Aung, P. N., & Hallinger, P. (2023). Research on sustainability leadership in higher education: A scoping review. *International Journal of Sustainability in Higher Education, 24*(3), 517–534.

Bandura, A. (1974). Behavior theory and the models of man. *American Psychologist, 29*(12), 859–869.

Bandura, A. (1986). *Social foundations of thought and action: A social cognitive theory*. Prentice Hall.

Bandura, A. (1997a). *The nature and structure of self-efficacy. Self-efficacy: The exercise of control* (pp. 37–78). WH Freeman and Company.

Bandura, A. (1997b). *Self-efficacy: The exercise of control*. W H Freeman/Times Books/Henry Holt.

Bandura, A. (2001). Social cognitive theory: An agentic perspective. *Annual Review of Psychology, 52*(1), 1–26.

Bandura, A. (2008). Observational learning. In W. Donsbach (Ed.), *The international encyclopedia of communication*. Wiley.

Bandura, A., Caprara, G. V., Barbaranelli, C., Gerbino, M., & Pastorelli, C. (2003). Role of affective self-regulatory efficacy in diverse spheres of psychosocial functioning. *Child Development, 74*(3), 769–782.

Banwo, A. O., & Du, J. (2019). Workplace pro-environmental behaviors in small and medium-sized enterprises: An employee level analysis. *Journal of Global Entrepreneurship Research, 9*(1), 34.

Boufounou, P., Moustairas, I., Toudas, K., & Malesios, C. (2023). ESGs and customer choice: Some empirical evidence. *Circular Economy and Sustainability, 3*(4), 1841–1874.

Bussey, K., & Bandura, A. (1999). Social cognitive theory of gender development and differentiation. *Psychological Review, 106*(4), 676.

Chen, S., Song, Y., & Gao, P. (2023). Environmental, social, and governance (ESG) performance and financial outcomes: Analyzing the impact of ESG on financial performance. *Journal of Environmental Management, 345*, 118829.

Çop, S., Olorunsola, V. O., & Alola, U. V. (2021). Achieving environmental sustainability through green transformational leadership policy: Can green team resilience help? *Business Strategy and the Environment, 30*(1), 671–682.

Dathe, T., Helmold, M., Dathe, R., & Dathe, I. (2024). Strategy implementation. In *Implementing environmental, social and governance (ESG) principles for sustainable businesses: A practical guide in sustainability management* (pp. 85–97). Springer.

Deng, W., Zhang, Z., & Guo, B. (2024). Firm-level carbon risk awareness and green transformation: A research on the motivation and consequences from government regulation and regional development perspective. *International Review of Financial Analysis, 91*, 103026.

Dey, M., Bhattacharjee, S., Mahmood, M., Uddin, M. A., & Biswas, S. R. (2022). Ethical leadership for better sustainable performance: Role of employee values, behavior and ethical climate. *Journal of Cleaner Production, 337*, 130527.

Dumont, J., Shen, J., & Deng, X. (2017). Effects of green HRM practices on employee workplace green behavior: The role of psychological green climate and employee green values. *Human Resource Management, 56*(4), 613–627.

Eccles, R. G., Ioannou, I., & Serafeim, G. (2014). The impact of corporate sustainability on organizational processes and performance. *Management Science, 60*(11), 2835–2857.

Fawehinmi, O., Yusliza, M. Y., Mohamad, Z., Noor Faezah, J., & Muhammad, Z. (2020). Assessing the green behaviour of academics: The role of green human resource management and environmental knowledge. *International Journal of Manpower, 41*(7), 879–900.

Hansmann, R., Laurenti, R., Mehdi, T., & Binder, C. R. (2020). Determinants of pro-environmental behavior: A comparison of university students and staff from diverse faculties at a Swiss university. *Journal of Cleaner Production, 268*, 121864.

Huang, D. Z. X. (2022). An integrated theory of the firm approach to environmental, social and governance performance. *Accounting & Finance, 62*, 1567–1598.

Khaireddine, H., Salhi, B., Aljabr, J., & Jarboui, A. (2020). Impact of board characteristics on governance, environmental and ethical disclosure. *Society and Business Review, 15*(3), 273–295.

Kumar, P., Singh, R. K., & Kumar, V. (2021). Managing supply chains for sustainable operations in the era of industry 4.0 and circular economy: Analysis of barriers. *Resources, Conservation and Recycling, 164*, 105215.

Lasrado, F., & Zakaria, N. (2020). Go green! Exploring the organizational factors that influence self-initiated green behavior in The United Arab Emirates. *Asia Pacific Journal of Management, 37*(3), 823–850.

Leeuwis, N., van Bommel, T., & Alimardani, M. (2022). A framework for application of consumer neuroscience in pro-environmental behavior change interventions. *Frontiers in Human Neuroscience, 16*, 886600.

Lei, X., & Yu, J. (2024). Striving for sustainable development: Green financial policy, institutional investors, and corporate ESG performance. *Corporate Social Responsibility and Environmental Management, 31*(2), 1177–1202.

Lüdeke-Freund, F. (2020). Sustainable entrepreneurship, innovation, and business models: Integrative framework and propositions for future research. *Business Strategy and the Environment, 29*(2), 665–681.

MacNeil, I., & Esser, I. M. (2022). From a financial to an entity model of ESG. *European Business Organization Law Review, 23*(1), 9–45.

Marrucci, L., Daddi, T., & Iraldo, F. (2024). Creating environmental performance indicators to assess corporate sustainability and reward employees. *Ecological Indicators, 158*, 111489.

Meng-tao, C., Da-peng, Y., Wei-qi, Z., & Qi-jun, W. (2023). How does ESG disclosure improve stock liquidity for enterprises—Empirical evidence from China. *Environmental Impact Assessment Review, 98*, 106926.

Muisyo, P. K., Qin, S., Julius, M. M., Ho, T. H., & Ho, T. H. (2021). Green HRM and employer branding: The role of collective affective commitment to environmental management change and environmental reputation. *Journal of Sustainable Tourism, 30*(8), 1897–1914.

Murphy, E., Guimaraes Da Costa, N., & Wong, C. Y. (2020). Decoding human intervention: Pathways to successful environmental management. *European Management Review, 17*(1), 247–265.

Nawaz Khan, A. (2023). Is green leadership associated with employees' green behavior? Role of green human resource management. *Journal of Environmental Planning and Management, 66*(9), 1962–1982.

Norton, T. A., Parker, S. L., Zacher, H., & Ashkanasy, N. M. (2015). Employee green behavior: A theoretical framework, multilevel review, and future research agenda. *Organization & Environment, 28*(1), 103–125.

Ones, D. S., & Dilchert, S. (2012). Environmental sustainability at work: A call to action. *Industrial and Organizational Psychology, 5*(4), 444–466.

Paillé, P., Boiral, O., & Chen, Y. (2013). Linking environmental management practices and organizational citizenship behaviour for the environment: A social exchange perspective. *The International Journal of Human Resource Management, 24*(18), 3552–3575.

Pham, N. T., Tučková, Z., & Jabbour, C. J. C. (2019). Greening the hospitality industry: How do green human resource management practices influence organizational citizenship behavior in hotels? A mixed-methods study. *Tourism Management, 72*, 386–399.

Rajesh, R., & Rajendran, C. (2020). Relating environmental, social, and governance scores and sustainability performances of firms: An empirical analysis. *Business Strategy and the Environment, 29*(3), 1247–1267.

Renwick, D. W., Jabbour, C. J., Muller-Camen, M., Redman, T., & Wilkinson, A. (2016). Contemporary developments in Green (environmental) HRM scholarship. *The International Journal of Human Resource Management, 27*(2), 114–128.

Salvador, R. O., & Burciaga, A. (2020). Organizational environmental orientation and employee environmental in-role behaviors: A cross-level study. *Business Ethics: A European Review, 29*(1), 98–113.

Santa-Maria, T., Vermeulen, W. J., & Baumgartner, R. J. (2022). How do incumbent firms innovate their business models for the circular economy? Identifying micro-foundations of dynamic capabilities. *Business Strategy and the Environment, 31*(4), 1308–1333.

Schunk, D. (2019). *Learning theories: An educational perspective* (8th ed.). Pearson.

Schutte, N. S., & Bhullar, N. (2017). Approaching environmental sustainability: Perceptions of self-efficacy and changeability. *The Journal of Psychology, 151*(3), 321–333.

Shah, S. H. A., Cheema, S., Al-Ghazali, B. M., Ali, M., & Rafiq, N. (2021). Perceived corporate social responsibility and pro-environmental behaviors: The role of organizational identification and coworker pro-environmental advocacy. *Corporate Social Responsibility and Environmental Management, 28*(1), 366–377.

Tahir, R., Athar, M. R., & Afzal, A. (2020). The impact of greenwashing practices on green employee behaviour: Mediating role of employee value orientation and green psychological climate. *Cogent Business & Management, 7*(1), 1781996.

Tian, H., Zhang, J., & Li, J. (2020). The relationship between pro-environmental attitude and employee green behavior: The role of motivational states and green work climate perceptions. *Environmental Science and Pollution Research, 27*(7), 7341–7352.

Torres, L., Ripa, D., Jain, A., Herrero, J., & Leka, S. (2023). The potential of responsible business to promote sustainable work–An analysis of CSR/ESG instruments. *Safety Science, 164*, 106151.

United Nations Global Compact. (2021). *Principles for responsible investment*. Retrieved from https://www.unpri.org/download?ac=10948

Xiaoyi, Z., Yang, H., Kumar, N., Bhutto, M. H., Kun, W., & Hu, T. (2023). Assessing Chinese textile and apparel industry business sustainability: The role of organization green culture, green dynamic capabilities, and green innovation in relation to environmental orientation and business sustainability. *Sustainability, 15*(11), 8588.
Yuriev, A., & Sierra-Barón, W. (2020). Exploring sustainability cross-culturally: Employees' beliefs on green behaviors. *Sustainable Development, 28*(5), 1199–1207.
Zhou, H., & Liu, J. (2023). Digitalization of the economy and resource efficiency for meeting the ESG goals. *Resources Policy, 86*, 104199.

Ethical and Work Green Climates: Drivers of Pro-Environmental Behavior in Public Sector Employees

Nurul Liyana Mohd Kamil and Wan Noor Azreen Wan Mohamad Nordin

Abstract This study explores the relationship between ethical climate, green work climate, environmental passion, and pro-environmental behavior (PEB) among public sector employees in Malaysia. The research aims to understand how ethical and green work climates influence PEB, with particular attention to the mediating influence of environmental passion. Drawing on social cognitive theory (SCT) and social exchange theory (SET), we utilized a cross-sectional multilevel design, surveying 407 public servants across 37 organizations. The findings reveal that, while ethical climate negatively impacts PEB and green work climate shows no direct effect, introducing environmental passion as a mediator transforms these relationships into positive outcomes. This highlights the critical role of environmental passion in aligning employee motivation with organizational sustainability goals. This research advances the theoretical understanding of public sector organizational behavior by demonstrating the pivotal role of environmental passion in overcoming the challenges posed by ethical and green work climates. These results offer valuable guidance for public sector organizations seeking to improve sustainability by fostering a work environment that nurtures environmental passion.

1 Introduction

Promoting sustainability is crucial for balancing economic growth with environmental protection. Over the past decade, governments worldwide have increasingly integrated environmental concerns into their policies (Ji & Darnall, 2021). In Malaysia, the 12th Malaysian Plan emphasizes the need for low-carbon, resilient development, and sustainable resource management, in line with the Sustainable Development Goals (SDGs). It also integrates environmental, social, and governance (ESG) principles into policy frameworks (Economic Planning Unit,

N. L. M. Kamil (✉) · W. N. A. W. M. Nordin
Faculty of Business and Economics, Universiti Malaya, Kuala Lumpur, Malaysia
e-mail: nurulliyana@um.edu.my; wannoorazreen@um.edu.my

M. Y. Yusliza et al. (eds.), *Employee Green Behavior*,
https://doi.org/10.1007/978-981-95-1422-9_5

2021). However, policy implementation alone is insufficient; human behavior plays a crucial role. Employees' pro-environmental behavior (PEB)—actions like recycling, conservation, and waste reduction—significantly impacts the success of environmental programs (Daily et al., 2009).

PEB refers to behavior that reduces the negative environmental impacts of workplace activities, contributing to sustainable organizational performance (Boiral & Paillé, 2012). Although various terms describe similar concepts, including workplace green behavior (Norton et al., 2017), ecological behavior (Noor Faezah et al., 2024), environmental pro-voluntary behavior (Renwick et al., 2024), and pro-environmental behavior (Yusliza et al., 2021), this study focuses on PEB due to its comprehensive nature. A significant obstacle exists in promoting PEB within a detrimental organizational climate, as an unpleasant moral or green work climate may result in resistance or only superficial compliance.

Organizational climate—encompassing ethical and green work climates—significantly influences employees' perceptions and behaviors. Ethical and green work climates are two important organizational-level constructs that have emerged in Western literature for decades (Martin & Cullen, 2006; Norton et al., 2014; Victor & Cullen, 1988). Although ethical climate and green work climate are distinct, they share overlapping characteristics. In general, ethical climate addresses ethical concerns such as fairness in compensation, employee morale, and managing turnover intentions, providing a framework for resolving issues (Fein et al., 2023). In contrast, a green work climate shapes employees' interpretations by highlighting the organization's environmental policies, procedures, and sustainable practices, fostering a shared perception of sustainability priorities (Sabbir & Taufique, 2022). Both climates have shown to influence PEB (Dahiya, 2020; Kamil & Nordin, 2025). However, limited research explores how these climates influence PEB through mediators such as environmental passion. Addressing this gap could provide valuable insights into the mechanisms linking organizational climates to employee behavior.

Therefore, we propose that environmental passion—a positive emotion motivating pro-environmental actions—has the potential to transform unsupportive climates into positive influcnces of PEB (Robertson & Barling, 2013). While the potential of environmental passion is compelling, empirical research on its mediating role is limited. Our study contributes to this area by examining how environmental passion can shift the dynamics within these relationships.

This study, based on social cognitive theory (SCT) and social exchange theory (SET), examines the influences of ethical and green work climates on employees' PEB. SCT suggests that individuals develop internal standards guiding their actions through social interactions (Bandura, 2001). SET posits that anticipated rewards drive behavior within social exchanges (Blau, 1964, p. 91). A multilevel approach is adopted to analyze how organizational-level ethical and green work climates influence individual-level environmental passion and PEB. This approach captures cross-level interactions, providing deeper insights into how organizational contexts shape individual outcomes.

Our paper is structured as follows: we begin with a detailed literature review and hypothesis development, followed by our research methodology and analysis results. We then discuss the findings, emphasizing their theoretical and practical implications, and conclude with main insights, research limitations, and suggestions future research.

2 Literature Review and Hypotheses Development

2.1 Social Cognitive Theory and Social Exchange Theory

To explore the connections between ethical climate, green work climate, environmental passion, and PEB, we draw on social cognitive theory (SCT) and social exchange theory (SET). On one hand, SCT (Bandura, 2014) examines how interactions among individuals, their environment, and their actions influence behaviors. Core elements of SCT include self-efficacy, task confidence, and anticipated outcomes. SCT suggests that fostering high environmental passion and self-efficacy can increase engagement in PEB. Thus, organizations should support employees in developing these traits to promote sustainability (Bandura, 2001).

On the other hand, social exchange theory (SET) offers a framework for comprehending how PEB develops in the workplace. According to SET, voluntary reciprocity and positive social interactions between employees and the organization lead to high-quality relationships (Cropanzano & Mitchell, 2005). A work environment that promotes positive perceptions can motivate employees to engage in PEB, as they reciprocate organizational support with a commitment to sustainability. By creating and nurturing supportive relationships, organizations can encourage employees to align with sustainability goals, enhancing overall environmental impact.

2.2 Ethical Climate, Work Green Climate, and PEB

The connection between ethical climate and PEB has attracted considerable interest in organizational studies, yet findings remain complex. Ethical climate pertains to the collective views within an organization regarding what constitutes ethical behavior, encompassing values such as fairness, integrity, and responsibility (Victor & Cullen, 1988). In contrast, PEB involves actions aimed at reducing environment impact, such as energy conservation, waste reduction, and participation in sustainability initiatives (Stern, 2000).

An ethical climate promotes positive organizational behaviors, although it does not consistently improve PEB. Organizations emphasizing fairness or compliance may neglect environmental issues, resulting in overlooking PEB. Social exchange theory suggests that employees' behaviors are influenced by the perceived reciprocity in their relationships with the organizations. Employees may experience moral licensing, where adherence to other ethical standards reduces motivation for environmental actions (Merritt et al., 2010).

Furthermore, resource allocation complicates the relationship. Ethical climates emphasizing compliance and interpersonal justice may divert attention and resources away from sustainability efforts, limiting employees' capacity or willingness to engage in PEB (Ramus & Steger, 2000). Additionally, employees could differ from organizational ethical responsibilities with pro-environmental actions in situations where environmental ethics are not explicitly incorporated into the ethical framework, resulting in the observed marginal effect (DesJardins, 1998).

Hypothesis 1 Ethical climate is negatively related to pro-environmental behavior.

The relationship between a green work climate and employees' PEB is multifaceted. While a green work climate typically results in positive outcomes, such as increased engagement in sustainability initiatives (Norton et al., 2014; Rubel et al., 2021), some studies have uncovered potential drawbacks. For example, strong environmental policies and high expectations may unintentionally cause stress, or they may be poorly implemented and result in employees' lack of support (Esty & Winston, 2009; Foster et al., 2022).

Social cognitive theory, which emphasizes learning through observation and environmental influences (Schunk & Usher, 2012), can help explain this. In a green work climate, if employees observe that the organization neglects or fails to follow through on environmental initiatives, they might perceive these efforts as sufficient, reducing their sense of personal responsibility. This perception diminishes their engagement in PEB. Additionally, employees who view a green work climate as insincere or symbolic may grow cynical, further lowering their participation in PEB.

In essence, while a green work climate intends to promote PEB, its effectiveness depends on employees' perceptions. When the climate is seen as genuine and well-supported, employees are more likely to adopt positive behaviors. Conversely, burdensome and insincere initiatives can reduce motivation. Social cognitive theory (Bandura, 1986) highlights the critical role of perception and interpretation in shaping employees' responses. Therefore, we hypothesize that:

Hypothesis 2 Green work climate is negatively related to PEB.

2.3 Ethical Climate, Work Green Climate, and Environmental Passion

An ethical climate fosters a supportive framework that motivates employees to adopt in environmentally responsible behaviors. By upholding ethical standards, organizations highlight the significance of environmental stewardship, which enhances employees' environmental passion. This environment motivates employees to support green initiatives, participate in sustainability programs, and advocate for eco-friendly practices, ultimately improving organizational performance in environmental sustainability (Robertson & Barling, 2013). Thus, an ethical climate is crucial for fostering and maintaining employees' environmental passion.

Studies have shown a positive link between ethical climate and environmental passion among employees. High ethical principles within an organization lead to higher acceptance of its ethical policies and practices. A workplace that prioritizes ethical behavior, transparency, and integrity helps align employees' values with organizational goals, thereby boosting their environmental motivation and emotional commitment to pro-environmental behavior (Boiral et al., 2015).

According to social cognitive theory (Bandura, 1986), employees develop attitudes by observing others and their experiences. In an ethical climate, employees observe ethical behaviors modeled by leaders and colleagues, which enhances their environmental passion. When employees see that their organization values ethical standards, their confidence in their ability to contribute environmental sustainability increases. Therefore, we describe that:

Hypothesis 3 Ethical climate positively influences environmental passion.

Research indicates a strong positive correlation between a green work climate and employees' environmental passion (Vallerand et al., 2014). An organization's commitment to sustainability and eco-friendly practices fosters a shared sense of responsibility toward environmental issues. When employees perceive their organization as genuinely committed to green practices, they develop a strong emotional connection and internal motivation for sustainability, which enhances their engagement in related initiatives.

Furthermore, social cognitive theory (Bandura, 1986) offers a nuanced understanding of how a green work climate influences environmental passion. This theory posits that individuals form their attitudes by observing and imitating behaviors in their environment. In a green work climate, employees observe eco-friendly behaviors from the organization, which strongly influences their actions. These observations reinforce employees' environmental values and increase their belief in the possibility of making a positive impact. When employees see their organization's commitment to green practices, they internalize these values, further increasing their environmental passion (Ali et al., 2023). Therefore:

Hypothesis 4 Green work climate positively influences environmental passion.

2.4 Environmental Passion as a Mediator Between Ethical Climate, Work Green Climate, and PEB

Organizational ethical dilemmas and conflicts have contributed to a negative trend in the relationship between ethical climate and PEB (Victor & Cullen, 1988). Such conflicts may hinder environmental initiatives and reduce employees' participation in PEB. However, environmental passion as a mediator can potentially transform this negative dynamic. Environmental passion, which is defined as a strong emotional commitment to environmental protection, improves PEB by reducing the negative effects of an ethical climate (De Bernardi & Pedrini, 2020). It is crucial to cultivate environmental passion within organizations because employees with strong environmental passion can transform disagreements into meaningful engagement (Afsar et al., 2016).

According to social exchange theory, employees are more inclined to adopt organizational values when reciprocal relationship exist between employees and their organization (Blau, 1964). Environmental passion reinforces this reciprocity by encouraging employees to overcome the challenges posed by a skeptical ethical workplace. Despite such challenges, employees who have a strong personal commitment to environmental protection are more likely to participate in PEB.

Furthermore, intrinsic motivation is essential for PEB, but it is not sufficient alone. The passion to prioritize collective benefits and environmental protection plays an equally, if not more, crucial role in encouraging PEB (Robertson & Barling, 2013). This theoretical framework indicates that environmental passion plays a crucial role in mediating the connection between ethical climate and PEB. Based on this, we assume that:

Hypothesis 5 Environmental passion mediates the relationship between ethical climate and PEB.

Research indicates that while a green work climate, characterized by organizational efforts to promote environmental sustainability, often aims to foster PEB, it can sometimes be associated with a negative relationship. This negative relationship may arise due to misalignment between organizational green policies and employees' values or perceptions, leading to conflicts or reduced motivation for PEB (Mouro et al., 2021). However, introducing environmental passion as a mediator can change this dynamic. Environmental passion, a deep emotional commitment to protecting the environment helps to connect a green work climate with PEB by aligning individual enthusiasm with organizational goals (Foster et al., 2022). This alignment transforms potential conflicts into opportunities, enhancing employees' motivation and participation in sustainability initiatives. As a result, environmental passion can turn the challenges associated with a green work climate into a driving force for meaningful and effective PEB.

From the perspective of social cognitive theory, environmental passion can amplify the effects of a green work climate on PEB by boosting employees' self-efficacy and behavioral intentions. According to Bandura (1986), self-efficacy is the belief in one's ability to accomplish particular goals, which plays main role in shaping motivation and behavior. Employees passionate about environmental issues are likely to have greater self-efficacy, boosting their ability to engage in PEB. This theoretical framework emphasizes the need to cultivate environmental passion in order to strengthen the positive outcomes of a green work climate on PEB, ultimately supporting a more sustainable organizational environment. Based on this, we propose the following hypothesis:

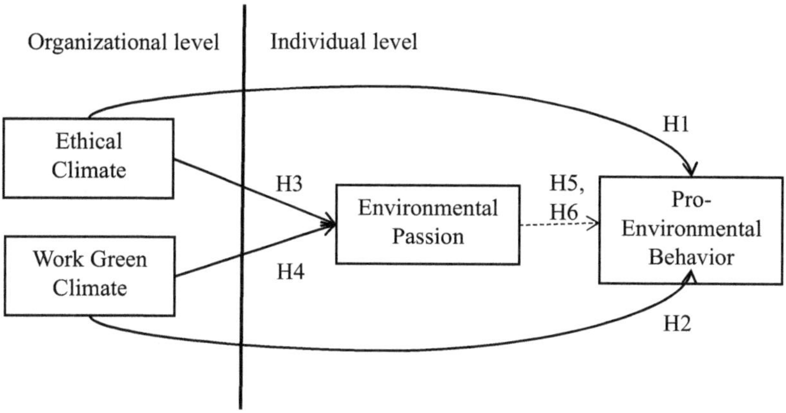

Fig. 1 Research model. Source: Author's own work

Hypothesis 6 Environmental passion mediates the relationship between green work climate and PEB.

Figure 1 depicts the research model for this study.

3 Methodology

3.1 Participants and Procedures

Our research utilized a cross-sectional multilevel design, gathering data from 407 civil servants representing 37 organizations in the Malaysian public sector. We focused on the public sector due to its proactive approach to promoting related pro-environmental behavior (Blok et al., 2015).

To recruit participants, we employed a combination of snowball sampling and professional networks, contacting Directors in the Service Division of various public sector organizations. Of the 55 organizations contacted, 37 agreed to participate, resulting in a response rate of 67%. The sample predominantly comprised 316 male participants (77.6%), with the largest age group being 31–40 years old (190 participants, 46.7%). Most participants identified as Malay (90.7%) and had diverse educational backgrounds, including secondary education (59%), bachelor's degrees (18.2%), master's degrees (15%), and diplomas (7.9%). Participants also varied in work experience, with 277 participants (68.1%) having 1–10 years of experience, and 130 participants (31.9%) with over 30 years of experience.

3.2 Measures

We translated all study instruments into Malay using Brislin's (1970) back-translation method. First, one individual translated the English items into Malay. Then, another translator, unfamiliar with the original English version, retranslated the Malay text back into English to ensure consistency. Both translators were fluent in Malay and English and were experts in organizational behavior.

Ethical climate was measured using 14 items adapted from Victor and Cullen (1988), such as "*The most important concern is the good of all the employees in the organization as a whole*," with reliability ranging from 0.50 to 0.81.

Green work climate was measured using eight items from Norton et al. (2014), including "*My organization is worried about its environmental impact*," showing strong reliability with a Cronbach's alpha of 0.92.

Environmental passion was measured with ten items from Robertson and Barling (2013), such as "*I am passionate about the environment*," with a Cronbach's alpha of 0.87.

Pro-environmental Behavior used seven items from Robertson and Barling (2013), like "*I print double-sided whenever possible*," with a reliability of 0.81.

3.3 Aggregation Procedures

We assessed whether ethical climate and green work climate should be evaluated at Level 2 by examining interrater agreement within groups. The mean r_{wg} values for ethical climate (0.97) and green work climate (0.94) exceeded the recommended threshold of 0.90, indicating substantial agreement within organizations (LeBreton & Senter, 2008). A one-way random effects ANOVA was conducted, revealing significant F-values for ethical climate (F(III) = 1.78, $p < 0.001$) and green work climate (F(III) = 2.37, $p < 0.001$), highlighting substantial differences between organizations. The intraclass correlation coefficient [ICC(I)] was calculated as 0.07 for ethical climate and 0.11 for green work climate, suggesting that 7% and 11% of the variance, respectively, could be attributed to random team effects. Given that ICC values between 0.05 and 0.20 are deemed sufficient for analysis (Bliese, 2000), these findings support the aggregation of ethical and green work climate measures at the organizational level.

3.4 Hypothesis Testing

In this study, hypothesis testing involved several steps to evaluate the relationships between variables. Initially, the null model was used to assess the extent of between-group variance in the measures (see Tables 3 and 4, Model 1). Hypothesis 1 tested the direct effect of ethical climate on PEB, while Hypothesis 2 focused on the relationship between green work climate and PEB. Hypothesis 3 and 4 explored the effects of environmental passion on PEB.

The study also tested mediation hypotheses (Hypotheses 5 and 6) using a cross-level mediation analysis, following Mathieu and Taylor's (2007) method. The analysis included evaluating the relationships between ethical climate, green work climate, and PEB, as well as between these climates and environmental passion. The mediation was tested by examining whether environmental passion acts as a mediator in the connection between ethical climate, green work climate, and PEB. Although the presence of a direct significant link between ethical climate, green work climate, and PEB is not necessary for mediation analysis, the study thoroughly tested the mediation paths using the Monte Carlo method (MCMAM). This approach calculated confidence intervals (CI) for indirect effects through 20,000 replications, providing detailed estimates of mediation effects.

3.5 Analysis Strategy

We analyzed the data using Hierarchical Linear Modeling (HLM) version 7.0. To ensure accuracy, we standardized the variables and organized them into two levels: Level 1 for individuals and Level 2 for the 37 organizations. Standardizing variables in multilevel modeling is crucial for obtaining precise estimates, especially when multiple predictors are involved.

To estimate cross-level effects, we reported several key metrics: the intercept values ($\gamma 00$), within-team variance at Level 1 (σ^2), intercept variance at Level 2 ($\tau 00$), slope variance at Level 2 ($\tau 11$), and intercept–slope variance at Level 2 ($\tau 01$). We also provided the log-likelihood value (−2 FIML), the number of estimated parameters, and the pseudo-R^2 value (Aguinis et al., 2013).

4 Results

4.1 Descriptive Analysis

The descriptive analysis (Table 1) showed significant relationships between the variables at both the individual level (Level 1) and the organizational level (Level 2), aligning with the study's expectation. Table 2 presents the results of confirmatory factor analysis (CFA) conducted with maximum likelihood estimation, which confirmed the distinctiveness of the variables. The four-factor model, consisting of ethical climate, green work climate, environmental passion and PEB, showed a superior fit compared to the two-factor and one-factor models. While some fit indices, such as CFI (0.792), NFI (0.735), NNFI (0.690), and RMSEA (0.08), were

Table 1 Summary of descriptive statistics and inter-correlations for study variables

Variables	M	SD	1	2	3	4
1. Ethical climate	3.31	0.50	(0.88)			
2. Green work climate	4.21	0.63	0.49***	(0.90)		
3. Environmental passion	4.23	0.56	0.45***	0.62***	(0.91)	
4. Pro-environmental behavior	3.92	0.69	0.14**	0.19***	0.42***	(0.82)

Note: *N* (individuals) = 407; *n* (organization) = 37. M = means, SD = standard deviation; values along the diagonal represent the reliability coefficients (Cronbach's alpha) for each construct; ***$p < 0.001$, *$p < 0.05$

Source: Author's own work

Table 2 Results of confirmatory factor analysis for different model comparison

Models	χ2 (df)	Δχ2 Δ(df)	CFI	NFI	NNFI	RMSEA
Four-factor model (EC, GWC, EP, and PEB)	2498.88 (696)	–	0.792	0.735	0.690	0.08
Two-factor model (combining GWC, EP, and PEB)	3850.40 (701)	1351.52 (5)***	0.637	0.591	0.560	0.105
Two-factor model (combining EC, EP, and PEB)	4153.00 (701)	1654.12 (5)***	0.602	0.559	0.529	0.110
One-factor model (combining all variables)	4763.49 (702)	2264.61 (6)***	0.532	0.495	0.469	0.119

Note: EC = ethical climate; GWC = green work climate; EP = Environmental passion; PEB = pro-environmental behavior; χ^2 = chi-square; df = degrees of freedom; $\Delta\chi^2$ = chi-square difference; CFI = Comparative Fit Index; NFI = Normed Fit Index; NNFI = Non-Normed Fit Index; RMSEA = root mean square error of approximation

Source: Author's own work

below the ideal threshold, the model still provided an acceptable fit, underscoring that, despite moderate correlations, these construct are distinct. Each construct captures a unique dimension of organizational climate and individual attitudes, contributing to understanding PEB.

4.2 *Hypothesis Testing*

The findings from the HLM analysis are presented in Tables 3 and 4. Hypothesis 1 posited a negative association between ethical climate and PEB. Model 3 in Table 3 confirms this hypothesis, showing a negative effect of ethical climate on PEB ($\gamma = -0.37$, SE = 0.17; $p < 0.05$). Hypothesis 2 proposed that a green work climate would negatively affect PEB. However, the results from Model 4 in Table 3 show no significant negative relationship with PEB ($\gamma = -0.14$; SE = 0.11; p = ns), indicating that the hypothesis is not supported.

Table 3 Multilevel analysis predicting pro-environmental behavior (PEB)

Level and Variable	Model 1	2	3	4
Level 1				
Intercept (γ_{00}) Environmental passion	3.92(0.03)	1.74(0.24) 0.52(0.06)***	2.89(0.56) 0.53(0.06)***	2.29(0.56) 0.52(0.08)***
Level 2				
Ethical climate Green work climate			−0.37(0.17)*	−0.14(0.11) ns
Variance components				
Within-team (L1) variance (σ^2)	0.47	0.38	0.38	0.38
Intercept (L2) variance (τ_{00})	0.0002	0.01	0.001	0.005
Additional information				
ICC(1) −2log likelihood (deviance)	0.0004 848.75	0.03 769.54	0.003 764.89	0.01 768.23
Number of estimated parameters	3	4	5	5
Pseudo R^2	0	0.19	0.19	0.19

Note: L1 = Level 1 (within-team); L2 = Level 2 (between-team). Standard errors are in parentheses. $^+p < 0.10$; $^*p < 0.05$; $^{**}p < 0.01$; $^{***}p < 0.001$. The models assess the impact of ethical climate (EC) and green work climate (GWC) on pro-environmental behavior (PEB) and environmental passion (EP), accounting for both within-team and between-team variances
Source: Author's own work

Table 4 Multilevel analysis for predicting environmental passion

Level and Variable	Model 1	2	3
Level 1			
Intercept (γ_{00})	4.23(0.02)	2.95(0.37)	3.39(0.43)
Level 2			
Ethical climate		0.39(0.11)***	
Green work climate			0.20(0.10)$^+$
Variance components			
Within-team (L1) variance (σ^2)	0.31	0.31	0.31
Intercept (L2) variance (τ_{00})	0.00004	0.00003	0.00003
Additional information			
ICC(1) −2log likelihood (deviance)	0.0001 680.83	0.0001 673.70	0.0001 676.90
Number of estimated parameters	3	4	4
Pseudo R^2	0	0	0

Note: L1 = Level 1 (within-team); L2 = Level 2 (between-team). Standard errors are in parentheses. $^+p < 0.10$; $^*p < 0.05$; $^{**}p < 0.01$; $^{***}p < 0.001$. The models assess the impact of ethical climate (EC) and green work climate (GWC) on pro-environmental behavior (PEB) and environmental passion (EP), accounting for both within-team and between-team variances
Source: Author's own work

Hypothesis 3 suggested a positive relationship between ethical climate and environmental passion. Model 2 in Table 4 shows that ethical climate significantly increases environmental passion ($\gamma = 0.39$; SE = 0.11; $p < 0.001$). Similarly, Hypothesis 4 predicted a positive effect of green work climate on environmental passion, which Model 3 in Table 4 supports with a significant positive relationship ($\gamma = 0.20$; SE = 0.10; $p < 0.10$).

Hypothesis 5 proposed that environmental passion mediates the effect of ethical climate on PEB. To test this, we analyzed two mediation paths: Path A (ethical climate to environmental passion) and Path B (environmental passion to PEB). The analysis in Model 3 of Table 3 shows a significant effect for Path B ($\beta = 0.53$; SE = 0.06; $p < 0.001$), with a 95% confidence interval ranging from 0.090 to 0.335, indicating significant mediation and thus supporting Hypothesis 5.

For Hypothesis 6, we proposed that environmental passion mediates the relationship between green work climate and PEB. Following a similar analysis approach, significant mediation effects observed for Path B (environmental passion to PEB) in Model 4 of Table 3 ($\beta = 0.52$; SE = 0.08; $p < 0.001$). The 95% confidence interval [0.003–0.217] excludes zero, thereby supporting Hypothesis 6.

5 Discussion

This study explores the impact of ethical climate and green work climate on PEB, with environmental passion serving as a mediator. Initially, a negative relationship was observed between ethical climate and PEB, while green work climate did not show a direct effect. However, introducing environmental passion as a mediator shifted both relationships to positive outcomes. The results emphasize the critical contribution of environmental passion in transforming the influence of organizational climates on employee behavior, offering valuable insights for fostering a more environmentally responsible workforce.

5.1 Theoretical Implications

Our study offers valuable insights into Malaysian public sector, addressing a significant gap in multilevel research. Grounded in social exchange theory, the findings reveal that an ethical climate can negatively influence PEB. Specifically, a strong ethical climate may hinder effective communication among employees, reducing their engagement in PEB. Social exchange theory (Blau, 1964) explains this as employees' behavior being shaped by the perceived reciprocity in organizational relationship. When organizations prioritize fairness over environmental concerns, employees may experience moral licensing, whereby fulfilling other ethical obligations diminishes their motivation for environmental action (Nordin et al., 2024).

Positively, our study emphasizes the beneficial effect of ethical and green work climates in promoting environmental passion. Organizations with clear ethical and environmental priorities inspire employees to develop a commitment to environmental causes. This passion fosters acceptance of ethical policies and active participation in sustainability efforts, thereby improving organizational environmental performance (Jia et al., 2018; Robertson & Barling, 2013). According to social cognitive theory (Bandura, 1986), the ethical and environmental values present in an organization help cultivate employees' environmental passion, leading to greater sustainability contribution.

Furthermore, environmental passion serves as mediator between ethical climate, green work climate, and PEB. This mediation aligns with both social exchange theory and social cognitive theory, indicating that passionate employees reciprocate organizational efforts by overcoming challenges and actively participate in sustainability initiatives (Afsar et al., 2016). By integrating these theories, our study provides a comprehensive perspective on how ethical and green work climates influence PEB through this mediating effect.

5.2 *Practical Implications*

This study provides practical insights for improving public sector employees' PEB. Creating a robust ethical and green work climate significantly increases employees' passion for environmental causes. Organizations could achieve this by establishing high ethical standards, committing to sustainability, and creating a motivational and supportive environment. Such a climate encourages employees to participate actively in environmental policies and practices.

Moreover, fostering PEB is closely linked to employees' environmental passion. Organizations can take advantages of this by putting in place focused environmental training programs that are intended to raise awareness and ignite a passion for sustainability. A multifaceted strategy is essential, combining strong leadership, employee involvement and clear communication. Leaders should model environmental values and prioritize sustainability. Additionally, communication strategies must involve employees in environmental decision-making, ensuring their voices contribute to organizational goals and initiatives.

5.3 *Limitations and Recommendation for Future Research*

This study offers valuable insights but has several limitations. First, self-report questionnaires have the potential to introduce common method bias. However, this bias is lessened by our within-group variance analysis and multilevel approach (Podsakoff et al., 2003). Second, the cross-sectional nature of the design limits the ability to infer causality. Collecting data at only one time point restricts the analysis of potential mediating pathways. A cause-and-effect relationship may be better understood through longitudinal research.

Third, the fit indices (CFI, NFI, NNFI) for the four-factor model do not meet ideal thresholds, suggesting a misalignment between the model and the data. Future research should refine the measurement instruments by revising existing items and developing new items to better align with theoretical constructs. Fourth, the focus on the public sector limits the generalizability of the findings. Future research could examine how ethical and green work climates affect PEB in other sectors, such as private or service industries.

Lastly, incorporating moderator variables could yield different findings. Factors like employee trust in leadership and their perceptions of organizational support for sustainability could influence the direct effect of a green work climate on PEB. Additionally, individual values like autonomy and self-efficacy could mediate employees' responses to organizational climate. Future research should explore these contextual factors to deepen our understanding of how ethical and green work climates influence PEB.

6 Conclusion

In conclusion, our multilevel study offers new insights into the relationships between ethical climate, green work climate, environmental passion, and PEB in the public sector. The findings indicate that although an ethical climate may hinder employees' PEB, environmental passion significantly mitigates this effect. Environmental passion also enhances PEB within green work climates. Despite limitations, including potential bias, cross-sectional design, and model fit concerns, the study lays a strong basis for future research to enhance ethical and green work climates, fostering employees' PEB.

Acknowledgments The authors would like to thank the Faculty of Business and Economics (Universiti Malaya) for its support in this research.

Funding This research was funded by the Universiti Malaya, under the Universiti Malaya Research Excellence Grant (UMReG) [Projectnumber: UMREG004-2025].

References

Afsar, B., Badir, Y., & Kiani, U. S. (2016). Linking spiritual leadership and employee pro-environmental behavior: The influence of workplace spirituality, intrinsic motivation, and environmental passion. *Journal of Environmental Psychology, 45*, 79–88. https://doi.org/10.1016/j.jenvp.2015.11.011

Aguinis, H., Gottfredson, R. K., & Culpepper, S. A. (2013). Best-practice recommendations for estimating cross-level interaction effects using multilevel modeling. *Journal of Management, 39*(6), 1490–1528. https://doi.org/10.1177/0149206313478188

Ali, M. A., Khan, A. Z., Azeem, M. U., & Inam, U. H. (2023). How does environmental corporate social responsibility contribute to the development of a green corporate image? The sequential mediating roles of employees' environmental passion and pro-environmental behavior. *Business Ethics, the Environment & Responsibility, 32*(3), 896–909. https://doi.org/10.1111/beer.12539

Bandura, A. (1986). *Social foundations of thought and action: A social cognitive theory*. Prentice Hall.

Bandura, A. (2001). Social cognitive theory: An agentic perspective. *Annual Review of Psychology, 52*(1), 1–26. https://doi.org/10.1146/annurev.psych.52.1.1

Bandura, A. (2014). Social cognitive theory of moral thought and action. In *Handbook of moral behavior and development* (pp. 45–103). Psychology Press.

Blau, P. M. (1964). *Exchange and power in social life*. John Wiley & Sons.

Bliese, P. D. (2000). Within-group agreement, non-independence, and reliability: Implications for data aggregation and analysis. In *Multilevel theory, research, and methods in organizations*. Jossey-Bass.

Blok, V., Wesselink, R., Studynka, O., & Kemp, R. (2015). Encouraging sustainability in the workplace: A survey on the pro-environmental behavior of university employees. *Journal of Cleaner Production, 106*, 55–67. https://doi.org/10.1016/j.jclepro.2014.07.063

Boiral, O., & Paillé, P. (2012). Organizational citizenship behavior for the environment: Measurement and validation. *Journal of Business Ethics, 109*, 431–445. https://doi.org/10.1007/s10551-011-1138-9

Boiral, O., Talbot, D., & Paillé, P. (2015). Leading by example: A model of organizational citizenship behavior for the environment. *Business Strategy and the Environment, 24*(6), 532–550. https://doi.org/10.1002/bse.1835

Brislin, R. W. (1970). Back-translation for cross-cultural research. *Journal of Cross-Cultural Psychology, 1*(3), 185–216. https://doi.org/10.1177/135910457000100301

Cropanzano, R., & Mitchell, M. S. (2005). Social exchange theory: An interdisciplinary review. *Journal of Management, 31*(6), 874–900. https://doi.org/10.1177/0149206305279602

Dahiya, R. (2020). Does organisational sustainability policies affect environmental attitude of employees? The missing link of green work climate perceptions. *Business Strategy & Development, 3*(3), 395–403. https://doi.org/10.1002/bsd2.110

Daily, G. C., Polasky, S., Goldstein, J., Kareiva, P. M., Mooney, H. A., Pejchar, L., et al. (2009). Ecosystem services in decision making: Time to deliver. *Frontiers in Ecology and the Environment, 7*(1), 21–28. https://doi.org/10.1890/080025

De Bernardi, C., & Pedrini, M. (2020). Entrepreneurial behavior: Getting eco-drunk by feeling environmental passion. *Journal of Cleaner Production, 256*, 120367. https://doi.org/10.1016/j.jclepro.2020.120367

DesJardins, J. (1998). Corporate environmental responsibility. *Journal of Business Ethics, 17*(8), 825–838. https://doi.org/10.1023/A:1005719707880

Economic Planning Unit. (2021). *Twelfth Malaysia plan 2021–2025: A prosperous, inclusive, sustainable Malaysia*. EPU, Malaysia.

Esty, D. C., & Winston, A. (2009). *Green to gold: How smart companies use environmental strategy to innovate, create value, and build competitive advantage*. John Wiley & Sons.

Fein, E. C., Tziner, A., & Vasiliu, C. (2023). Perceptions of ethical climate and organizational justice as antecedents to employee performance: The mediating role of employees' attributions of leader effectiveness. *European Management Journal, 41*(1), 114–124. https://doi.org/10.1016/j.emj.2021.11.003

Foster, B., Muhammad, Z., Yusliza, M. Y., Faezah, J. N., Johansyah, M. D., Yong, J. Y., et al. (2022). Determinants of pro-environmental behavior in the workplace. *Sustainability, 14*(8), 4420. https://doi.org/10.3390/su14084420

Ji, H., & Darnall, N. (2021). All are not created equal: Assessing local governments' strategic approaches towards sustainability. In *Sustainable public management* (pp. 159–180). Routledge.

Jia, J., Liu, H., Chin, T., & Hu, D. (2018). The continuous mediating effects of GHRM on employees' green passion via transformational leadership and green creativity. *Sustainability, 10*(9), 3237. https://doi.org/10.3390/su10093237

Kamil, N. L. M., & Nordin, W. N. A. W. M. (2025). How does ethical climate attract employees' low-carbon behavior? The role of environmental passion and green mindfulness. *Journal of Management & Organization, 31*(1), 103–121. https://doi.org/10.1017/jmo.2024.38

LeBreton, J. M., & Senter, J. L. (2008). Answers to 20 questions about interrater reliability and interrater agreement. *Organizational Research Methods, 11*(4), 815–852. https://doi.org/10.1177/1094428106296642

Martin, K. D., & Cullen, J. B. (2006). Continuities and extensions of ethical climate theory: A meta-analytic review. *Journal of Business Ethics, 69*, 175–194. https://doi.org/10.1007/s10551-006-9084-7

Mathieu, J. E., & Taylor, S. R. (2007). A framework for testing meso-mediational relationships in organizational behavior. *Journal of Organizational Behavior: The International Journal of Industrial, Occupational and Organizational Psychology and Behavior, 28*(2), 141–172. https://doi.org/10.1002/job.436

Merritt, A. C., Effron, D. A., & Monin, B. (2010). Moral self-licensing: When being good frees us to be bad. *Social and Personality Psychology Compass, 4*(5), 344–357. https://doi.org/10.1111/j.1751-9004.2010.00263.x

Mouro, C., Lomba, V., & Duarte, A. P. (2021). Pro-environmental behaviors at work: The interactive role of norms and attitudinal ambivalence. *Sustainability, 13*(21), 12003. https://doi.org/10.3390/su132112003

Noor Faezah, J., Yusliza, M. Y., Wan Kassim, W. Z., Muhammad, Z., & Imaningsih, E. S. (2024). Exploring corporate social responsibility research in the era of sustainable development goals in Malaysia. In M. Y. Yusliza & D. Renwick (Eds.), *Green human resource management*. Springer. https://doi.org/10.1007/978-981-99-7104-6_10

Nordin, W. N. A. W. M., Kamil, N. L. M., & Govindaraju, V. C. (2024). Multilevel study of transformational leadership and work behavior: Job autonomy matters in public service. *Management Research Review, 47*(10), 1684–1701. https://doi.org/10.1108/MRR-08-2023-0596

Norton, T. A., Zacher, H., & Ashkanasy, N. M. (2014). Organisational sustainability policies and employee green behavior: The mediating role of work climate perceptions. *Journal of Environmental Psychology, 38*, 49–54. https://doi.org/10.1016/j.jenvp.2013.12.008

Norton, T. A., Zacher, H., Parker, S. L., & Ashkanasy, N. M. (2017). Bridging the gap between green behavioral intentions and employee green behavior: The role of green psychological climate. *Journal of Organizational Behavior, 38*(7), 996–1015. https://doi.org/10.1002/job.2178

Podsakoff, P. M., MacKenzie, S. B., Lee, J. Y., & Podsakoff, N. P. (2003). Common method biases in behavioral research: A critical review of the literature and recommended remedies. *Journal of Applied Psychology, 88*(5), 879–903. https://doi.org/10.1037/0021-9010.88.5.879

Ramus, C. A., & Steger, U. (2000). The roles of supervisory support behaviors and environmental policy in employee "Ecoinitiatives" at leading-edge European companies. *Academy of Management Journal, 43*(4), 605–626. https://doi.org/10.5465/1556357

Renwick, D. W. S., Brint, A., De Camargo Fiorini, P., de Sousa, L., Jabbour, A. B., & Chiappetta Jabbour, C. J. (2024). Investigating voluntary workplace green behaviors in Brazil. In M. Y. Yusliza & D. Renwick (Eds.), *Green human resource management*. Springer. https://doi.org/10.1007/978-981-99-7104-6_7

Robertson, J. L., & Barling, J. (2013). Greening organizations through leaders' influence on employees' pro-environmental behaviors. *Journal of Organizational Behavior, 34*(2), 176–194. https://doi.org/10.1002/job.1820

Rubel, M. R. B., Kee, D. M. H., & Rimi, N. N. (2021). Green human resource management and supervisor pro-environmental behavior: The role of green work climate perceptions. *Journal of Cleaner Production, 313*, 127669. https://doi.org/10.1016/j.jclepro.2021.127669

Sabbir, M. M., & Taufique, K. M. R. (2022). Sustainable employee green behavior in the workplace: Integrating cognitive and non-cognitive factors in corporate environmental policy. *Business Strategy and the Environment, 31*(1), 110–128. https://doi.org/10.1002/bse.2877

Schunk, D. H., & Usher, E. L. (2012). Social cognitive theory and motivation. *The Oxford Handbook of Human Motivation, 2*, 11–26.

Stern, P. C. (2000). New environmental theories: Toward a coherent theory of environmentally significant behavior. *Journal of Social Issues, 56*(3), 407–424. https://doi.org/10.1111/0022-4537.00175

Vallerand, R. J., Houlfort, N., & Forest, J. (2014). Passion for work: Determinants and outcomes. In *The Oxford handbook of work engagement, motivation, and self-determination theory* (pp. 85–105). Oxford University Press.

Victor, B., & Cullen, J. B. (1988). The organizational bases of ethical work climates. *Administrative Science Quarterly, 33*(1), 101–125. https://doi.org/10.2307/2392857

Yusliza, M. Y., Faezah, J. N., Mat, N. H. N., Saputra, J., Muhammad, Z., Muhamad, A. S., & Ramayah, T. (2021). Modelling pro-environmental behavior in the workplace: A preliminary study. In *Proceedings of the International Conference on industrial engineering and operations management, July* (pp. 3953–3963).

Optimizing Remote Work for a Sustainable Workplace: Strategies to Promote Employee Green Behavior

Nelvin XeChung Leow, Woon Leong Lin, Jing Yi Yong, Siong Choy Chong, Chuen Khee Pek, and Feranita Feranita

Abstract This chapter examines the relationship between remote work policies and employee green behavior (EGB) in sustainable workplaces, offering unique contributions by integrating evidence-based strategies tailored to remote and hybrid work environments. Remote work presents notable environmental benefits, such as reducing commuting emissions and lowering office energy consumption, while fostering a transition to paperless workflows. These shifts contribute to a reduced organizational carbon footprint and align with sustainability goals. However, increased energy use in home offices and the rise in electronic waste (e-waste) present challenges that can offset these environmental gains. To address these challenges, the chapter highlights novel strategies, including structured e-waste recycling programs, guidance on energy-efficient home office setups, and incentives for green commuting during occasional office visits. For instance, encouraging the use of energy-saving devices, promoting cloud-based solutions, and supporting sustainable practices at home can help mitigate the environmental trade-offs of remote work. Organizations are positioned as key enablers, with policies that foster EGB through resources, incentives, and education, thereby promoting a culture of environmental responsibility. The chapter introduces a multidimensional framework combining organizational policies with employee engagement to maximize the sustainability benefits of remote work. This framework emphasizes proactive digital waste reduction, energy-efficient behaviors, and eco-friendly practices that employees can adopt both at home and in hybrid work scenarios. By aligning individual

N. X. Leow (✉) · W. L. Lin · J. Y. Yong · C. K. Pek
School of Management and Marketing, Faculty of Business and Law, Taylor's University, Subang Jaya, Selangor, Malaysia
e-mail: NelvinXeChung.Leow@taylors.edu.my; Woonleong.lin@taylors.edu.my; Serene.Yong@taylors.edu.my; Eugene.Pek@taylors.edu.my

S. C. Chong
Faculty of Business and Law, Taylor's University, Subang Jaya, Selangor, Malaysia
e-mail: eddy.chong@taylors.edu.my

F. Feranita
Nottingham University Business School China, University of Nottingham, Ningbo, China
e-mail: feranita.feranita@nottingham.edu.cn

M. Y. Yusliza et al. (eds.), *Employee Green Behavior*,
https://doi.org/10.1007/978-981-95-1422-9_6

actions with corporate sustainability goals, remote work emerges as a powerful tool for achieving environmental stewardship. These insights bridge theoretical frameworks, such as the Theory of Planned Behavior, with practical applications, offering a roadmap for organizations to optimize remote work for sustainability. This chapter advances the discourse on EGB by linking workplace policies with actionable strategies, positioning remote work as a transformative opportunity for environmental impact.

1 Introduction

The rise of remote work, accelerated by the COVID-19 pandemic, has reshaped traditional office environments and work dynamics globally. This shift offers organizations a unique opportunity to integrate employee green behavior (EGB) into their sustainability strategies (Galanti et al., 2021). Remote work can significantly reduce environmental impacts by decreasing the need for office resources such as energy, water, and paper, while also cutting commuting-related carbon emissions. As employees become less dependent on daily transportation, greenhouse gas emissions and urban air pollution are reduced (Hook et al., 2020). Additionally, the absence of a centralized office allows organizations to reassess their natural resource consumption and transition to more sustainable, digital-first business models (Asatiani & Norström, 2023). However, the environmental benefits of remote work must be carefully managed to avoid unintended sustainability challenges. For instance, while reducing commutes leads to lower carbon emissions, the shift to working from home often increases household energy consumption. This increased consumption can offset some of the environmental gains, especially if homes are not optimized for energy efficiency. Furthermore, the growing reliance on technology in remote work environments contributes to e-waste, exacerbating environmental concerns.

Despite a growing body of literature highlighting the environmental benefits of remote work, there is limited research on strategies to address the challenges it poses, such as increased home energy use and e-waste (Hook et al., 2020; Zacher & Rudolph, 2022). Additionally, few studies explore how organizations can actively foster EGB to balance these trade-offs (Asatiani & Norström, 2023). This chapter builds on prior research by offering a holistic framework for promoting EGB tailored to remote and hybrid work environments. By addressing these gaps, it extends the discourse on workplace sustainability and provides actionable strategies to maximize the environmental potential of remote work.

To fully realize the environmental benefits of remote work, organizations must adopt strategies that promote EGB both in the office and at home. These strategies include providing resources to help employees improve the energy efficiency of their home offices, adopting cloud-based solutions to reduce digital waste, and establishing e-waste recycling programs. Encouraging such behaviors is essential not only for reducing an organization's environmental footprint but also for engaging employees and fostering a culture of sustainability (Zacher & Rudolph, 2022).

EGB refers to voluntary actions that employees take to support environmental sustainability within their organizations. These behaviors include reducing energy consumption, minimizing waste, and promoting the use of eco-friendly resources (Norton et al., 2014). In remote work environments, fostering EGB requires deliberate policies and resources that guide employees toward adopting eco-friendly habits at home (Blok et al., 2015). By encouraging green behavior among employees, especially in remote settings, organizations can not only achieve their sustainability goals but also improve employee engagement and well-being by creating a sense of purpose (Zacher & Rudolph, 2022).

2 Problem Statement

Remote work, a trend accelerated by the COVID-19 pandemic, has demonstrated potential environmental benefits, primarily through reductions in commuting emissions and office energy consumption. However, despite its advantages, remote work also introduces new environmental challenges, such as increased household energy consumption and the growing problem of e-waste. This study focuses on optimizing remote work to maximize sustainability benefits while addressing these challenges, particularly in Malaysia—a country with rapidly growing digital infrastructure and diverse industries transitioning to remote work.

The primary research question explored is: How can organizations in Malaysia leverage remote work policies to promote EGB and achieve sustainable workplace practices while mitigating environmental drawbacks such as increased home energy usage and e-waste? This study distinguishes itself by concentrating on Malaysia's unique socioeconomic and cultural context, which encompasses both urban and rural work settings. It emphasizes industries with significant digital operations, such as technology and services. Unlike studies focusing on Western countries or large corporate environments, this research considers the localized implications of remote work on sustainability, including challenges posed by Malaysia's uneven adoption of green technologies and public transport infrastructure.

2.1 Commuting Emissions: From Reduction to Optimization

One of the most notable environmental advantages of remote work is the reduction in commuting emissions. Traditional commuting is a significant contributor to global greenhouse gas (GHG) emissions, particularly in urban settings where traffic congestion is prevalent. According to the International Energy Agency (IEA, 2021), transportation contributes to approximately 29% of global GHG emissions, with personal vehicles being a primary source. Remote work reduces the need for daily commuting, leading to decreased fuel consumption and lower carbon emissions. During the COVID-19 pandemic, this shift translated into substantial environmental gains. For instance, in cities such as Los Angeles and London, CO_2 emissions dropped by 35% and 22%, respectively, due to reduced vehicular traffic (Liu et al., 2020).

However, for these gains to be sustained, remote work must be optimized to address residual impacts. While full-time remote work eliminates daily commuting, hybrid models still necessitate occasional office visits, which can contribute to transportation emissions. To mitigate this, organizations should encourage eco-friendly commuting options for the days employees travel to the office. Providing incentives for public transportation, carpooling, or cycling can help further reduce commuting-related emissions, even in hybrid work settings (Onnis, 2019).

2.2 *The Hidden Energy Costs of Remote Working*

Although remote work reduces energy consumption in commercial office spaces, it presents challenges related to increased home energy usage. Office buildings typically benefit from energy-efficient technologies and centralized heating and cooling systems, making them more energy-efficient than individual households. However, remote workers may consume more electricity to power devices and maintain comfortable home environments throughout the day. This increase in household energy consumption can negate some of the environmental benefits of reduced office operations. For example, Zhang et al. (2021) highlighted that the rise in home energy use during remote work could offset up to 50% of the emissions savings from reduced commuting.

To mitigate the environmental impact of increased home energy use, organizations can implement policies that support energy-efficient home office setups. This can involve providing financial incentives or guidance on energy-saving measures, such as using energy-efficient appliances, smart thermostats, and LED lighting (Richardson & Kelliher, 2015). In some cases, companies have begun offering stipends or discounts on energy-efficient equipment to encourage employees to adopt more sustainable practices at home (Orzeł & Wolniak, 2022).

2.3 *The Shift to a Paperless, Yet Digital-Dependent Workplace*

Another environmental benefit of remote work is the shift toward paperless operations. Traditional office environments rely heavily on paper for communication and documentation, contributing to deforestation and waste generation. Remote work, on the other hand, promotes the use of digital tools, such as cloud-based solutions and electronic signatures, reducing the need for physical documents.

However, the increased reliance on digital tools also presents a new environmental challenge: e-waste. The demand for personal electronic devices, such as laptops, tablets, and smartphones, has surged as remote workers rely on these tools to stay connected and productive. The frequent need to replace or upgrade devices exacerbates the growing e-waste problem, which poses significant environmental risks. E-waste contains hazardous materials like lead, mercury, and cadmium, which can pollute the environment if not properly disposed of (Awasthi et al., 2022b).

Without adequate e-waste management practices, the environmental and health impacts of discarded electronics could worsen. To tackle this issue, organizations should implement comprehensive e-waste recycling programs. By partnering with certified e-waste recyclers, companies can ensure that obsolete devices are disposed of in an environmentally responsible manner (Zacher & Rudolph, 2022). Additionally, some organizations have introduced buy-back programs or refurbishing initiatives, where employees can return old devices for repair or resale, extending their lifespan and reducing e-waste.

2.4 *Balancing Remote Work and Sustainability*

To fully capitalize on the environmental benefits of remote work, organizations must adopt a holistic approach that addresses the associated challenges. This includes reducing home energy consumption, managing e-waste responsibly, and encouraging sustainable commuting options for hybrid work setups. Promoting EGB is a key strategy in achieving these goals. Companies can provide training on energy-efficient practices, offer subsidies for renewable energy adoption, and facilitate e-waste recycling programs (Zacher & Rudolph, 2022).

Moreover, remote work policies should be flexible enough to adapt to long-term sustainability objectives. Hybrid work models, for instance, can provide a balanced solution by allowing employees to work remotely while still accessing energy-efficient office resources when necessary. By adopting eco-friendly commuting options and promoting energy-saving practices in both home and office environments, organizations can optimize the sustainability potential of remote work.

2.5 *Challenges of Remote Work for Sustainability*

Despite the significant environmental advantages of remote work, such as reduced commuting and lower office energy consumption, it also presents new sustainability challenges. Increased household energy consumption and the growing issue of e-waste are pressing concerns that could counterbalance the environmental gains of remote work. Studies have shown that remote work can lead to a 10% increase in residential electricity usage, as employees spend more time at home utilizing electronic devices and heating or cooling systems (Cicala, 2020). This rise in household energy consumption can offset the reductions achieved by decreased office energy use. Additionally, the shift to remote work has led to a surge in demand for electronic devices, contributing to a 30–40% increase in e-waste generation (eSmart Recycling, 2024). E-waste contains hazardous materials that pose significant environmental and health risks if not properly managed.

To mitigate these challenges, organizations can adopt several strategies. Providing employees with energy-efficient equipment and offering guidance on reducing home energy consumption can help lower the increased residential energy use associated with remote work (Richardson & Kelliher, 2015). Implementing e-waste recycling programs and promoting the responsible disposal of electronic devices are also crucial steps in addressing the environmental impact of increased device usage (Awasthi et al., 2022b).

By fostering a culture of sustainability through targeted policies, organizations can leverage remote work as a tool for achieving long-term environmental objectives. Encouraging energy-efficient home office setups, supporting e-waste recycling, and promoting eco-friendly commuting are essential steps toward creating a sustainable remote work environment.

3 Literature Review

This section reviews the key variables relevant to the study of EGB in promoting a sustainable workplace, particularly in the context of remote work. The variables include Energy-Efficient Practices, Digital Waste Reduction, Green Commuting Intentions, Organizational Support, and EGB. Each variable plays a crucial role in understanding how remote work can contribute to environmental sustainability within organizations.

3.1 Contradictions and Gaps in the Literature

While the literature highlights the potential of energy-efficient practices to reduce the environmental impact of remote work, there is a lack of consensus on its net benefits. For example, some studies argue that home energy consumption may offset energy savings from reduced office usage, particularly in less energy-efficient households (Hook et al., 2020). Conversely, others emphasize that the flexibility to adopt energy-saving behaviors at home outweighs these drawbacks (Zhang et al., 2021). This contradiction underscores the need for more granular studies that assess the net environmental impact across diverse household settings.

Similarly, while digital waste reduction is frequently promoted as a key sustainability goal, there is limited research on practical implementation within remote work contexts. Studies highlight the growing problem of e-waste, but few explore how organizations can provide effective support for responsible disposal or recycling (Awasthi et al., 2022a). The lack of standardized guidelines for digital waste management creates a gap between awareness and actionable practices.

Regarding green commuting intentions, the literature agrees broadly on the benefits of incentivizing eco-friendly transportation options during hybrid work scenarios. However, gaps persist in understanding how these intentions translate into long-term behavior, particularly when employees face logistical barriers such as inadequate public transport infrastructure (Tao et al., 2024).

Finally, while organizational support is widely recognized as a critical driver of EGB, inconsistencies arise in defining the most effective types of support. For instance, some studies prioritize financial incentives for energy-efficient tools, while others advocate for cultural shifts that promote sustainability as a core value (Zacher & Rudolph, 2022; Blok et al., 2015). These differing approaches highlight the need for integrated models that combine tangible resources with cultural transformation.

3.2 *Underpinning Theory*

The Theory of Planned Behavior (TPB), developed by Icek Ajzen (1991), is a psychological model that helps explain and predict human behavior based on three primary components: attitude toward the behavior, subjective norms, and perceived behavioral control. This theory is particularly relevant for understanding behaviors in contexts where individuals can exercise control, such as promoting EGB within remote work environments. In this setting, TPB provides a framework to explore how employees' intentions and actions align with organizational sustainability goals and directly informs the development of hypotheses by identifying specific behavioral drivers.

The attitude toward behavior refers to an individual's positive or negative evaluation of a particular action. In the context of EGB, employees' attitudes toward environmentally friendly practices, such as conserving energy or reducing digital waste, play a critical role. Hypotheses based on this component can examine how employees' perceptions of the benefits of green practices—such as reducing environmental impact and enhancing organizational sustainability—motivate pro-environmental actions. Research suggests that when employees believe their actions contribute to meaningful environmental outcomes, they are more inclined to adopt green practices (Ajzen, 1991). This highlights the importance of fostering positive attitudes through awareness campaigns and demonstrating the tangible impact of EGB initiatives.

Subjective norms refer to the social pressures employees experience regarding whether they should engage in certain behaviors. These norms are shaped by colleagues, supervisors, and organizational culture. Hypotheses can explore the influence of a pro-environmental workplace culture and visible leadership support on employees' likelihood of adopting EGB. For instance, organizations that prioritize sustainability and actively encourage green practices through recognition programs or shared success stories can strengthen social norms, leading to higher compliance among employees. The role of subjective norms underscores the need to embed sustainability values within the organization to foster widespread green behaviors.

Perceived behavioral control is the employee's belief in their ability to successfully perform the behavior. This component is critical for understanding barriers to EGB, such as a lack of resources, knowledge, or opportunities to act sustainably. Hypotheses derived from this element could investigate how providing employees with resources—such as energy-efficient tools, training on sustainable practices, and access to e-waste recycling programs—enhances their capacity to engage in green behavior. Employees who feel equipped and supported are more likely to translate their intentions into actions.

Recent studies have applied TPB to explore EGB in various contexts. For instance, a meta-analysis by Katz et al. (2022) examined the relationship between TPB components and EGB, highlighting the significance of attitudes, subjective norms, and perceived behavioral control in predicting green behaviors among employees. Additionally, research by Bouarar and Mouloudj (2021) utilized TPB to investigate employees' intentions to implement green practices, emphasizing the role of environmental knowledge and organizational support in shaping these intentions.

3.3 Energy-Efficient Practices

Energy-efficient practices are essential in reducing the environmental impact of remote and hybrid work models. Homes are typically less energy-efficient than office buildings, which often use centralized, energy-efficient systems. To mitigate this issue, employees must adopt energy-saving behaviors such as using energy-efficient lighting (e.g., LED bulbs), smart thermostats, and energy-saving office equipment.

Recent empirical studies have examined the impact of hybrid work models on sustainability efforts. For instance, Choudhury et al. (2022) highlighted that hybrid work arrangements—where employees split time between home and office—allow for significant commuting reductions while maintaining productivity. Additionally, hybrid models reduce energy use in office spaces by lowering occupancy rates. However, these benefits are offset by increased household energy consumption, as employees rely more on home utilities and personal devices (Cicala, 2020).

Despite these findings, gaps remain in the literature regarding the environmental implications of hybrid work, particularly concerning digital waste management. A recent study by Awasthi et al. (2022a, b) indicated that the rise in remote and hybrid work has contributed to a 30–40% increase in e-waste globally. E-waste, if improperly disposed of, poses severe environmental risks due to hazardous materials such as lead and mercury. Current organizational policies often lack adequate measures for managing e-waste, highlighting the need for structured recycling programs and employee education initiatives.

To address these challenges, organizations can influence employee behaviors by providing resources and incentives. Richardson and Kelliher (2015) found that companies offering energy-efficient devices or subsidizing home office equipment saw significant reductions in employees' home energy consumption. Moreover, energy-efficient behaviors at home can help reduce the overall carbon footprint of remote work, aligning with corporate sustainability goals.

Promoting responsible e-waste management practices is also crucial. A recent review by Mehta et al. (2023) (Sharma et al., 2023) explored how digitalization can play a pivotal role in waste recycling by enabling better tracking and management of electronic devices. Implementing e-waste recycling programs, partnering with certified recyclers and educating employees on proper disposal methods can significantly reduce the environmental impact of increased electronic device usage in hybrid work setups.

3.4 Digital Waste Reduction

Digital waste reduction refers to practices aimed at minimizing unnecessary data storage, reducing paper usage, and properly handling e-waste. As remote work relies heavily on digital platforms and tools, the amount of digital waste generated has increased significantly. E-waste, including discarded electronic devices like laptops and smartphones, poses a major environmental challenge due to the hazardous materials these devices contain, such as lead, mercury, and cadmium.

The rise of cloud computing and digital-first solutions has created opportunities for organizations to reduce their digital waste footprint. Cloud-based platforms can improve efficiency by reducing the duplication of files, which often leads to unnecessary storage and digital clutter (Richardson & Kelliher, 2015). Moreover, recent studies have found that adopting cloud-based solutions can reduce energy consumption by up to 40% compared to traditional data storage systems, highlighting their environmental benefits (Maji et al., 2023).

Organizations also play a crucial role in encouraging proper e-waste disposal. Awasthi et al. (2022b) emphasized that companies implementing structured e-waste recycling programs could reduce the environmental risks posed by discarded electronic devices by over 30%. These programs are particularly critical as the demand for electronic devices increases due to hybrid work models.

Recent empirical studies have also examined how hybrid work models affect sustainability efforts. Choudhury et al. (2022) found that hybrid work can reduce overall commuting emissions by up to 27%, but it may lead to higher residential energy use and e-waste generation. The study highlighted that employees working from home frequently upgrade electronic devices, exacerbating the e-waste problem.

Despite these efforts, gaps remain in the literature regarding sustainable digital waste management practices. For example, Sharma, Metha, and Sharma et al. (2023) pointed out inconsistencies in how organizations across industries approach digital waste reduction, particularly in hybrid work setups. Many lack clear policies for tracking and managing digital waste effectively.

To address these challenges, organizations can implement policies promoting responsible digital waste practices. Providing employees with guidance on data management, introducing file lifecycle policies to avoid redundant storage, and offering incentives for responsible device recycling are actionable strategies identified in recent research (Singh & Pandey, 2023). Additionally, promoting employee education on e-waste disposal and partnering with certified recycling facilities can help mitigate the environmental impact of hybrid work models.

3.5 Green Commuting Intentions

Although remote work reduces daily commuting, many employees still travel to the office for hybrid work or occasional meetings. Green commuting refers to sustainable transportation methods, including public transport, carpooling, cycling, and walking. Green commuting intentions measure employees' willingness to adopt these environmentally friendly commuting methods when necessary.

The transportation sector is one of the largest contributors to greenhouse gas emissions, accounting for approximately 29% of global emissions (IEA, 2021). Remote work significantly reduces these emissions by limiting the need for daily commutes. However, employees' commuting habits during hybrid workdays can still contribute to the overall environmental impact. Encouraging green commuting can therefore help organizations reduce their carbon footprint.

Blok et al. (2015) found that organizations offering incentives, such as public transportation subsidies or bicycle parking, foster higher levels of engagement in sustainable commuting practices. Moreover, Tao et al. (2024) argue that creating a corporate culture that values sustainability can lead to higher adoption rates of green commuting behaviors. For instance, companies that promote environmental values through internal communications and provide tools for sustainable travel see better outcomes in terms of employee participation in green commuting.

Employees' personal values and awareness of environmental issues also play a role in their commuting behaviors. Hook et al. (2020) found that employees, who are more environmentally conscious, are likely to adopt green commuting practices, even without significant organizational support. This highlights the importance of aligning organizational initiatives with employee values to drive more sustainable commuting choices.

Recent empirical studies have examined the impact of hybrid work models on sustainability efforts. Choudhury et al. (2022) conducted a field experiment that found an intermediate number of days in the office resulted in increased communication and novelty of work products, suggesting that hybrid work can offer the "best of both worlds" by providing greater work–life balance without isolating employees from colleagues.

However, gaps remain in the literature regarding the environmental implications of hybrid work, particularly concerning energy consumption and digital waste management. While some studies suggest that hybrid work can reduce commuting emissions, the increased use of electronic devices at home may lead to higher residential energy consumption and increased e-waste generation. For example, the shift to remote work has led to a surge in demand for electronic devices, contributing to a 30–40% increase in e-waste generation (Awasthi et al., 2022b).

Additionally, promoting responsible e-waste management practices is crucial. Rossitto et al. (2022) highlighted the importance of digital environmental stewardship, emphasizing that implementing e-waste recycling programs and educating employees on proper disposal methods can mitigate the environmental impact of increased electronic device usage.

3.6 *Organizational Support*

Organizational support is a key factor in promoting EGB, as it reflects the degree to which an organization provides the resources, incentives, and culture necessary for employees to adopt sustainable practices. Organizational support includes providing energy-efficient office tools, promoting paperless work environments, offering e-waste recycling programs, and incentivizing green commuting behaviors (Zacher & Rudolph, 2022).

Research indicates that organizations with strong environmental policies are more likely to foster pro-environmental behaviors among their employees (Norton et al., 2014). Organizational support can take many forms, such as providing training on sustainability practices, offering financial incentives for adopting green behaviors, and integrating environmental goals into the corporate mission. When employees perceive that their organization values sustainability, they are more likely to engage in EGB, both in and out of the office.

Moreover, this sense of support enhances employee engagement, leading to better retention rates and higher levels of job satisfaction. By embedding sustainability into their corporate culture, organizations can drive lasting behavioral change and achieve their environmental goals more effectively.

3.7 *Employee Green Behavior (EGB)*

Employee green behavior (EGB) refers to voluntary actions taken by employees to reduce their environmental impact in the workplace. EGB is a critical component of organizational sustainability initiatives, as employees' day-to-day actions directly influence the organization's overall environmental performance (Zacher & Rudolph, 2022).

Several factors influence EGB, including personal values, organizational culture, and the availability of resources. Employees, who hold strong pro-environmental values are more likely to engage in EGB, especially when they feel that their actions align with organizational goals (Blok et al., 2015). Additionally, employees are more likely to adopt green behaviors when they perceive that their efforts are supported and valued by the organization (Norton et al., 2014).

The transition to remote work has created new opportunities and challenges for fostering EGB. While remote work reduces certain environmental impacts, such as commuting emissions, it also requires employees to take greater responsibility for their environmental footprint at home. Despite the growing interest in EGB, several underexplored areas remain in the context of remote work.

The role of digital waste reduction—such as managing e-waste and optimizing cloud storage for environmental benefits—has not been extensively studied. Implementing green computing practices, including efficient data management and responsible disposal of electronic devices, is essential to minimize the environmental impact of increased digital reliance (Sharon, 2016). Secondly, the interplay between hybrid work models and EGB is another area lacking sufficient research. Hybrid work combines remote and office-based tasks, creating variability in employees' environmental impacts. Studies are needed to understand how to optimize green behaviors across both settings. Recent findings suggest that hybrid workers, who work from home 2–4 days/week can reduce their carbon footprint by 11–29%, indicating potential benefits that require further exploration (Tao et al., 2023). Moving on, the effectiveness of incentives such as stipends for sustainable commuting or energy-efficient home office equipment has been documented in some cases but requires more empirical evidence to establish best practices across industries. Providing resources and support for employees to adopt sustainable practices at home can enhance EGB, yet the optimal strategies for various organizational contexts remain underexplored (Lossin et al., 2016).

4 Proposition Development

In the context of remote and hybrid work environments, organizations are increasingly focusing on fostering EGB to align with their sustainability goals. Employee engagement in pro-environmental actions, such as reducing energy consumption, minimizing digital waste, and opting for sustainable commuting practices, is influenced by various factors both at the individual and organizational level. This study examines the relationships between key independent variables—Energy-Efficient Practices, Digital Waste Reduction, Green Commuting Intentions, and Organizational Support—and their impact on Employee Green Behavior. The following hypotheses are proposed to explore these relationships, based on previous research and theoretical frameworks in sustainability and employee behavior.

4.1 Proposition 1

Employees, who engage in energy-efficient practices at home, such as optimizing energy use with smart thermostats, energy-efficient appliances, and actively minimizing energy waste, are more likely to exhibit broader pro-environmental behaviors in their professional roles. Research supports the notion that EGB is often influenced by habits formed in personal environments, especially when those behaviors are encouraged by organizational policies. Employees who consciously manage their home energy use not only reduce their carbon footprint but also demonstrate a commitment to sustainability that extends into other areas of their lives, including the workplace.

Moreover, when organizations provide the necessary tools and support, such as subsidies for energy-efficient appliances or guidance on energy conservation, employees are more likely to integrate these practices into their daily routines (Blok et al., 2015; Al-Ghazali & Afsar, 2021). A study by Al-Ghazali and Afsar (2021) found that employees, who felt supported by their organizations in adopting energy-saving technologies, displayed a higher level of engagement in broader environmental initiatives within the workplace, such as reducing waste, participating in recycling programs, and advocating for sustainable policies. This highlights the importance of organizational encouragement in shaping employees' attitudes toward energy conservation.

Recent findings also emphasize that the adoption of energy-efficient practices at home is often a reflection of employees' overall environmental consciousness (Zacher & Rudolph, 2022). The integration of smart home technologies, for instance, not only reduces household energy consumption but also increases awareness of resource management, which can influence behavior at work. For organizations aiming to promote EGB, encouraging home energy efficiency serves as a foundational step in fostering a more environmentally responsible workforce.

Proposition 1: Energy-efficient practices will have a positive effect on EGB.

4.2 Proposition 2

Digital waste management, including practices such as minimizing e-waste, efficient data management, and reducing unnecessary printing, plays a critical role in fostering environmentally responsible behavior. As organizations increasingly rely on digital tools for remote and hybrid work, managing digital waste has become an important aspect of sustainability. Employees who actively reduce digital waste—by managing files efficiently, avoiding redundant printing, and responsibly disposing off outdated electronic devices—are more likely to contribute to the organization's broader sustainability goals (Hameed et al., 2021).

Furthermore, the practice of digital decluttering, which involves organizing and optimizing digital storage, has been linked to a reduction in energy consumption, as efficient data storage requires less energy for maintenance (Wu et al., 2023). Employees, who actively engage in such practices, demonstrate a commitment to sustainability that often extends beyond the digital realm, influencing their behavior in other environmental initiatives, such as resource conservation and recycling.

Proposition 2: Digital Waste Reduction will have a positive effect on EGB.

4.3 Proposition 3

Although remote work significantly reduces the need for commuting, hybrid work models still necessitate periodic travel to the office. Employees who express strong intentions to engage in green commuting behaviors—such as using public transportation, cycling, or carpooling—are more likely to adopt other pro-environmental practices both at home and in the workplace. The decision to commute sustainably often reflects a deeper commitment to environmental stewardship, which can extend to other aspects of daily life (Galanti et al., 2021).

Research by Tao et al. (2024) demonstrates that employees with pro-environmental attitudes are more likely to participate in green commuting, which positively correlates with their engagement in other green behaviors, such as energy conservation and waste reduction. These findings are supported by Blok et al. (2015), who argue that individuals who adopt sustainable transportation methods often develop a broader awareness of their environmental impact, leading them to practice sustainability in other areas of their professional and personal lives.

Additionally, green commuting practices can reinforce the adoption of organizational sustainability goals. For example, employees, who commute using environmentally friendly modes of transportation, are more likely to advocate for energy-efficient technologies or participate in corporate sustainability programs. The link between green commuting and broader EGB is critical for organizations aiming to reduce their overall carbon footprint and promote a culture of sustainability.

Proposition 3: Green commuting intentions will have a positive effect on EGB.

4.4 Proposition 4

Organizational support is a crucial factor in promoting EGB. Employees are more likely to engage in sustainable practices when they perceive that their organization is committed to environmental responsibility. This support may come in various forms, including the provision of resources such as energy-efficient technologies, digital waste reduction programs, and incentives for adopting sustainable commuting practices (Afsar & Umrani, 2020).

Galanti et al. (2021) highlight the importance of a strong organizational culture that emphasizes sustainability. Organizations that actively promote environmental initiatives—such as e-waste recycling programs, paperless offices, and energy-saving technologies—are more likely to see employees adopt green behaviors in their everyday work activities. This is supported by Norton et al., who found that organizations that offer incentives for sustainable behavior, such as financial rewards for energy savings or recognition for green practices, foster higher levels of employee engagement in sustainability efforts.

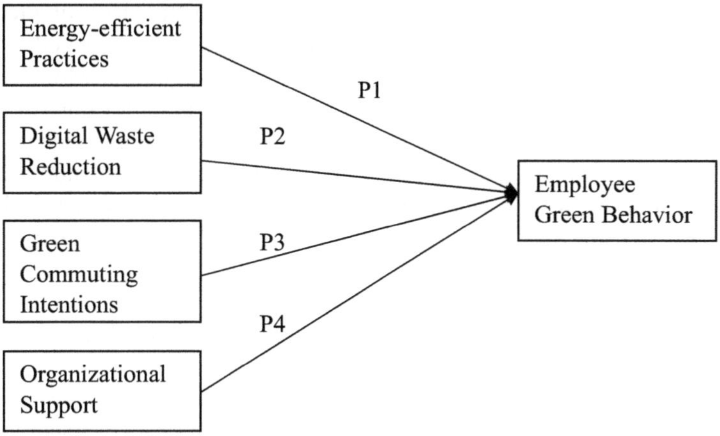

Fig. 1 Conceptual framework. Source: Author's own work

Furthermore, organizational support plays a significant role in shaping employees' attitudes toward environmental issues. Employees who perceive their workplace as environmentally responsible are more likely to engage in pro-environmental behaviors, both at work and at home. A study by Zacher and Rudolph (2022) demonstrates that employees in organizations with strong sustainability cultures are more motivated to adopt green behaviors, driven by a sense of shared responsibility for the organization's environmental impact.

Providing organizational support for environmental initiatives also helps to address potential barriers to GB. For instance, employees may be more willing to adopt energy-efficient practices or reduce digital waste if they receive the necessary training, resources, or incentives from their employers. This support not only promotes sustainability but also reinforces the organization's commitment to its environmental goals (Fig. 1).

Proposition 4: Organizational support will have a positive effect on EGB.

5 Research Methodology

This study employs a quantitative research methodology to explore the relationship between remote work policies and EGB in fostering sustainable workplace practices. A quantitative approach is chosen for its ability to provide objective measurements, allowing for the collection of data that can be statistically analyzed to draw conclusions about the population of interest. The study follows a deductive approach, testing hypotheses derived from existing literature on EGB and workplace sustainability, aligning with the positivist paradigm that underpins this research (Saunders et al., 2019).

The data collection process involves a survey administered to remote workers across various industries. Given the complex nature of behaviors related to environmental sustainability, Partial Least Squares Structural Equation Modeling (PLS-SEM) is used for the analysis. PLS-SEM is particularly suited for this study, as it is capable of handling complex models with multiple constructs, such as energy-efficient home office practices, digital waste reduction behaviors, and the adoption of green commuting solutions. Moreover, PLS-SEM is robust when dealing with non-normally distributed data and is effective with smaller sample sizes compared to covariance-based SEM, making it ideal for analyzing employee behavior in specific organizational settings (Hair et al., 2019).

Cochran's sample size formula is applied to ensure statistical reliability and validity, guiding the minimum sample size needed to achieve accurate population representations (Cochran, 1977). To account for potential nonresponses and incomplete data, the target sample size is set at 500 respondents. This oversampling strategy aims to improve the robustness of the results by ensuring adequate data for analysis even in the case of unusable responses.

The study uses convenience sampling, focusing on employees working remotely across various sectors, particularly those in knowledge-intensive industries. This sampling technique is appropriate given the time and resource constraints, although the study acknowledges potential biases. The survey aims to assess a representative group of employees, who are actively engaged in remote work and whose organizations have implemented sustainability initiatives.

The research instrument includes a set of questions designed to measure each variable identified in the literature review. These variables include energy efficiency, waste reduction practices, and green commuting behaviors. The survey items are adapted from validated scales used in prior research to ensure the reliability and validity of the constructs being measured. A sample of the questionnaire sources is provided in Table 1.

The survey data will be analyzed using PLS-SEM to test the proposed hypotheses regarding the influence of organizational support on EGB and the effectiveness of remote work policies in promoting sustainable behaviors. This method allows for the simultaneous estimation of both measurement and structural models, capturing the mediating effects of organizational support and the moderating role of employee engagement in sustainability initiatives. Model fit will be assessed using criteria such as composite reliability, goodness-of-fit index, and the average variance extracted (Hair et al., 2019).

Table 1 Questionnaire source

Variable	Cronbach's alpha	Authors
Energy-efficient practices	0.871	Chen and Chen (2021)
Digital waste reduction	0.959	Malokani et al. (2023)
Green commuting intentions	0.956	Liao et al. (2023)
Organizational support	0.915	Temminck et al. (2015)
Employee green behavior	0.805	Cheng et al. (2022)

Source: Author's own work

In addition, bootstrap resampling techniques will be applied to evaluate the statistical significance of the path coefficients in the model, enhancing the reliability of the results (Chin, 1998). This method generates multiple subsamples, providing confidence intervals for the relationships between variables and ensuring the robustness of the analysis. Sensitivity analyses will be conducted to account for various assumptions and verify that methodological choices do not disproportionately affect the outcomes.

Although this paper is conceptual in nature, several key statistical analyses will be considered for future empirical validation. These include descriptive statistics of respondents to summarize demographic and other relevant characteristics and a correlation matrix to identify relationships between key variables.

Ethical considerations are rigorously adhered to throughout the research process. All participants are required to provide informed consent, and their responses are anonymized to ensure confidentiality. The study should comply with the ethical guidelines set forth by universities and relevant regulatory bodies, ensuring transparency and integrity in data collection and analysis.

6 Conclusion

In conclusion, integrating remote work into organizational sustainability strategies offers a unique opportunity to foster EGB and achieve broader environmental goals. Remote work can significantly reduce commuting emissions, office energy consumption, and reliance on paper-based operations. However, these benefits must be balanced against challenges such as increased home energy use and the rising issue of e-waste.

To address these challenges, organizations should leverage the TPB to understand and influence EGB. By fostering positive attitudes toward sustainability, strengthening social norms, and enhancing employees' perceived control, companies can drive meaningful behavior change. Actionable steps include providing energy-saving guidance for home offices, implementing e-waste recycling programs, and incentivizing eco-friendly commuting options for hybrid workers.

While the study focuses primarily on remote work, many of these findings can extend to hybrid and in-office work environments. For example, encouraging energy-efficient practices in physical office spaces—such as using energy-efficient lighting or centralized heating and cooling systems—aligns with strategies for home offices. Similarly, green commuting initiatives, such as providing subsidies for public transportation or prioritizing bike-friendly infrastructure, are equally relevant for hybrid and in-office employees. Embedding sustainability into workplace design, such as creating paperless workflows and promoting shared digital resources, can further support environmentally conscious behaviors across all work models.

To integrate these findings into long-term sustainability strategies, organizations must embed energy-efficient and waste-reduction policies within their corporate culture. This begins with clear communication of sustainability goals, ensuring employees understand how their individual actions contribute to broader environmental objectives. Leadership commitment is crucial, as visible support from the management reinforces the importance of green initiatives. For instance, leaders can model eco-friendly behaviors, such as using energy-efficient devices or participating in recycling programs, to inspire similar actions among employees.

Incorporating sustainability into employee training programs can also normalize energy-saving- and waste-reduction practices. Regular workshops or e-learning modules on eco-friendly habits, such as managing digital waste or optimizing home energy use, can help sustain long-term behavioral change. Additionally, integrating sustainability metrics into performance evaluations or corporate reporting can keep environmental objectives at the forefront of organizational priorities. Recognizing and rewarding employees, who actively engage in green practices further reinforces their importance.

Limitations in this study, such as the reliance on conceptual frameworks and the potential bias of self-reported measures in future empirical investigations, should be acknowledged. Self-reports may overestimate pro-environmental behaviors, while demographic factors such as age, role, or industry might influence EGB adoption. Future research should aim to include diverse populations and validate findings through longitudinal or experimental studies to provide a more comprehensive understanding of EGB across different work environments.

If managed effectively, remote work aligns individual behaviors with organizational sustainability goals, creating a win–win scenario for employees and the planet. By adopting a holistic approach, organizations can transform remote work into a powerful tool for reducing their environmental footprint and contributing to a sustainable future.

Appendix: Question Items

Employee Green Behavior

1. I avoid printing and use digital alternatives whenever possible.
2. I recycle household waste, including paper, plastics, and electronics.
3. I turn lights off when not in use.
4. I take part in environmentally friendly programs.
5. I suggest eco-friendly practices to my organization that could improve environmental performance.

Organizational Support

1. My organization values my environmental contributions while working remotely.
2. The organization fails to appreciate any of my environmental efforts (r).
3. My organization would ignore any complaint, from me, relating to their harmful environmental practices (r).

4. My organization cares about my views on sustainability and environmental issues.
5. Even if I did the best environmentally friendly job possible, my organization would fail to notice (r).
6. The organization shows very little concern for my environmental opinions (r).

Green Commuting Intention

1. I am willing to use public transport, even if it is inconvenient some of the time.
2. If it is close, I would like to take public transport or walk.
3. When I need to travel to the office, I prefer to use public transport or other eco-friendly options (e.g., cycling, carpooling).
4. I would like to participate in activities related to low-carbon travel.

Digital Waste Reduction

1. We minimize resource consumption in our business processes.
2. We consider the use of sustainable materials in our business processes.
3. Optimization of processes is used to reduce energy consumption.
4. We consider water efficiency in our projects.

Energy-efficient Practices

1. When I leave the workplace for a long time, I will turn off all unnecessary lighting and air conditioners.
2. I will take the initiative to turn off the lighting of the stairwell when the sunshine is sufficient during the daytime.
3. When the indoor air conditioner is running, I will close the door spontaneously.
4. When I see that a colleague is wasting resources or electricity, I will remind and advise him/her to change such behavior.

References

Afsar, B., & Umrani, W. A. (2020). Corporate social responsibility and pro-environmental behavior at the workplace: The role of moral reflectiveness, employee commitment, and organizational support. *Corporate Social Responsibility and Environmental Management, 27*(1), 109–125.

Ajzen, I. (1991). The theory of planned behavior. *Organizational Behavior and Human Decision Processes, 50*(2), 179–211.

Al-Ghazali, B. M., & Afsar, B. (2021). Green human resource management and employee green behavior: The role of pro-environmental psychological climate and employee green values. *Corporate Social Responsibility and Environmental Management, 28*(5), 1448–1464.

Asatiani, A., & Norström, L. (2023). Information systems for sustainable remote workplaces. *The Journal of Strategic Information Systems, 32*(3), 101789.

Awasthi, A. K., Cucchiella, F., D'Adamo, I., Li, J., & Zeng, X. (2022a). Global perspectives on e-waste management: From waste to resource recovery. *Waste Management, 150*, 321–329.

Awasthi, A. K., Zeng, X., Li, J., & Singh, R. (2022b). Sustainable e-waste management: An emerging challenge. *Journal of Hazardous Materials, 423*, 127893.

Blok, V., Wesselink, R., Studynka, O., & Kemp, R. (2015). Encouraging sustainability in the workplace: A survey on the pro-environmental behaviour of university employees. *Journal of Cleaner Production, 106*, 55–67.

Bouarar, A. C., & Mouloudj, K. (2021). Using the theory of planned behavior to explore employees' intentions to implement green practices. *Dirassat Journal Economic Issue, 12*(1), 641–659.

Chen, C. H. V., & Chen, Y. C. (2021). Assessment of enhancing employee engagement in energy-saving behavior at workplace: An empirical study. *Sustainability, 13*(5), 2457.

Cheng, Z., Wu, B., Deng, X., & Li, W. (2022). The impact of employees' pro-environmental behaviors on corporate green innovation performance: The mediating effect of green organizational identity. *Frontiers in Psychology, 13*, 984856.

Chin, W. W. (1998). The partial least squares approach to structural equation modeling. In G. A. Marcoulides (Ed.), *Modern methods for business research* (pp. 295–336). Lawrence Erlbaum Associates Publishers.

Choudhury, P., Khanna, T., Makridis, C. A., & Schirmann, K. (2022). *Is hybrid work the best of both worlds? Evidence from a field experiment*. Harvard Business School Working Paper No. 22-063.

Cicala, S. (2020). *Powering work from home*. National Bureau of Economic Research Working Paper No. 27937. Retrieved from https://www.nber.org/system/files/working_papers/w27937/w27937.pdf

Cochran, W. G. (1977). *Sampling techniques* (3rd ed.). John Wiley & Sons.

eSmart Recycling. (2024). *eSmart Recycling in 2024: Leaders in electronic recycling and sustainability in Tampa*. Retrieved from https://esmartrecycling.com/blog/esmart-recycling-in-2024-leaders-in-electronic-recycling-and-sustainability-in-tampa

Galanti, T., Guidetti, G., Mazzei, E., Zappalà, S., & Toscano, F. (2021). Work from home during the COVID-19 outbreak: The impact on employees' well-being and productivity. *Frontiers in Psychology, 12*, 632225.

Hair, J. F., Hult, G. T. M., Ringle, C. M., & Sarstedt, M. (2019). *A primer on partial least squares structural equation modeling (PLS-SEM)* (2nd ed.). SAGE Publications.

Hameed, Z., Khan, I. U., Islam, T., Sheikh, Z., & Naeem, R. M. (2021). Do green HRM practices influence employees' environmental performance? The role of green knowledge sharing and pro-environmental behavior. *Corporate Social Responsibility and Environmental Management, 28*(5), 1056–1064.

Hook, A., Court, V., Sovacool, B. K., & Sorrell, S. (2020). A systematic review of the energy and climate impacts of teleworking. *Environmental Research Letters, 15*(9), 093003.

International Energy Agency (IEA). (2021). *The role of energy efficiency in post-pandemic economic recovery*. IEA Reports.

Katz, I. M., Rauvola, R. S., Rudolph, C. W., & Zacher, H. (2022). Employee green behavior: A meta-analysis. *Corporate Social Responsibility and Environmental Management, 29*(5), 1146–1157.

Liao, C., Huang, Y., Zheng, Z., & Xu, Y. (2023). Investigating the factors influencing urban residents' low-carbon travel intention: A comprehensive analysis based on the TPB model. *Transportation Research Interdisciplinary Perspectives, 22*, 100948.

Liu, Z., Ciais, P., Deng, Z., et al. (2020). Near-real-time monitoring of global CO2 emissions reveals the effects of the COVID-19 pandemic. *Nature Communications, 11*, 5172.

Lossin, F., Kozlovskiy, I., Sodenkamp, M., & Staake, T. (2016). *Incentives to go green: An empirical investigation of monetary and symbolic rewards to motivate energy savings*. Research Papers. 70.

Maji, D., Pfaff, B., Vipin, P. R., Sreenivasan, R., Firoiu, V., Iyer, S., Josephson, C., Pan, Z., & Sitaraman, R. K. (2023). Bringing carbon awareness to multi-cloud application delivery. In *Proceedings of the 2nd workshop on sustainable computer systems (HotCarbon '23), July 9, 2023, Boston, MA, USA*. ACM. https://doi.org/10.1145/3604930.3605711

Malokani, D. K. A. K., Qureshi, A. A., Inayat, A., & Mahar, Q. (2023). Impact of reduction of waste, green employee behavior, and reduction of resource use on environmental strategy:

Manufacturing companies' top management's perception. *Journal of Social Sciences Review, 3*(2), 27–39.
Norton, T. A., Zacher, H., & Ashkanasy, N. M. (2014). Organisational sustainability policies and employee green behaviour: The mediating role of work climate perceptions. *Journal of Environmental Psychology, 38*, 49–54.
Onnis, L. A. (2019). Human resource management policy choices, management practices, and health workforce sustainability: Remote Australian perspectives. *Asia Pacific Journal of Human Resources, 57*(1), 3–23.
Orzeł, B., & Wolniak, R. (2022). Digitization in the design and construction industry—Remote work in the context of sustainability: A study from Poland. *Sustainability, 14*(3), 1332.
Richardson, H., & Kelliher, C. (2015). Managing energy consumption in hybrid offices: The role of flexible working. *Journal of Corporate Sustainability, 12*(3), 125–140.
Rossitto, C., Comber, R., Tholander, J., & Jacobsson, M. (2022). Towards digital environmental stewardship: The work of caring for the environment in waste management. In *Proceedings of the 2022 CHI Conference on human factors in computing systems* (pp. 1–15).
Saunders, M. N. K., Lewis, P., & Thornhill, A. (2019). *Research methods for business students* (8th ed.). Pearson.
Sharma, S., Mehta, A., & Sharma, S. (2023). A bibliometric analysis of electronic waste management: Issues and challenges. *Current World Environment, 18*(3).
Sharon, T. (2016). Big data, big waste? A reflection on the environmental sustainability of big data initiatives. *Science, Technology, & Human Values, 41*(5), 692–712.
Singh, R., & Pandey, S. (2023). Strategies for effective digital waste management in organizations. *Journal of Environmental Management, 250*, 109456.
Tao, Y., Yang, L., Jaffe, S., Amini, F., Bergen, P., Hecht, B., & You, F. (2023). Climate mitigation potentials of teleworking are sensitive to changes in lifestyle and workplace rather than ICT usage. *Proceedings of the National Academy of Sciences, 120*(39), e2304099120.
Tao, Y., You, S., Zhu, J., & You, F. (2024). Energy, climate, and environmental sustainability of the trend toward occupational-dependent hybrid work: Overview, research challenges, and outlook. *Journal of Cleaner Production, 442*, 141083.
Temminck, E., Mearns, K., & Fruhen, L. (2015). Motivating employees towards sustainable behaviour. *Business Strategy and the Environment, 24*(6), 402–412.
Wu, H., Han, M., & Shen, Y. (2023). Technology-driven energy revolution: The impact of digital technology on energy efficiency and its mechanism. *Frontiers in Energy Research, 11*, 1242580.
Zacher, H., & Rudolph, C. W. (2022). Employee green behavior as the core of environmentally sustainable organizations. *Annual Review of Organizational Psychology and Organizational Behavior, 9*, 223–249.
Zhang, B., Yang, L., Cheng, X., & Chen, F. (2021). How does employee green behavior impact employee well-being? An empirical analysis. *International Journal of Environmental Research and Public Health, 18*(4), 4169.

Promoting Environmental Sustainability in the Workplace: Strategies for Business Organizations and Employee Well-being

Shoeb Ahmad, Abdulaziz Abdulmohsen Alfalih, and Chetan Sharma

Abstract Awareness of human activity's harmful impact on the environment has made environmental sustainability a major issue in modern civilization. It has become one of the most crucial ethical imperatives for businesses, which focus more and more on encouraging employees to act in the greenest and most environmentally sustainable ways when within the organization. Green human resource management (HRM) approaches, incorporating environmental considerations, are being used to address this. HRM, rewards, and motivation with green HRM practices are key to environmental sustainability in the company of today. HRM's involvement in supporting sustainable practices has grown as organizations tackle environmental issues. Organizations can motivate employees to embrace environmentally friendly practices and contribute to organizational sustainability goals by linking rewards and recognition to green habits and performance. As a consequence, sustainability in the workplace has become a crucial topic for scholars to study. In this study, the authors considered a corpus from the Scopus database from 1984 to 2024. They performed bibliometric analysis using VOSviewer to provide insights into year-wise publication, leading authors, dominating journals, and keyword analysis. The chapter highlights various steps businesses can take to promote workplace sustainability, which can benefit both the business and employee well-being.

S. Ahmad (✉)
Department of Management, College of Business & Management, Fahad Bin Sultan University, Tabuk, Kingdom of Saudi Arabia
e-mail: sahmad@fbsu.edu.sa

A. A. Alfalih
Department of Business Administration, College of Business & Economics, Qasim University, Buraidah, Kingdom of Saudi Arabia
e-mail: afalh@qu.edu.sa

C. Sharma
PW-Institute of Innovation, PhysicsWallah Limited, Lucknow, India
e-mail: chetan.sharma@pw.live

M. Y. Yusliza et al. (eds.), *Employee Green Behavior*,
https://doi.org/10.1007/978-981-95-1422-9_7

1 Introduction

The importance of environmental sustainability has grown significantly in contemporary society, as people are becoming more conscious of the detrimental effects of human activities on the environment (Uralovich et al., 2023). As the world heads toward multiple climate crises, staying sustainable has become essential, especially for the corporate world. Many organizations have environmental sustainability goals as a significant part of their success strategy, evaluating their contribution toward overall organizational goals (Malik et al., 2021). An organization can only accomplish its environmental sustainability goals through the proper actions of its employees at different levels (Unsworth et al., 2021). Employee green behavior (EGB) entails behavior under the employee's control that is measurable in terms of quality and frequency (Ansari et al., 2021). The impact of their performance on the natural environment can be quickly evaluated. Employees' sustainable behavior differs from nonwork pro-environmental behavior. Even though the level of relevancy can differ depending on an employee's position and formal job role, EGB still contributes to the overall success of the organization's sustainability goals (Aboramadan, 2022). Green behavior can also be discretionary or include particular levels of adaptability (Al-Swidi et al., 2021). Employees' sustainable behavior can be shown as part of their work roles.

At different hierarchal levels, from the bottom to top managers, employees' green and sustainable behavior can be classified along three main dimensions (Chang & Hung, 2021). First, sustainable behavior can be carried out per the role end tasks. Second, it can be direct or indirect. For example, employees may take actions to directly ensure environmental benefits. Third, EGB can range from low-intensity to high-intensity efforts, which include more risk (Chaudhary, 2020).

The formation and implementation of EGB within a corporation can be understood as illustrated in Fig. 1. The whole process begins with individual characteristics, such as values and attitudes regarding the environment, which are subsequently reinforced by organizational elements, including culture, leadership, and human resource policy. When these occurrences transpire, employees engage in environmentally beneficial actions, resulting in EGB. This produces perceptible outcomes such as reduced expenses and an enhanced environmental reputation.

Ensuring sustainable and green practices in the workplace helps an organization achieve its overall sustainability goal. It pushes employees toward environmental sustainability within and outside the organization (Al-Swidi et al., 2021). In response to this problem, firms have begun using green human resource management (HRM) methods, which aim to include environmental factors in HRM procedures (Saeed et al., 2019). According to Podgorodnichenko et al. (2020), as businesses try to figure out how to deal with environmental issues, HRM's role in encouraging eco-friendly actions has grown. Natural maintainability is being integrated into HR technologies. Green administration of personnel is rationalized by extensive research. These researches might aid hiring managers in implementing an environmentally friendly employee relations system. To safeguard and grow firms, it may

Fig. 1 Employee green behaviors. Author's own work

encourage staff to adopt environmentally friendly methods of management. Further investigation may reveal fresh information.

This study uniquely contributes to understanding EGB in the workplace by examining the intersection of green HRM practices, green leadership, and employee behavior in fostering workplace sustainability. By integrating a bibliometric analysis of publications from 1984 to 2024, the study provides a comprehensive overview of how green behavior influences organizational goals and sustainability strategies. It highlights critical components, such as green training, empowerment, and motivational tools, while demonstrating the role of transformational leadership in aligning employees' actions with environmental objectives. Furthermore, the study emphasizes the necessity of embedding environmental sustainability into HRM policies and organizational cultures to stimulate pro-environmental behaviors. By linking these practices to performance evaluations, compensation systems, and green innovations, it offers actionable insights for creating eco-conscious workplaces. This approach not only contributes to advancing theoretical frameworks but also provides practical strategies to enhance employee engagement and ecological sustainability within organizations.

Stevan Hobfoll (1989) proposed the conservation of resources (COR) Theory in 1989. It is a motivational theory that examines how people seek to acquire, maintain, and defend their prized resources. COR Theory's key ideas include the importance of resources such as time, energy, personal attributes, social support, items, and settings. People become stressed when they are threatened with losing resources, lose resources, or do not obtain resources despite working hard.

Resource gain: You must invest money in resources to protect them from loss, recover them after they have been lost, or acquire more of them. Implementing EGB

COR theory discusses how and why employees choose to do (or not do) green behaviors in the context of EGB, based on their perceptions of the costs and benefits: Employees may refuse to do green things if they believe it will cost them time, energy, or performance (for example, if they must take extra measures to recycle or commute in an environmentally friendly manner).

Resource gain spiral: When employers foster EGB by providing employees with rewards, recognition, or a lighter workload, these actions may be viewed as opportunities to obtain additional resources.

Stress reduction: When green practices are integrated into workflows and supported by infrastructure, employees are less likely to consider them a waste of resources. Resource conservation theory, which embraces the preservation of resources and their proper use, is instrumental in encouraging a green stance in organizations.

Deci and Ryan (1985) developed the Self-Determination Theory (SDT), which was further expanded by Ryan and Deci (2000). This established psychological paradigm elucidates the rationale for human behavior through the fulfillment of three fundamental psychological needs which is elaborated as Three Fundamental Needs in Self-Determination Theory:

Autonomy—the desire to perceive oneself as the architect of one's own actions.
Competence—the necessity to perceive oneself as capable of performing tasks proficiently.
Relatedness—the necessity to experience respect and connection with others.

Although there is significant literature on consumer green behavior, the cross-fertilization of social exchange theory and self-determination theory is an important avenue that will enhance understanding of green behavior within the business environment. This paper conceptualizes that self-organization adopts social exchange theory, in which employees return the investment in green training and leadership practices provided by their organizations with pro-environmental behaviors due to self-interest and trust. On the other hand, self-determination theory talks about internal motivation, stating that employees are likely to participate in sustainability voluntarily if they can exercise a high level of autonomy, competence, and perceived relatedness in their organization's chosen sustainability goals. In combination, these frameworks present a broader vision of how green practice, green leadership, and HRM can lead to sustainability culture, with sustainability as a prized and planned organizational value proposition.

2 Methodology

This research is subjective and relies on comprehensive assessments of the existing literature on green HRM. Records inspection is widely embraced due to its contribution to the development of research and information in the field. The research aimed to employ effective methodologies to conduct a thorough literature audit via printed books, personal journals, electronic paper, and online portals. A broad range of databases, such as Web of Science, Pubmed, and Scopus, are used for bibliometric analysis. The study employed the VOSviewer software, which offers a comprehensive range of bibliometric tools that allow researchers to enhance and capitalize on the advantages while addressing the limits of traditional reviews. An exhaustive and efficient inquiry was carried out on the gathered materials.

3 Literature Review

Considering today's environmental issues, companies are quickly making green concepts an essential component of their operational strategies (Al-Swidi et al., 2021). Based on resource conservation theory, the green behavior of employees explores green management tools called green HRM practices and green transformational leadership. The green behavior of employees can be understood as employee activities that mainly focus on conserving natural resources, maintaining the ecological environment, and working to reduce environmental degradation and improve environmental quality (Tsai et al., 2017). HRM practices are necessary to manage and regulate green behavior in organizational policies (Sharma et al., 2022a). In exploring the antecedents of EGB, Norton et al. (2015) proposed a multilevel pathway model emphasizing the role of perceived sustainability policy and green work climate perceptions. Their model (Fig. 2) distinguishes between two key mediators—perceptions of the organization's green climate and perceptions of coworkers' green climate—each playing a unique role in shaping different forms of green behavior.

Factors that promote environmentally responsible conduct among employees in the workplace EGB are influenced by personal convictions as well as the organization's facilitation and promotion of sustainability. There are two important contributing factors:

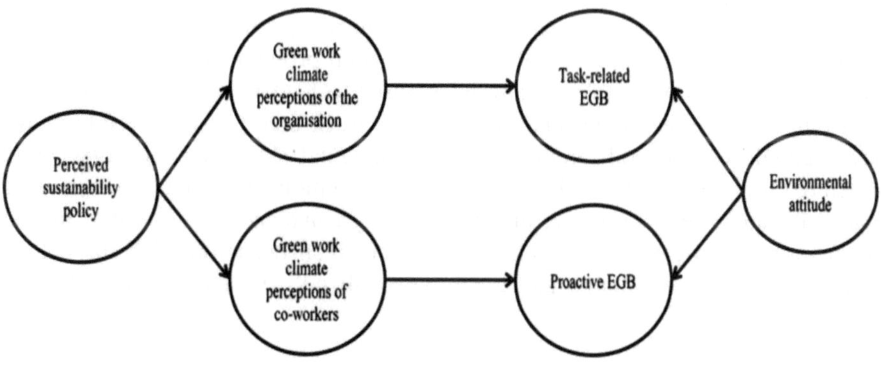

Fig. 2 Adapted from Norton et al. (2015).

Green HRM—it refers to the alignment of human resource practices with environmental objectives. It helps EGB with: green recruitment and training, performance evaluations with goals for sustainability, and incentives for participating in ecologically beneficial activities. Establishing a culture of sustainability in the workplace. Green HRM practices are manifested by introducing green issues into recruitment, training, evaluation of employees' overall performance, and compensation based on sustainability in employees' performance (Dahiya, 2020). Employees are then likely to demonstrate knowledge acquisition in resource conservation, waste utilization, environmental preservation, and sustainable work within the workplace (Zihan et al., 2024). Companies should recruit green-oriented staff and follow green practices to attract and bring in environmentally conscious employees. Such employees will, by default, show green behavior after taking their place within the workplace. In one study, Yang et al. (2018) contended that by organizing environmental skills training, companies can focus on enriching environmental knowledge and train employees to solve environmental problems so they are more capable of implementing green behavior. By introducing green behavior and exposing employees to green issues in the workplace, companies can ensure that employees are showing green behavior in their work (Darvishmotevali & Altinay, 2022). By including employees who promote eco-friendly behavior in performance reviews and pay, employees can better understand their roles and receive fair awards for being proactive and supporting sustainability at work (Tsai et al., 2017).

4 Green Leadership

Green leadership refers to leaders focusing on and motivating employees to achieve green goals and making them demonstrate green behavior beyond the bare minimum. The characteristics of green transformational leadership can impact the coordinated relationship between organizational strategy and HRM (Tsai et al., 2017). This includes influencing the motivation and intellectual stimulation of employees. Green motivation is a tool that organizations use to motivate their employees to go beyond their short-term self-interest and work harder to achieve green goals. Green intellectual stimulation refers to leaders who inspire employees to challenge old ideas and use new methods to solve environmental problems more creatively and sustainably. Green personalized care is essential to leaders who focus on valuing their employees' green contributions, regularly track them, and analyze how they have helped develop both employees' environmental skills and the organization (Edeh & Okwurume, 2019). Green influence, green motivation, and green transformational leaders help employees understand that green behavior is essential and that organizations can easily encourage and expect green behavior. With the help of green intellectual stimulation and green personalized care, many sustainable and environment-related issues can be resolved. Promoting environmental knowledge in the workplace and helping employees develop environmental skills enhances their ability to demonstrate green behaviors. The effectiveness of green initiatives through HRM practices can be tracked by the organization's leadership, by seeing how motivated employees are toward sustainability (Sharma et al., 2022b). When employees perceive a higher level of green HRM practice, and leaders show a high level of green transformation leadership behavior, there are more chances that the employees will be able to perceive both sides and stay consistent, which eventually contributes to the acceleration of recognition and adoption of green ideas in the workplace (Ababneh, 2021). Leaders have a critical role in developing the framework for sustainability. Environmentally minded leaders showcase environmentally mindful actions to serve as an example for others, communicate their vision for sustainability clearly, encourage employee participation, and acknowledge their contributions. Assistance is required to reduce resistance in accordance with COR Theory.

Green HRM and green leadership work together to provide a cultural and institutional basis that supports and encourages effective EGB. These are resources responsible for improving the overall functioning of resource management practices like green training and green empowerment of employees (Singh et al., 2023).

Chen and Wu (2022) proposed the EGB Model within the framework of Chinese culture, as illustrated in Fig. 3. Their study concluded that green transformational leadership and green HRM practices positively influence employees' green behavior, supporting the notion that leadership behavior and human resource management practices function synergistically.

Employees can quickly obtain more consistent indications from their work environment, and can focus on the importance of green behavior and believe that green HRM and sustainability in the workplace can first be implemented within the department, and later on spread across the organization (Norton et al., 2015).

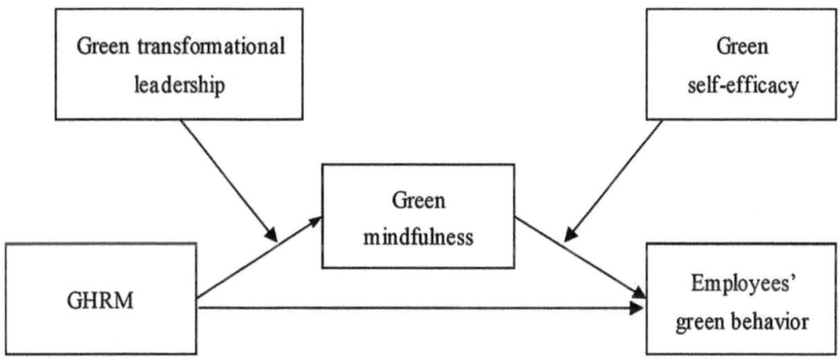

Fig. 3 Adapted from Chen and Wu (2022)

5 Key Factors Impacting Sustainable Talent Development

The impact on the environment of the negligence of multiple organizations around the world in the past few decades has necessitated organizational sustainability. Growing environmental concerns and the demand for sustainable practices have led to a requirement for industrial enterprises to adopt environmentally, commercially, and socially sustainable processes. These days, all organizations—especially those that function in highly regulated and competitive environments—have a need to balance their economic, social, and environmental performance. According to Renwick et al. (2013), striking a balance between these variables is seen as a difficult undertaking that can even be contentious at times. Improving an organization's environmental, social, and economic performance while upholding cultural norms and incorporating eco-friendly practices into daily operations becomes exceedingly challenging when environmental management systems are implemented (Al-Ghazali & Afsar, 2021). The cross-functional dispersion of green ideology may help to overcome these issues (Dost et al., 2019). According to Roscoe et al. (2019), researchers looked into how green management might be implemented in two crucial company areas: innovation and human resources.

6 Primary Difficulties of Implementing Sustainable Management of Personnel

Environmental issues have only become worse in the past 10 years, although being given unprecedented prominence across the globe. The plan for human resources is creating guidelines for supervising people, but it also poses challenges to its implementation. According to Deshwal (2015), green HRM challenges are:

- Changing the behavior of workers within a small time frame is a difficult task.
- Not all employees are equally inclined to engage with the implementation of green HRM techniques inside the firm.
- Developing a green HRM lifestyle throughout the firm can be a cumbersome and time-consuming effort.
- Recruiting and compensating environmentally conscious employees with suitable rewards is difficult.
- It requires an extremely high level of prediction at the beginning and a rate of return close to the average.
- Assessing the suitability of green HR policies for employee behavior is difficult.

A study conducted by Noor Faezah et al. (2022) found that the rising recognition of employees' sustainable and green behavior within the workplace provides many new conceptualizations and questions regarding what green employee behavior in the workplace should be. Some scholars consider the green behavior part of HRM, which focuses on green practices in the workplace. Others claim that employee behavior must go far beyond formal requirements to qualify as green. For a clearer understanding of employees' green behavior, there is a requirement for the development of context-specific and psychological processes that motivate multiple behaviors for organizational sustainability (Pham et al., 2019).

Problems caused by climate change result in consequences like natural disasters and degradation of the overall ecosystem, which have become a concern for every sector and industry. Since the 1980s, demand for green practices within organizations has increased, due to the rapid depletion of natural resources and the increase in pollution caused by unethical organizational activities, especially economic activities related to industrial production. Many organizations now include environmental performance goals in their triple-bottom-line strategy, along with social and economic goals. In the words of Fawehinmi et al. (2020), not only are organizations accountable to their shareholders, but they are also entirely responsible for the safety and well-being of others where the organization is functioning. As per the triple bottom line, the organizations are sponsors who must protect their surroundings and limit economic activities that might impact the environment.

Figure 4 presents a comprehensive multilevel theoretical framework for understanding the antecedents and outcomes of EGB, adapted from Zhang et al. (2022). The model integrates influences from the intrapersonal, interpersonal, organizational, and societal levels, illustrating how EGB is shaped and manifested in the workplace.

When the company is building a culture, it has to invest in all members, from senior leadership to the other employees of the organization, and this cannot be done by dictating to the stakeholders, but by engaging them within the organization's sustainable activities. Staff participation in sustainability entails

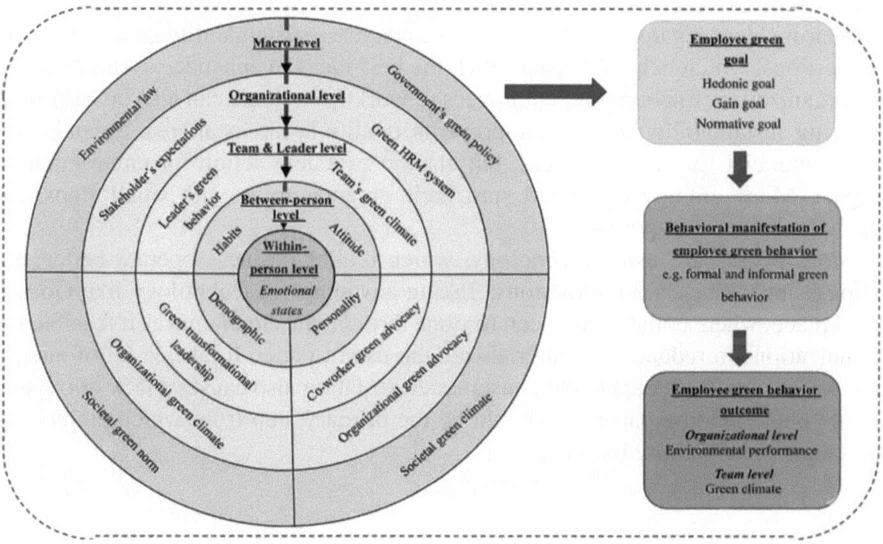

Fig. 4 Adapted from Tang et al. (2023).

interactive discussions on how the organization may eliminate harmful behaviors and embrace sustainability in day-to-day work sessions. Employees have the responsibility of formulating policies and executing them in order to achieve sustainability, which is the overarching environmental objective of the firm. Updating by the leader of the mission statement for the employees is an excellent initiative because it creates a sustainable statement, which is essential to show employees how serious green issues are. The leader needs to be very clear and concise regarding their sustainability statement and explain how staying green matters to the business and employees' personal growth, why its accomplishment is essential, and how it impacts the organization's overall success criteria.

It is essential to use the organization's company newsletter, as one of the most critical internal communication media, to update all employees within the workplace regarding the sustainability push. The leader must ensure that staff know what needs to be achieved, document the efforts to ensure fundamental changes are being made, and make employees understand specifically what targets are essential and need to be achieved first regarding sustainability. Sharing goals and aims with organization members increases the chances of achieving them faster. It inspires employees to set up their own green working practices to contribute to the organization's sustainability.

Another way to promote workplace sustainability is by running an employee awareness campaign. The recruited employees must know about sustainable work practices before implementing the strategy. Running a campaign raises awareness of global practices that can be implemented and the time limit within which sustainable changes must be achieved. Another great way of achieving sustainability in the green workplace is taking constant advice from recruited employees. The organization's employees are a vast pool of ideas, with multiple backgrounds, diversity, and experiences combined.

Sharing ideas regarding green activities to increase sustainability generates many suggestions, which helps the leader pick the best ideas to enhance sustainability in the organization. Encouraging employees to work from home can also be helpful in ensuring sustainability, as it can improve air quality in the local area. It can be not only beneficial for the employees, but also an excellent action for environmental impact. Many employees should start their green journey with small steps, for example, by going paperless.

More people are using technology, which makes business operate better and helps reduce their carbon emissions. Taking advantage of technology like Google Workplace, where collaboration can be done through the cloud, makes it feasible for organizations to reduce or even eradicate the use of paper. It takes a lot of energy and water to produce paper, and constant deforestation also causes a lack of oxygen on the planet. Going paperless should be the primary step from which employees start their sustainability journey.

Conserving water resources is another factor that needs to be focused on (Pham et al., 2019). It is essential to ensure that wasting water is unnecessary, as the impact of climate change is becoming ever more pronounced. Encouraging sustainable transportation in the workplace can be beneficial to start the green revolution (Ojo et al., 2022). Cutting down the amount of employee travel for work with different carbon-emitting sources like airplanes can be a huge step toward sustainability. The organization's leader should encourage sharing transport between employees to ensure the green path is followed. Cutting down transportation, promoting hybrid practices, and holding online meetings instead of traveling to other places are hugely sustainable ways of functioning within the workplace (Arshad et al., 2020).

7 Performance Analysis

In this research, the author collected data from the Scopus database, which is the most reliable and extensive database (Dash et al., 2023). The authors used the following string to extract the database.

(TITLE-ABS-KEY(("GREEN EMPLOYEE" OR "green behavior" OR "sustainable workplace"OR "green human resource management")

In the first pass, the authors received 1726 articles. In the second phase, the authors implied inclusion/exclusion, in which only English-written articles were considered for this study, as English is a global language, which can benefit readers. Finally, the authors considered 1709 articles for this meta-analysis (Singh et al., 2023). The authors used the VOSviewer tool to analyze and present the work.

7.1 Year-Wise Growth in Publication

According to Fig. 5, there has been a consistent increase in the duration of publications in the field of green technology over time. Although there was a lack of scholarly publications during the early stages of the green revolution, with only one paper published in 1984, there is a noticeable tendency toward a substantial increase. Despite experiencing minor oscillations in publication rates, such as a decrease to 198 publications in 2021, the highest point in 2023 was reached with 446 articles, accounting for 26.14% of the corpus. This significant increase reflects the growing interest in this particular field. Recognizing the potential for further publications after the authors' time frame is crucial, indicating continuous progress in investigating and developing environmentally friendly technologies.

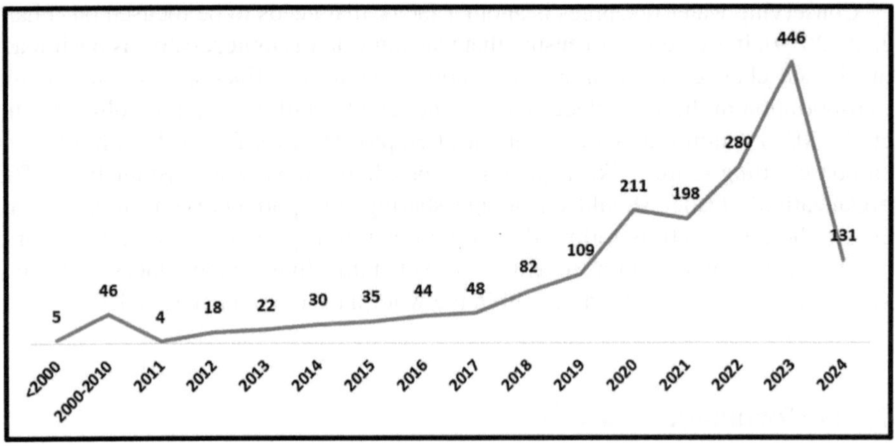

Fig. 5 Year-wise publication analysis. Source: Author's own work

7.2 *Subject Area-Wise Analysis*

A total of 1709 articles have surfaced within the academic literature, spanning many domains. Business, Management, and Accounting emerge as the most prominent, comprising 863 articles. This strongly emphasizes organizational strategy, financial management, and company operations.

The field of Environmental Science is actively monitored, with 593 publications dedicated to the critical analysis and resolution of environmental issues. The field of Social Sciences makes a substantial contribution, as evidenced by the publication of 563 papers that explore a wide range of societal challenges, human behavior, and cultural phenomena. As shown by 304 articles on renewable energy sources, sustainability, and energy regulations, Energy Research is a noteworthy study area. The fields of Engineering and Economics, Econometrics, and Finance have significant content, with 280 and 276 publications, respectively, indicating the interdisciplinary character of these domains. In addition, Computer Science, Decision Sciences, and Psychology have made significant contributions by producing 220, 129, and 124 papers, respectively. These articles highlight the convergence of technology, decision-making processes, and human behavior. Figure 6 represents the subject area-wise analysis.

7.3 *Source Type-Wise Analysis*

Figure 7 represents the analysis based on source type. Unsurprisingly, research was published in various magazines and conference journals. The analysis reveals that MDPI was responsible for publishing the most significant number of publications, which totaled 139. These articles were published in Sustainability. The Journal of

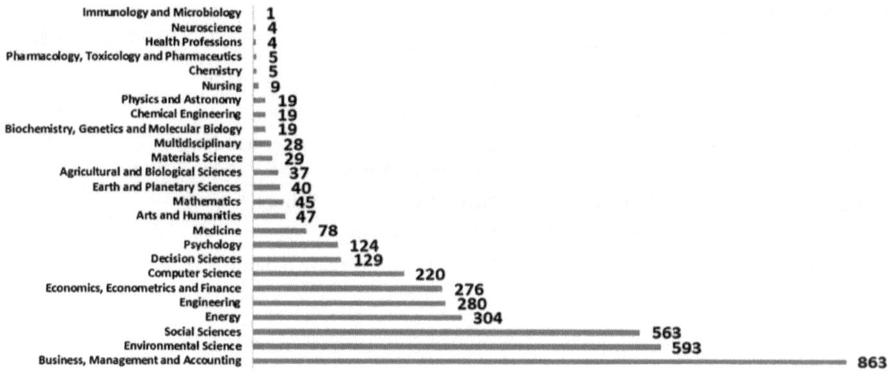

Fig. 6 Subject area-wise analysis. Source: Author's own work

Fig. 7 Source-wise analysis. Source: Author's own work

Cleaner Production was in second position, with 72 articles. Environmental Science and Pollution Research published 39 articles, placing them third. Sustainability has an H-index of 136, and the Journal of Cleaner Production has an H-index of 268.

7.4 Author-Wise Analysis

Within the acknowledged community of scholars who have contributed substantially to their respective areas, some notable authors have made a lasting impact in their respective domains. Table 1 presents the authors contributing the most articles in this field of research.

M.Y. Yusliza, a Malaysian individual linked with Universiti Malaysia Terengganu, has made a significant contribution by publishing 26 scholarly works. C.J.C. Jabbour, who represents France and is connected with NEOMA Business School, has contributed to Business and Management studies through the authorship of 17 works.

Table 1 Author analysis

Author	Count	Country	Affiliation	H-Index
Yusliza, M.Y.	26	Malaysia	Universiti Malaysia Terengganu	39
Jabbour, C.J.C.	17	France	NEOMA Business School	89
Tang, G.	14	China	School of Management, Shandong University, Jinan	NA
Paillé, P.	13	France	NEOMA Business School	41
Nisar, Q.A.	12	Australia	RMIT University	27
Renwick, D.W.S.	12	United Kingdom	Nottingham Business School	23

Source: Author's own work

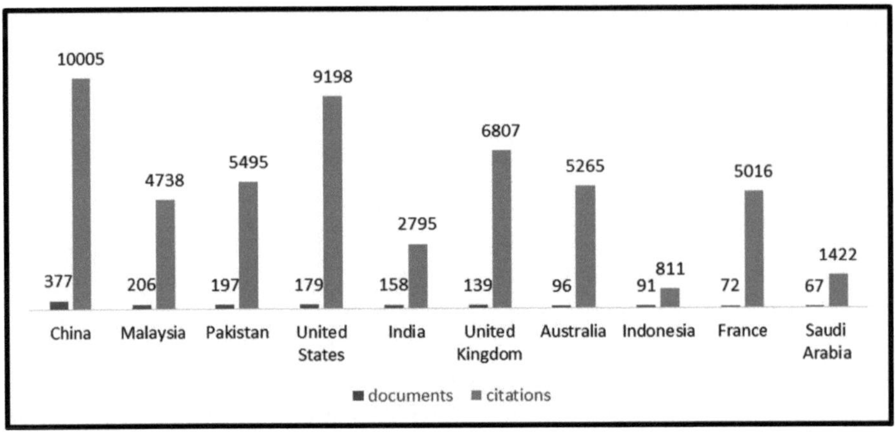

Fig. 8 Countries' contribution based on publication and citation. Source: Author's own work

Furthermore, G. Tang, a scholar from China affiliated with the School of Management at Shandong University in Jinan, has authored 14 articles.

7.5 *Country-Wise Analysis*

Figure 8 represents the analysis based on countries contributing to research regarding the green revolution. The study showed that China leads with 377 articles and 10,005 citations. China contributed 22.06% of the total corpus. Malaysia follows China in terms of published documents. They published 206 articles and had 4738 citations. Meanwhile, the United States follows China in terms of citations. Pakistan, the United Kingdom, and India are also among the top five countries with 197, 179, and 158 articles, respectively.

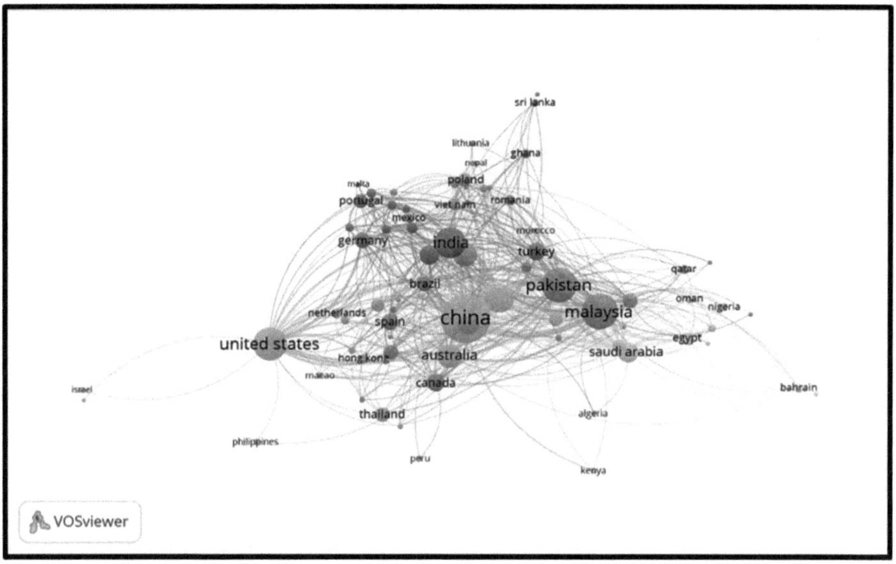

Fig. 9 Countries' contribution based on publication and citation. Source: Author's own work

Figure 9 represents the network graph of countries connected with each other via their research publications. China is leading with 306 link strengths, while Malaysia has 239 link strengths for 206 published articles.

8 Keyword Analysis

The keywords presented in Fig. 10 highlight the increasing focus on sustainability and environmental issues in HRM. Terms such as "Green Human Resource Management" and "Sustainability" indicate an increased emphasis on incorporating environmental factors into human resource operations. Furthermore, incorporating concepts such as "Sustainable Development" and "Environmental Management" signifies the overarching dedication of organizations to achieving enduring ecological equilibrium. Significantly, the inclusion of terminology such as "Human Resource" and "Human Resource Management" in conjunction with "Green Behavior" signifies an integration of conventional HR operations with environmentally aware practices inside the organizational setting.

Moreover, incorporating the "Corporate Social Responsibility" concept underscores firms' growing involvement in tackling environmental concerns. The substantial inclusion of China in the keyword list indicates the country's acknowledgment of the significance of green HRM and environmental conservation within its

Fig. 10 Word cloud of keyword analysis. Source: Author's own work

industrial and organizational context. The themes mentioned above illustrate a transition toward a comprehensive approach to HRM, wherein sustainability principles are incorporated to promote organizational achievement and environmental responsibility.

9 Implications

This study can help manufacturing companies set environmental strategy goals that are connected to certain parts of their HRM and innovation processes. Employees will likely play a big role in making environmental laws because of this (Mousa & Othman, 2020). There is evidence that businesses should switch to an environmental management system that both human resource managers and operations managers like. Organizations should actively cultivate interest in environmental management and strive to establish an ecologically friendly image to attract like-minded job seekers.

In order to stimulate employees' enthusiasm and creativity toward environmental sustainability, companies should establish a supportive work environment by providing financial assistance, implementing advanced environmental management techniques, and creating a framework that encourages employees' innovative ideas for green initiatives.

Organizations are guided to cultivate sustainable practices and establish an eco-friendly atmosphere. This in turn provides the necessary resources and support to generate ideas that are beneficial for the organization's sustainable development. Training on sustainability should not solely focus on enhancing workers' environmental security skills, but should also strive to communicate the organization's environmental management goals and values to individuals.

In the early stages of the development of green HRM, innovators can apply transformational practices to establish benchmarks for cooperation, sustainability, and growth. They can also provide a certain amount of organizational resources to generate environmentally conscious employee enthusiasm and creativity. Transformational administration can enhance enterprises' environmental creativity by influencing green HRM and stimulating workers' inventions.

In order to improve green behavior in conjunction with the practical implications of organizational structure, organizations can implement specific strategies and follow specific steps. For instance, integrating environmental objectives into HRM activities, including selection, training, appraisal, and performance incentive measures, may foster green behaviors.

One real-world example is Google, which promoted car sharing to work by offering financial bonuses. Likewise, IKEA has implemented green HRM by involving its employees in sustainability through training and making the company's individual and organizational goals green.

Specific recommendations involve the introduction of green training, conducting environmental awareness activities, and establishing green incentives, which include cash rewards for green practices and a trophy for green ideas. Additionally, the application of technology to eradicate paper use and encourage working from home decreases an organization's carbon footprint. Another way, using the pluralistic approach, is making major organizational decisions in relation to the green policy through all-staff brainstorming meetings, so that commitment is inherent among all employees. Organizations can build collective responsibility through environmental culture and employee engagement.

10 Conclusion

A sustainable workplace where employees are motivated to perform green practices has become a basic need for organizations in the world of climate change and crisis. Sustainability is already crucial globally and in almost every sector. Because of the extra generation of waste by different industries, it has become mandatory for organizations to meet sustainability obligations and encourage their employees to focus on green activities within the workplace. Unstable businesses usually share the same problem, which is a need for sustainability practices.

Sustainability in the workplace consists of business practices that focus on promoting conservation and encouraging socially and economically responsible behavior toward the environment. Environmental issues directly impact the everyday activity of business functions, and ecological sustainability includes addressing aspects of the workplace, like how negativity in society as well as in the workplace impacts the planet and the behavior of employees.

EGB within organizations not only benefits the environment, but also spreads to different sections of society and the economy. Economic sustainability helps organizations create a sustainable workplace and promotes business practices that endorse economic conditions for a more comfortable working life. Elements of financial sustainability (such as active ownership, sustainable projects, and green bonds) are the fundamental mainstay for promoting employees' green behavior in the workplace. Adopting sustainable practices within the workplace helps promote a healthier workplace and reduces overall operational costs.

COVID-19 has played a massive role in promoting workplace sustainability and motivating employees to focus more on green behavior. Unintentionally, the pandemic forced people to work from home, which was an excellent initiative for organizations looking for sustainability management by promoting hybrid working spaces. Extant literature surveys mentioned in the above sections has revealed that remote working has turned out to be more economically and environmentally effective than traditional working methods.

Green practices should be promoted in all types of organizations, and they should be engrained into employee behavior. Existing employees are responsible for motivating their juniors and newly joined workforce to focus on keeping an organization sustainable and contributing to its overall sustainability. Performing green activities in the workplace will help employees not only to become better workers, but also to become better people in their personal lives and be more careful regarding the impact they have on the environment.

References

Ababneh, O. M. A. (2021). How do green HRM practices affect employees' green behaviors? The role of employee engagement and personality attributes. *Journal of Environmental Planning and Management, 64*(7), 1204–1226.

Aboramadan, M. (2022). The effect of green HRM on employee green behaviors in higher education: The mediating mechanism of green work engagement. *International Journal of Organizational Analysis, 30*(1), 7–23.

Al-Ghazali, B. M., & Afsar, B. (2021). Retracted: Green human resource management and employees' green creativity: The roles of green behavioral intention and individual green values. *Corporate Social Responsibility and Environmental Management, 28*(1), 536. https://doi.org/10.1002/csr.1987

Al-Swidi, A. K., Gelaidan, H. M., & Saleh, R. M. (2021). The joint impact of green human resource management, leadership and organizational culture on employees' green behaviour and organisational environmental performance. *Journal of Cleaner Production, 316*, 128112. https://doi.org/10.1016/j.jclepro.2021.128112

Ansari, N. Y., Farrukh, M., & Raza, A. (2021). Green human resource management and employees pro-environmental behaviors: Examining the underlying mechanism. *Corporate Social Responsibility and Environmental Management, 28*(1), 229–238.

Arshad, M., Abid, G., & Khan, M. (2020). Impact of employee's environmental concern on ecological green behaviour: Mediation mechanism of employee-customer oriented OCB and organisational commitment. *International Journal of Innovation, Creativity and Change, 14*(12), 614–633.

Chang, T. W., & Hung, C. Z. (2021). How to shape the employees' organization sustainable green knowledge sharing: Cross-level effect of green organizational identity effect on green management behavior and performance of members. *Sustainability, 13*(2), 626.

Chaudhary, R. (2020). Green human resource management and employee green behavior: An empirical analysis. *Corporate Social Responsibility and Environmental Management, 27*(2), 630–641.

Chen, T., & Wu, Z. (2022). How to facilitate employees' green behavior? The joint role of green human resource management practice and green transformational leadership. *Frontiers in Psychology, 13*, 906869. https://doi.org/10.3389/fpsyg.2022.906869

Dahiya, R. (2020). Does organisational sustainability policies affect the environmental attitude of employees? The missing link of green work climate perceptions. *Business Strategy & Development, 3*(3), 395–403.

Darvishmotevali, M., & Altinay, L. (2022). Green HRM, environmental awareness and green behaviors: The moderating role of servant leadership. *Tourism Management, 88*, 104401.

Dash, G., Sharma, C., & Sharma, S. (2023). Sustainable marketing and the role of social media: An experimental study using natural language processing (NLP). *Sustainability, 15*(6), 5443.

Deci, E. L., & Ryan, R. M. (1985). *Intrinsic motivation and self-determination in human behavior*. Plenum Press.

Deshwal, P. (2015). Green human resource management: An organizational strategy of greening people. *International Journal of Applied Research, 1*(13), 176–181.

Dost, M., Pahi, M. H., Magsi, H. B., & Umrani, W. A. (2019). Effects of sources of knowledge on frugal innovation: Moderating role of environmental turbulence. *Journal of Knowledge Management, 23*(7), 1245–1259. https://doi.org/10.1108/JKM-01-2019-0035

Edeh, F. O., & Okwurume, C. N. (2019). Green human resource management and organisational sustainability of deposit money banks in Nigeria. *Journal of Management Sciences, 3*(1), 224–235. https://doi.org/10.2139/ssrn.3504961

Fawehinmi, O., Yusliza, M. Y., Wan Kasim, W. Z., Mohamad, Z., & Sofian Abdul Halim, M. A. (2020). Exploring the interplay of green human resource management, employee green behaviour, and personal moral norms. *SAGE Open, 10*(4), 2158244020982292.

Hobfoll, S. E. (1989). Conservation of resources: A new attempt at conceptualizing stress. *American Psychologist, 44*, 513–524.

Malik, S. Y., Hayat Mughal, Y., Azam, T., Cao, Y., Wan, Z., Zhu, H., & Thurasamy, R. (2021). Corporate social responsibility, green human resources management, and sustainable performance: Is organizational citizenship behavior towards environment the missing link? *Sustainability, 13*(3), 1044.

Mousa, S. K., & Othman, M. (2020). The impact of green human resource management practices on sustainable performance in healthcare organisations: A conceptual framework. *Journal of Cleaner Production, 243*, 118595. https://doi.org/10.1016/j.jclepro.2019.118595

Noor Faezah, J., Yusliza, M. Y., Azlina, Y. N., Saputra, J., & Zulkifli, W. K. W. (2022). Developing a conceptual model to implement the employee ecological behavior in organisations. *Journal of Environmental Management & Tourism, 13*(3), 746–755.

Norton, T. A., Parker, S. L., Zacher, H., & Ashkanasy, N. M. (2015). Employee green behavior: A theoretical framework, multilevel review, and future research agenda. *Organization & Environment, 28*(1), 103–125. https://doi.org/10.1177/1086026615575773

Ojo, A. O., Tan, C. N. L., & Alias, M. (2022). Linking green HRM practices to environmental performance through pro-environment behaviour in the information technology sector. *Social Responsibility Journal, 18*(1), 1–18.

Pham, N. T., Tučková, Z., & Phan, Q. P. T. (2019). Greening human resource management and employee commitment toward the environment: An interaction model. *Journal of Business Economics and Management, 20*(3), 446–465.

Podgorodnichenko, N., Edgar, F., & McAndrew, I. (2020). The role of HRM in developing sustainable organizations: Contemporary challenges and contradictions. *Human Resource Management Review, 30*(3), 100685.

Renwick, D. W. S., Redman, T., & Maguire, S. (2013). Green human resource management: A review and research agenda. *International Journal of Management Review, 15*(1), 1–14. https://doi.org/10.1111/j.1468-2370.2011.00328.x

Roscoe, S., Subramanian, N., Jabbour, C. J. C., & Chong, T. (2019). Green human resource management and the enablers of green organisational culture: Enhancing a firm's environmental performance for sustainable development. *Business Strategy and the Environment, 28*(5), 737–749. https://doi.org/10.1002/bse.2277

Ryan, R. M., & Deci, E. L. (2000). Self-determination theory and the facilitation of intrinsic motivation, social development, and well-being. *American Psychologist, 55*(1), 68–78.

Saeed, B. B., Afsar, B., Hafeez, S., Khan, I., Tahir, M., & Afridi, M. A. (2019). Promoting employee's proenvironmental behavior through green human resource management practices. *Corporate Social Responsibility and Environmental Management, 26*(2), 424–438.

Sharma, C., Ahmad, S., & Singh, S. (2022a). Impact of human resource practices on individual and organization growth. In *2022 International Conference on decision aid sciences and applications (DASA)* (pp. 937–941). IEEE.

Sharma, C., Sakhuja, S., & Nijjer, S. (2022b). Recent trends of green human resource management: Text mining and network analysis. *Environmental Science and Pollution Research, 29*, 84916–84935. https://doi.org/10.1007/s11356-022-21471-9

Singh, S., Sharma, C., Bali, P., Sharma, S., & Shah, M. A. (2023). Making sense of glass ceiling: A bibliometric analysis of conceptual framework, intellectual structure and research publications. *Cogent Social Sciences, 9*(1), 2181508.

Tang, G., Ren, S., Wang, M., Li, Y., & Zhang, S. (2023). Employee green behaviour: A review and recommendations for future research. *International Journal of Management Reviews, 25*(2), 297–317.

Tsai, H. L., Lu, C. S., & Chang, C. C. (2017). The influence of organisational green climate on employees' green behaviors: Evidence from the Eco Port of Kaohsiung. *International Journal of Shipping and Transport Logistics, 9*(6), 696–723.

Unsworth, K. L., Davis, M. C., Russell, S. V., & Bretter, C. (2021). Employee green behaviour: How organizations can help the environment. *Current Opinion in Psychology, 42*, 1–6.

Uralovich, K. S., Toshmamatovich, T. U., Kubayevich, K. F., Sapaev, I. B., Saylaubaevna, S. S., Beknazarova, Z. F., & Khurramov, A. (2023). A primary factor in sustainable development and environmental sustainability is environmental education. *Caspian Journal of Environmental Sciences, 21*(4), 965–975.

Yang, L., Manika, D., & Bowen, F. (2018). Organisational and employee symbolic environmental behaviors: An integrated multi-level framework. In *Research handbook on employee pro-environmental behaviour* (pp. 229–264). Edward Elgar Publishing.

Zhang, Y. L., Zhao, Z. Y., Yang, Y. Y., Fu, B. J., Ma, R. M., Lue, Y. H., & Wu, X. (2022). Identifying ecological security patterns based on the supply, demand and sensitivity of ecosystem service: A case study in the Yellow River Basin, China. *Journal of Environmental Management, 315*, 115158.https://doi.org/10.1016/j.jenvman.2022.115158

Zihan, W., Makhbul, Z. K. M., & Alam, S. S. (2024). Green human resource management in practice: Assessing the impact of readiness and corporate social responsibility on organizational change. *Sustainability, 16*(3), 1153.

The Impact of Perceived Organizational Support for the Environment on Hotel Employee Green Behavior in the Workplace: Evidence from Vietnam

Hung Trong Hoang, Nga Thi Thuy Ho, Nhat Tan Pham, and Lan Thi Kieu Mai

Abstract Extant research on the linkage between perceived organizational support for the environment (POSE) and hotel employee green behavior (EGB) in the workplace, especially studies that investigate the boundary conditions of this effect, is very limited. Using social exchange theory, this study aims to fill this research gap. Data were collected from a two-wave survey of 316 four- and five-star hotel employees in Vietnam, and structural equation modeling (SEM) was applied to analyze the data. The findings show that, in addition to being an important driver of hotel EGB, POSE also significantly influences it through employee green passion. Furthermore, employees' collectivist orientation constitutes a boundary condition that enhances the effect of employee green passion on hotel EGB. This research extends existing literature on the POSE–EGB relationship by explicating both mediation and moderation effects and provides managerial implications for promoting EGB.

H. T. Hoang
Faculty of Business Administration, University of Economics, Hue University, Hue City, Vietnam
e-mail: hthung@hueuni.edu.vn

N. T. T. Ho (✉)
Faculty of Accounting and Finance, University of Economics, Hue University, Hue City, Vietnam
e-mail: hothuynga@hueuni.edu.vn

N. T. Pham
International University, Vietnam National University, Ho Chi Minh City, Vietnam
e-mail: ptnhat@hcmiu.edu.vn

L. T. K. Mai
Dalat University, Dalat City, Vietnam

School of Hospitality and Tourism, Hue University, Hue City, Vietnam
e-mail: lanmtk@dlu.edu.vn; maithikieulan@hueuni.edu.vn

M. Y. Yusliza et al. (eds.), *Employee Green Behavior*,
https://doi.org/10.1007/978-981-95-1422-9_8

1 Introduction

The increasing pressure to address environmental challenges such as climate change and resource depletion has forced organizations to change, placing sustainability goals at the forefront of their development (Pham et al., 2019). This pressure comes from different stakeholders, including governments, investors, and consumers, requiring businesses to demonstrate their commitment to environmental accountability (Pham et al., 2023b). While organizations often adopt advanced green technologies and strategies, the success of these initiatives heavily relies on the actions and behaviors of their employees. This highlights the critical role of employee green behavior (EGB), which is defined as the voluntary or discretionary actions by employees that contribute to environmental sustainability at work (Norton et al., 2015). Understanding and fostering such behaviors is vital to bridging organizational strategies with tangible environmental outcomes.

EGB is not merely a by-product of organizational policies but an essential driver of sustainability efforts (Zacher et al., 2023). Employees' actions, whether conserving energy, reducing waste, or participating in green initiatives, directly influence the success of environmental policies. However, organizations often encounter challenges in ensuring employee alignment with sustainability goals. Despite considerable investment in green infrastructure and processes, the human dimension, including how employees perceive, engage with, and enact green initiatives, remains insufficiently explored (Norton et al., 2017). Studying EGB provides a nuanced understanding of these dynamics, offering insights into how organizations can effectively promote environmentally responsible behaviors among their workforce (Katz et al., 2022; Zacher et al., 2023).

In addition to addressing organizational needs, exploring EGB provides critical benefits for employees themselves. Engaging in green behaviors often fosters a sense of purpose and meaningfulness, contributing to their well-being and career satisfaction (Ahmed et al., 2020). Research demonstrates that employees who perceive their work as aligned with environmental values are more likely to feel engaged and committed to their organizations (Pham et al., 2024a). This dual benefit enhances environmental outcomes while improving employee experiences, underscoring the strategic value of studying EGB in various organizational contexts. Moreover, as industries worldwide adapt to stricter environmental regulations, understanding the factors that drive employee engagement in green practices becomes a necessity for maintaining competitive advantage (Zacher et al., 2023).

The growing body of research on EGB highlights the significant role of some organizational practices in fostering EGB (Dumont et al., 2017; Pham et al., 2023b; Yusliza et al., 2017). These practices, which include green training programs, recognition of eco-friendly efforts, and leadership-driven support for sustainability, create a foundation for embedding green behaviors into the workplace. Nevertheless, our understanding of what drives employee green behavior (EGB) in the hotel sector in developing countries such as Vietnam remains incomplete. Specifically, little is known about the extent to which perceived organizational support for the environment (POSE), taking into account the influence of employees' cultural orientation, drives the EGB of hotel employees. POSE is a specialized construct derived from the broader framework of perceived organizational support (POS), which refers to employees' beliefs about the extent to which their organization appreciates their efforts and prioritizes their well-being (Eisenberger et al., 1986). POSE narrows this focus to the environmental domain, emphasizing how much employees perceive their organization as valuing and supporting their sustainability efforts. It provides a lens through which employees assess organizational commitment to environmental sustainability, which, in turn, influences their behaviors and attitudes at work (Lamm et al., 2015). Further, there is still much to uncover about the mechanism through which POSE influences individual EGB (Khalid et al., 2022; Zacher et al., 2023).

By using social exchange theory, which explains how the exchange quality between organizations and employees could result in reciprocity among employees (Blau, 1968), this chapter advances existing knowledge on the antecedents of EGB by linking it to both organizational and cultural characteristics. First, we advance knowledge on how POSE influences EGB through the mediation of employee green passion. Second, we examine how employees' collectivist orientation plays as a cultural factor on the influence of employee green passion on EGB. By doing so, our study offers a better understanding of the mechanism of how POSE impacts EGB. In particular, we draw from the experience of Vietnam, an Asian emerging market for several reasons. First, Vietnam's rapidly growing tourism sector, particularly its expanding hospitality industry, makes it an ideal setting for examining EGB in hotels (Hoang et al., 2025). Second, the country has placed increasing importance on sustainability, with many hotels, especially four- and five-star establishments, adopting eco-friendly policies and practices (Luu, 2019). This offers a unique opportunity to investigate the impact of these practices on employee behavior. In line with this, the Vietnamese government introduced the "Strategy on Green Growth for the period 2011–2020 and Vision to 2050" in 2011 to promote sustainable economic development, compelling hotels to adhere to environmental regulations that are closely monitored by local authorities (Pham et al., 2019). Third, Vietnam's rich cultural background provides an opportunity to explore how local values, such as collectivism (Ho et al., 2016), influence EGB in the workplace and offer insights into employee motivation. Finally, many Vietnamese hotels comply with international environmental standards, such as ISO 14,001, offering a structured framework to study employee participation in green initiatives (Hoang et al., 2025).

2 Literature Review

2.1 Employee Green Behavior in the Workplace

EGB is a concept that reflects the willingness of employees to get involved and participate in activities that help the environment (Scherbaum et al., 2008). According to Bissing-Olson et al. (2013), EGB includes not only completing job tasks in an eco-friendly manner but also taking proactive steps beyond job expectations to benefit the environment. This type of behavior seeks to minimize harm to the environment or even contribute positively to it. Examples include actions such as saving energy, avoiding disposable cups, supporting green strategies in the workplace, cycling to work, reducing waste, and initiating projects to combat environmental damage. These behaviors, often discretionary, include conserving resources, participating in green initiatives, and advocating for sustainable practices, contributing significantly to organizational sustainability goals (Lamm et al., 2015; Paillé & Meija-Morelos, 2019). By exploring the drivers of EGB, researchers and practitioners can design strategies that align organizational practices with environmental objectives (Dumont et al., 2017; Zhang et al., 2024). This alignment is crucial for organizations aiming to achieve sustainability goals while fostering a workforce that is both engaged and committed to meaningful environmental contributions.

2.2 Perceived Organizational Support for the Environment

As indicated earlier, POSE refers to the extent to which employees perceive their organization appreciating and supporting their sustainability efforts. Empirical studies highlight the positive outcomes of fostering POSE. For instance, Lamm et al. (2015) demonstrated that POSE is strongly associated with organizational citizenship behaviors toward the environment (OCB-E), which are voluntary actions employees undertake to benefit the environment. Additionally, POSE enhances job satisfaction, organizational identification, and psychological empowerment, while reducing turnover intentions (Karatepe et al., 2022; Lamm et al., 2015). These findings underscore the dual benefits of POSE that advances sustainability while enhancing employee well-being and commitment.

To foster POSE, organizations must integrate sustainability into their human resource management practices. Green human resource management (HRM) practices, including green training, green performance appraisals, and green recognition, have been identified as critical enablers of POSE (Aboramadan & Karatepe, 2021). Furthermore, leadership plays a pivotal role in shaping employees' perceptions of support by consistently communicating and demonstrating the organization's green commitment (Paillé et al., 2023).

In summary, POSE represents a critical concept that links organizational strategies with employee behaviors to promote sustainability (Zhang et al., 2022). By fostering a supportive environment for sustainable practices, organizations can not only improve their environmental performance but also cultivate a more motivated, engaged, and committed workforce. As environmental challenges intensify, understanding and leveraging POSE will be critical for organizations seeking to embed sustainability into their core values and operations.

3 Theoretical Foundation and Research Hypotheses

Social exchange theory explains how reciprocal relationships between organizations and employees are formed (Cropanzano & Mitchell, 2005). When employees perceive that their organization supports their efforts and fosters their skills and knowledge, it leads employees to form positive perceptions about how much their organization appreciates their contributions and prioritizes their well-being (Rhoades & Eisenberger, 2002). This, in turn, results in employees' favorable work outcomes. In the context of green management and sustainability, previous studies (Lamm et al., 2015; Paillé et al., 2023) have shown that POSE could generate reciprocity among employees. When organizations provide training, rewards, and leadership support, allowing employees to have more capabilities to facilitate their work and understand the environmental issues, such kind of support will make employees strive for better behavior toward the environment. Thus, to better understand the mechanism of how POSE influences EGB, based on social exchange theory, this chapter will explore the effect of POSE on EGB through the mediation and moderation effects of employee green passion and employee collectivist orientation.

3.1 Perceived Organizational Support for the Environment and Employee Green Passion

Passion is a positive emotion that an individual can control, driving them to engage in activities or relationships tied to what they are passionate about (Vallerand et al., 2007). In the green and sustainability context, employee green passion refers to the emotional drive motivating employees to participate in environmental initiatives within their organization (Robertson & Barling, 2013).

There are several ways that organizations can show commitment and provide resources for employees to perform their tasks. For example, providing employees with the necessary tools and resources to further enhance their capacity to contribute effectively in an environmentally friendly manner, or fostering a collaborative and inclusive work environment for employees to engage in green practices. When organizations implement environmental practices and offer support, it can inspire employees to have more concerns and awareness about environmental issues. This kind of support can also spark a desire and passion within them to actively care for and address these issues in their workplace. Employees are more likely to feel a deeper connection to the company's mission and be motivated to adopt sustainable behaviors. Based on this argument, the following hypothesis is proposed:

Hypothesis 1 POSE has a positive effect on employee green passion.

3.2 Perceived Organizational Support for the Environment and Employee Green Behavior in the Workplace

In the context of sustainability in hotels, actions from organizations such as green training, eco-friendly reward systems, and visible leadership support for sustainability signal to hotel employees that their organization prioritizes environmental values (Rhoades & Eisenberger, 2002). When employees perceive that their organization provides the necessary resources to do so (Paillé & Meija-Morelos, 2019), this will cultivate an environment where employees feel empowered and valued (Karatepe et al., 2022). Thus, employees feel a stronger sense of obligation and motivation to reciprocate and are more inclined to engage in environmentally friendly behaviors related to their tasks to give back to their organization (Paillé et al., 2023; Pham et al., 2024b). In other words, the support activities of organizations will encourage hotel employees to adopt more responsible environmental behaviors in their tasks, and the following hypothesis is proposed:

Hypothesis 2 POSE has a positive effect on EGB.

3.3 The Mediation Role of Employee Green Passion on the Relationship Between Organizational Support for the Environment and Employee Green Behavior in the Workplace

Hotel employees who exhibit green passion are more inclined to put in greater effort in performing their tasks in a greener way. Further, passionate and engaged employees tend to perform better than those who are not, as they show more enthusiasm and satisfaction in addressing related tasks (Peng et al., 2021). According to social exchange theory, employees who receive environmental support from their organization are likely to feel a stronger sense of motivation, trust, and passion and give back to their organization by supporting their hotel's green objectives and trying to complete their tasks in a green manner (Buil et al., 2019).

The arguments mentioned above imply the partial mediation role of employee green passion on the relationship between POSE and EGB. POSE positively affects employee green passion, which in turn affects EGB in the workplace. POSE also directly influences EGB. From this, the following hypothesis is proposed.

Hypothesis 3 Employee green passion positively mediates the relationship between POSE and EGB.

3.4 The Moderating Role of Employees' Collectivist Orientation

Employees' collectivist orientation refers to the extent to which individuals see themselves as separate (individualistic) or interconnected (collectivistic) with their in-group members (Triandis, 2001; Triandis et al., 1985). Research by Farh et al. (2007) and Hoang et al. (2018) highlights that cultural diversity in terms of individual orientation is common within any given country. Caprar (2011) also observed that employees in multinational corporations (MNCs) in developing markets display a variety of cultural values. Prior studies (e.g., Farh et al., 2007; Newman & Butler, 2014) found that the influences of employee attitudes on employee behavior tend to be stronger among those with lower individualism or higher collectivism. Specifically, employees with high collectivism are more likely to align with norms of support organizational objectives. Thus, we expect that the influence of employee green passion on EGB will be more significant for individuals who score lower on individualism or higher on collectivism. We propose the following:

Hypothesis 4 Employees' collectivist orientation strengthens the positive relationship between employee green passion and EGB (Fig. 1).

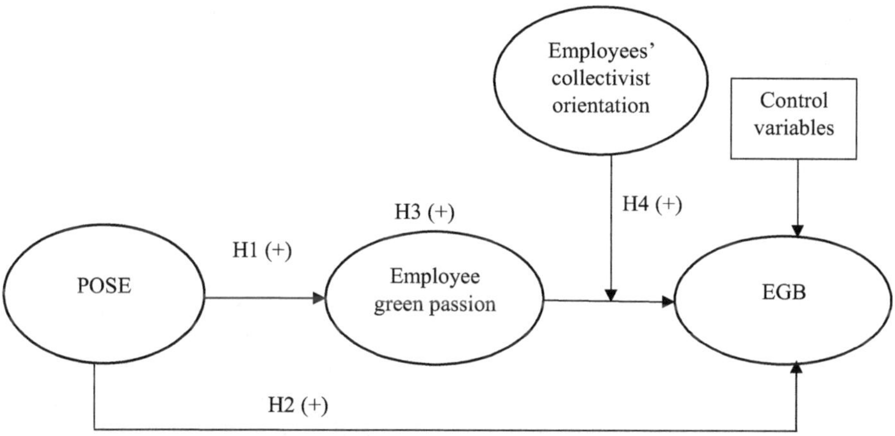

Fig. 1 Research framework explaining hotel employee green behavior in the workplace. Source: Author's own work

4 Methodology

4.1 Sample and Data Collection

This study focused on employees from four- and five-star hotels in Vietnam, which play a significant role in tourism development while adopting environmental sustainability practices (Luu, 2019). These hotels followed environmental standards like ISO:14,001 and TCVN: 4391–2015 and emphasized eco-friendly practices such as creating green teams, training staff on environmental protection, reducing waste in food services, and using sustainable materials in guest rooms (Pham et al., 2023a). The selection of these hotels aligned with previous research on sustainability in the hospitality industry (e.g., Darvishmotevali & Altinay, 2022; Hoang et al., 2025).

We contacted the human resource departments of 34 four- to five-star hotels, explained the study's purpose, and sought permission to conduct a survey with their employees. Of the 34 hotels, 30 agreed to participate. To ensure valid responses, participants needed to have working experience at the hotel for at least a year and be involved in sustainable practices. Employees from different departments who met these criteria were invited to participate, with their participation being voluntary, anonymous, and for research purposes only.

Data collection was conducted from August 2021 to February 2022. The study used a two-wave time-lagged design to reduce common method bias. In the first wave, participants provided demographic information and answered questions about green passion, POSE, collectivism orientation, and individual information (e.g., gender, tenure). Four weeks later, they completed a follow-up survey on workplace green behavior. The questionnaire was translated into Vietnamese using back-translation to ensure accuracy. A pilot study with five hotel employees helped refine the survey before the final version was distributed.

4.2 Sample Description

Out of 316 respondents, 56.6% were female. Regarding age, the majority of participants (67.1%) fell into the 30–49 years age group, followed by 28.2% of individuals who were under 30 years of age. The 50+ years group was significantly smaller, with only 4.7% of participants. In terms of department, the largest groups were found in the front office (23.7%) and food and beverage (22.5%) departments. These were followed by the maintenance (18.4%) and housekeeping (12.7%) departments, with the administration (13.9%) and others (8.9%) making up the remaining representation. As regards job tenure, a substantial proportion of individuals (42.4%) had worked between 3 and 5 years in their roles, followed by 32.9% with more than 5 years of experience. Meanwhile, 24.7% of participants had been in their current positions for <3 years. As for management positions, the data revealed that a majority of the sample (70.6%) did not hold a management role, while 29.4% were in managerial positions. Finally, when it came to hotel star rating, the majority of participants worked in 4-star hotels (64.2%), while 35.8% were employed at 5-star hotels.

4.3 Measurements

We used multi-item, five-point Likert scales ranging from (1) "strongly disagree" to (5) "strongly agree."

POSE was adopted from Saifulina and Carballo-Penela (2017). Example items include "My organization makes an active effort to help employees be environmentally pro-active" and "My organization provides its employees with useful information they need to behave in an environmentally friendly way."

The concept of employee green passion was derived from Robertson and Barling (2013). Sample items include "I enthusiastically discuss environmental issues with others" and "I passionately encourage others to be more environmentally responsible."

EGB was adopted from Robertson and Barling (2013) and Zhang and Huang (2019), including six items. Example items include "I print double-sided whenever possible" and "I take part in environmentally friendly programs."

The measure for employees' collectivist orientation was adopted from Yoo and Donthu (2002). An example item includes: "Individuals should stick with the group even through difficulties."

Following prior studies (Hoang et al., 2025; Lamm et al., 2015), personal information of hotel employees (e.g., gender, job tenure, management position, hotel star rating) was used as control variables.

4.4 Data Analysis

Data analysis was conducted in the following steps. First, a confirmatory factor analysis (CFA) was applied to check the reliability and validity of the scales and the measurement model. Then, structural equation modeling (SEM) in AMOS 27.0 was used to test the proposed hypotheses. The indirect effect of POSE on hotel EGB, mediated by employee green passion, was tested by using the bootstrap method. Finally, to test the moderation effects, Hayes (2013)'s PROCESS macro for SPSS with 5000 bootstrap resamples was applied.

5 Findings

5.1 Measurement Model

The measurement components included POSE, employee green passion, EGB, and employees' collectivist orientation. Table 1 indicates that all values of composite reliability (CR) were higher than the threshold of 0.7. All average variance extracted (AVE) values were higher than 0.50, and all factor loadings were >0.5, achieving convergent reliability. The square roots of AVE were higher than interconstruct correlations, achieving discriminant validity. All alpha coefficients were higher than 0.70, suggesting that the items in the related constructs had relatively high internal consistency.

Table 2 shows the results of the CFA models. The four-factor model shows a good fit ($\chi^2 = 172.77$, df = 97, χ^2/df = 1.89, CFI = 0.97, TLI = 0.96, RMSEA = 0.047).

Table 1 Descriptive statistics, correlations, validity and reliability of the constructs involved in the study

	Mean	SD	Cronbach's alpha	CR	AVE	1	2	3	4
1. POSE	4.52	0.470	0.83	0.861	0.608	**0.780**			
2. Employee green passion	4.76	0.399	0.89	0.891	0.732	0.276***	**0.856**		
3. EGB	4.57	0.387	0.86	0.811	0.519	0.219***	0.376***	**0.720**	
4. Employees' collectivist orientation	4.19	0.443	0.82	0.830	0.507	0.261***	0.115**	0.202***	**0.712**

Notes: **$p < 0.05$; ***$p < 0.01$; The square root of the AVE for each construct is along the diagonal (in bold). *SD* Standard deviation, *AVE* Average variance extracted, *CR* Composite reliability
Source: Author's own work

Table 2 Comparison of measurement models

Measurement model	χ2 (df)	χ2/df	CFI	TLI	RMSEA	Δ χ2(Δdf)
Four-factor model	172.77 (97)	1.89	0.97	0.96	0.047	
Three-factor model[a]	851.91 (99)	8.61	0.73	0.68	0.139	679.14(2)***
Two-factor model[b]	1376.29 (101)	13.63	0.55	0.46	0.179	1203.52(4)***
One-factor model[c]	1705.19 (102)	16.72	0.43	0.33	0.199	1532,42(5)***

Notes: [a]POSE and employee green passion were combined into one factor; [b]POSE, employee green passion and employees' collectivist orientation were combined into one factor; [c]All items were combined into one factor. ***$p < 0.01$; $\chi 2$ Chi-square, *df* Degrees of freedom, *CFI* Comparative fit index, *TLI* Tucker-Lewis index, *RMSEA* Root mean square error of approximation
Source: Author's own work

We compared this model with the alternative models, including the three-factor model, the two-factor model, and the one-factor model. The results show that the former model was a better fit (*p*-value of $\Delta\ \chi2 < 0.01$). Similarly, the four-factor model was a better fit than the other models. Thus, the discriminant validity of all constructs was satisfied.

5.2 Common Method Bias

Since the data were collected from only hotel employees, which is single source, there was a potential risk of common method bias (Delcourt et al., 2013). To mitigate this risk, we carefully designed the questionnaire with clear and straightforward scale items, encouraged honest responses, and instructed participants to complete the survey in two waves. Additionally, we assessed common method bias by using a common latent factor approach (MacKenzie & Podsakoff, 2012). We compared the factor loadings of a CFA model with and without the common latent factor, and found that the difference was below the 0.2 threshold (Cohen, 2013). Therefore, common method bias does not appear to be a significant concern in the current study.

5.3 Hypothesis Testing

The hypothesized model demonstrated a good fit to the data ($\chi^2 = 153.41$, df = 81, $\chi^2/df = 1.89$, CFI = 0.96, TLI = 0.95, RMSEA = 0.053). Table 3 and Fig. 2 indicate the completely standardized structural effects on EGB. Table 4 and Fig. 3 show the moderation effect of employees' collectivist orientation.

Table 3 Direct, indirect, and total effects

	Direct effects	Indirect effects	Total effects
POSE	0.16***	0.08***	0.24***
Employee green passion	0.29***		0.29***
Gender	-0.03^{ns}		-0.03^{ns}
Job tenure	0.06^{ns}		0.06^{ns}
Hotel star rating	0.01^{ns}		0.01^{ns}
Management position	0.05^{ns}		0.05^{ns}

Notes: ^{ns}nonsignificant; ***$p < 0.01$
Source: Author's own work

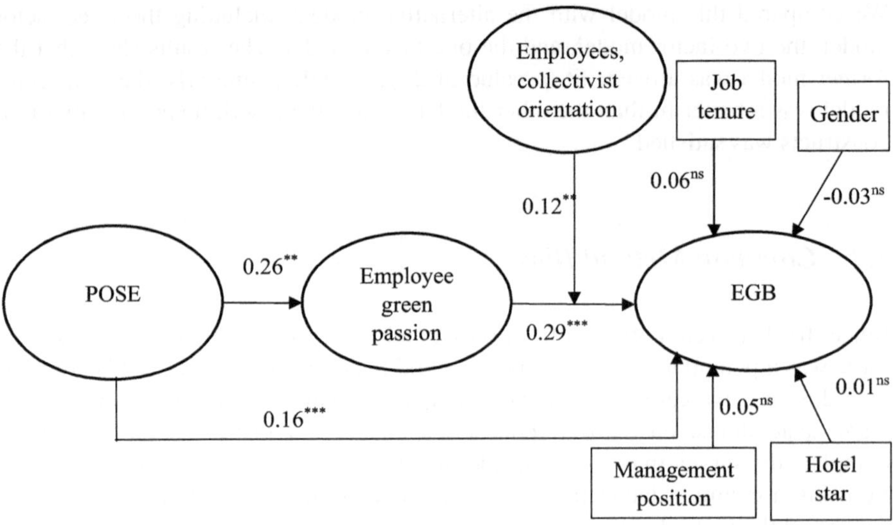

Fig. 2 Completely standardized structural coefficients for the theoretical model. Note: nsnonsignificant, $^{**}p < 0.05$; $^{***}p < 0.01$. Source: Author's own work

Table 4 Moderation effects

Variables	Level of moderator	Conditional effect	Std. error	t-values	Bootstrapping 95%CI	
					Lower	Upper
Employee green passion X Employees' collectivist orientation	High collectivist orientation	0.33***	0.06	5.060	0.20	0.46
	Low collectivist orientation	0.23***	0.05	5.011	0.14	0.32

Notes: $^{***}p < 0.01$
Source: Author's own work

5.3.1 Direct Linkages

H1 stated that POSE is positively associated with employee green passion. The results in Fig. 2 show that POSE was significantly positively related to employee green passion ($\beta = 0.26$, $p < 0.01$). Thus, H1 is supported.

H2 stated that POSE is positively associated with EGB. The results in Fig. 2 show that POSE positively affected EGB ($\beta = 0.16$, $p < 0.01$). Therefore, H2 is supported.

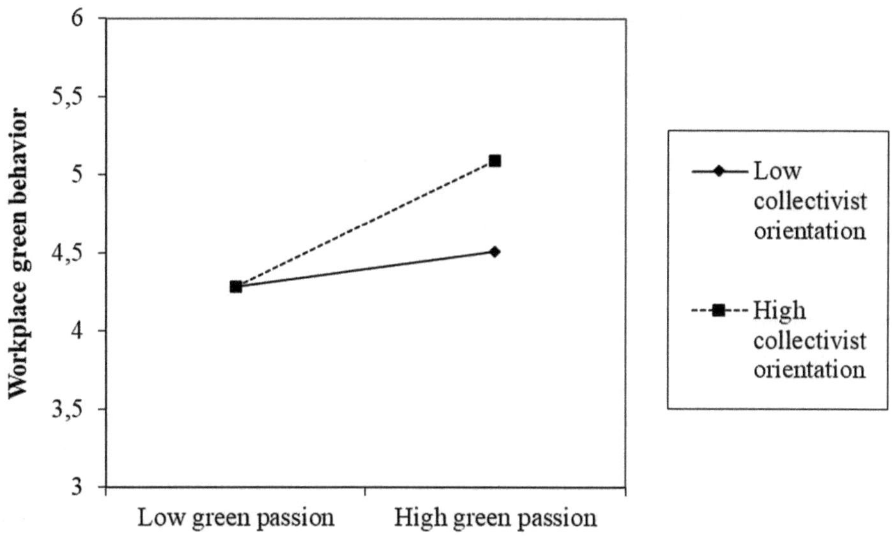

Fig. 3 Moderation effect of employees' collectivist orientation. Source: Author's own work

5.3.2 Mediation Linkages

To test the mediation effects, we conducted bootstrapping with 2000 resamples to determine the significance of the indirect effects (Cheung & Lau, 2008). If the independent variable has a significant effect on the mediator, and the mediator, in turn, significantly influences the dependent variable, and the indirect effect is significant, we can confirm the existence of mediation. Figure 2 shows that POSE was significantly associated with employee green passion ($\beta = 0.26$, $p < 0.01$), and employee green passion was significantly related to EGB ($\beta = 0.29$, $p < 0.01$). Table 3 indicates the direct effect of POSE on EGB via employee green passion was significant ($\beta = 0.08$, $p < 0.01$), supporting H3.

5.3.3 Moderation Effects

Table 4 shows that employee green passion was more positively associated with EGB when employees' collectivist orientation was high (conditional effect of 0.33 with a 95% confidence interval (CI) of [0.20, 0.46]) than when it was low (conditional effect of 0.23 with a 95% CI of [0.14, 0.32]), supporting H4. Figure 3 graphically shows the moderating effect of employees' collectivist orientation.

6 Discussion of Implications

6.1 Discussion of Theoretical Implications

The present study advances current research on EGB in different ways. First, although prior research (Lamm et al., 2015; Zhang et al., 2022) has identified the relationship between POSE and EGB, limited studies have identified the motivational process that could explain EGB. Further, no study explores the impact of cultural factors in the process of driving EGB. Our study contributes to the literature by investigating how and when POSE impacts EGB in the hotel industry context in Vietnam. The study found that in addition to POSE constituting an important driver of hotel EGB, POSE also significantly influences hotel EGB through employee green passion. The research also identifies that employees with collectivist orientations, those who value group goals and social harmony, are more likely to exhibit green behaviors in the workplace when they are passionate about environmental sustainability. In other words, a collectivist mindset strengthens the positive impact of green passion on green behavior in the workplace.

In addition to advancing prior research, the results of this study reinforce the application of social exchange theory (Blau, 1968) in the context of EGB. Our study indicates that when an organization provides support for employees to implement environmental practices, this will increase their passion for green initiatives, and thus employees feel obligated to reciprocate by engaging in green behaviors in their workplace. Our study also extends this theory in the context of an Asian emerging market, Vietnam, by confirming that the social exchange theory is more applicable with a high level of employees' collectivist orientation.

6.2 Discussion of Managerial Implications

Enhancing organizational support to foster green behavior requires an integrated approach that leverages sustainable practices, inclusive leadership, and supportive workplace environments. By embedding sustainability into the core of organizational values and daily operations, organizations can create a culture where employees feel empowered and motivated to contribute to environmental goals. Promoting organizational support to enhance green behavior requires a comprehensive and strategic approach that integrates sustainable HRM practices, such as green HRM, SRHRM. These frameworks provide mechanisms to align employee behaviors with organizational sustainability goals while fostering a culture of environmental stewardship and social responsibility.

A critical step in this process is ensuring that employees perceive their contributions to environmental initiatives as valued and impactful. This begins with clear communication of the organization's sustainability objectives and how employees' efforts align with these goals. Leaders play a crucial role in setting this tone by modeling green behaviors and consistently reinforcing the importance of sustainability. When leaders visibly support environmental initiatives, it sends a strong message that these actions are integral to the organization's success, inspiring employees to follow suit.

Providing employees with the necessary tools, resources, and training to engage in green practices further enhances their capacity to contribute effectively. For example, providing recycling bins, energy-efficient technologies, or guidance on resource conservation could demonstrate the organization's commitment to sustainability. These actions not only facilitate green behavior but also reinforce employees' belief that their organization prioritizes environmental stewardship. Another important strategy is fostering a collaborative and inclusive work environment where employees feel encouraged to participate in decision-making related to sustainability. This can be achieved through initiatives such as sustainability committees, regular feedback mechanisms, or involving employees in the design and implementation of green policies.

Recognition and rewards are also powerful motivators for promoting green behavior. Employees are more likely to engage in pro-environmental actions when their efforts are acknowledged and celebrated. Organizations can implement both formal and informal recognition systems, such as awards for sustainability achievements, public acknowledgment of contributions, or career advancement opportunities tied to green initiatives. These practices highlight the value of employees' contributions and reinforce the organization's commitment to supporting them.

Hotel managers should also consider the impact of employees' collectivist orientation on promoting EGB in the workplace. For employees with a higher collectivist orientation, green passion could have a stronger positive influence on their EGB. Therefore, it is recommended that hotel managers assess the collectivist orientation of hotel employees during the hiring process. Additionally, creating an environment that fosters collaboration among hotel staff and departments could be beneficial.

7 Future Directions

This study relied on a small sample of hotel employees in Vietnam and used self-reported data. To enhance the robustness of the findings, future research could incorporate data from multiple sources, including input from coworkers and supervisors. For example, green passion and EGB could be collected from employees' coworkers or their supervisors. While this study focuses on Vietnam, validating these results in other countries would broaden our understanding of the model. Further research could also explore other factors that could affect EGB, such as leadership, or other employees' cultural orientations, such as power distance and uncertainty avoidance.

8 Conclusion

In conclusion, POSE is a crucial factor in promoting hotel EGB. This study extended the current knowledge of hotel EGB by investigating the effect of POSE on EGB through the mediation and moderation effects of employee green passion and employee collectivist orientation. The study also provided implications to help hotel managers promote POSE and EGB in their workplace.

Acknowledgments The authors would like to thank the editors and anonymous reviewers for their insightful comments and suggestions for the improvement of this chapter.

References

Aboramadan, M., & Karatepe, O. M. (2021). Green human resource management, perceived green organizational support and their effects on hotel employees' behavioral outcomes. *International Journal of Contemporary Hospitality Management, 33*(10), 3199–3222. https://doi.org/10.1108/IJCHM-12-2020-1440

Ahmed, M., Zehou, S., Raza, S. A., Qureshi, M. A., & Yousufi, S. Q. (2020). Impact of CSR and environmental triggers on employee green behavior: The mediating effect of employee well-being. *Corporate Social Responsibility and Environmental Management, 27*(5), 2225–2239. https://doi.org/10.1002/csr.1960

Bissing-Olson, M. J., Iyer, A., Fielding, K. S., & Zacher, H. (2013). Relationships between daily affect and pro-environmental behavior at work: The moderating role of pro-environmental attitude. *Journal of Organizational Behavior, 34*(2), 156–175. https://doi.org/10.1002/job.1788

Blau, P. M. (1968). Social exchange. In D. L. Sills (Ed.), *International encyclopedia of the social sciences* (Vol. 7, pp. 452–457). MacMillan.

Buil, I., Martínez, E., & Matute, J. (2019). Transformational leadership and employee performance: The role of identification, engagement and proactive personality. *International Journal of Hospitality Management, 77*, 64–75. https://doi.org/10.1016/j.ijhm.2018.06.014

Caprar, D. V. (2011). Foreign locals: A cautionary tale on the culture of MNC local employees. *Journal of International Business Studies, 42*, 608–628. https://doi.org/10.1057/jibs.2011.9

Cheung, G. W., & Lau, R. S. (2008). Testing mediation and suppression effects of latent variables bootstrapping with structural equation models. *Organizational Research Methods, 11*(2), 296–325. https://doi.org/10.1177/1094428107300343

Cohen, J. (2013). *Statistical power analysis for the behavioral sciences*. Routledge.

Cropanzano, R., & Mitchell, M. S. (2005). Social exchange theory: An interdisciplinary review. *Journal of Management, 31*(6), 874–900. https://doi.org/10.1177/0149206305279602

Darvishmotevali, M., & Altinay, L. (2022). Toward pro-environmental performance in the hospitality industry: Empirical evidence on the mediating and interaction analysis. *Journal of Hospitality Marketing & Management, 31*(4), 431–457. https://doi.org/10.1080/19368623.2022.2019650

Delcourt, C., Gremler, D. D., van Riel, A. C., & van Birgelen, M. (2013). Effects of perceived employee emotional competence on customer satisfaction and loyalty: The mediating role of rapport. *Journal of Service Management, 24*(1), 5–24. https://doi.org/10.1108/09564231311304161

Dumont, J., Shen, J., & Deng, X. (2017). Effects of green HRM practices on employee workplace green behavior: The role of psychological green climate and employee green values. *Human Resource Management, 56*(4), 613–627. https://doi.org/10.1002/hrm.21792

Eisenberger, R., Huntington, R., & Sava, D. (1986). Perceived organisational support. *Journal of Applied Psychology, 71*(3), 500–507.

Farh, J.-L., Hackett, R. D., & Liang, J. (2007). Individual-level cultural values as moderators of perceived organizational support–employee outcome relationships in China: Comparing the effects of power distance and traditionality. *Academy of Management Journal, 50*(3), 715–729. https://doi.org/10.5465/amj.2007.25530866

Hayes, A. F. (2013). *Introduction to mediation, moderation, and conditional process analysis: A regression-based approach.* Guilford Publications.

Ho, N. T. T., Seet, P.-S., & Jones, J. (2016). Understanding re-expatriation intentions among overseas returnees–an emerging economy perspective. *The International Journal of Human Resource Management, 27*(17), 1938–1966. https://doi.org/10.1080/09585192.2015.1088884

Hoang, H. T., Pham, N. T., Seet, P.-S., Jones, J., & Ho, N. T. T. (2025). Top management green commitment and employee in-role green performance: An emerging economy study. *Sustainable Futures, 9*, 100432. https://doi.org/10.1016/j.sftr.2025.100432

Hoang, H. T., Rao Hill, S., Lu, V. N., & Freeman, S. (2018). Drivers of service climate: An emerging market perspective. *Journal of Services Marketing, 32*(4), 476–492. https://doi.org/10.1108/JSM-06-2017-0208

Karatepe, O. M., Hsieh, H., & Aboramadan, M. (2022). The effects of green human resource management and perceived organizational support for the environment on green and non-green hotel employee outcomes. *International Journal of Hospitality Management, 103*, 103202. https://doi.org/10.1016/j.ijhm.2022.103202

Katz, I. M., Rauvola, R. S., Rudolph, C. W., & Zacher, H. (2022). Employee green behavior: A meta-analysis. *Corporate Social Responsibility and Environmental Management, 29*(5), 1146–1157. https://doi.org/10.1002/csr.2260

Khalid, B., Shahzad, K., Shafi, M. Q., & Paille, P. (2022). Predicting required and voluntary employee green behavior using the theory of planned behavior. *Corporate Social Responsibility and Environmental Management, 29*(5), 1300–1314. https://doi.org/10.1002/csr.2270

Lamm, E., Tosti-Kharas, J., & King, C. E. (2015). Empowering employee sustainability: Perceived organizational support toward the environment. *Journal of Business Ethics, 128*, 207–220. https://doi.org/10.1007/s10551-014-2093-z

Luu, T. T. (2019). Building employees' organizational citizenship behavior for the environment: The role of environmentally-specific servant leadership and a moderated mediation mechanism. *International Journal of Contemporary Hospitality Management, 31*(1), 406–426. https://doi.org/10.1108/IJCHM-07-2017-0425

MacKenzie, S. B., & Podsakoff, P. M. (2012). Common method bias in marketing: Causes, mechanisms, and procedural remedies. *Journal of Retailing, 88*(4), 542–555. https://doi.org/10.1016/j.jretai.2012.08.001

Newman, A., & Butler, C. (2014). The influence of follower cultural orientation on attitudinal responses towards transformational leadership: Evidence from the Chinese hospitality industry. *The International Journal of Human Resource Management, 25*(7), 1024–1045. https://doi.org/10.1080/09585192.2013.815250

Norton, T. A., Parker, S. L., Zacher, H., & Ashkanasy, N. M. (2015). Employee green behavior: A theoretical framework, multilevel review, and future research agenda. *Organization & Environment, 28*(1), 103–125. https://doi.org/10.1177/1086026615575773

Norton, T. A., Zacher, H., Parker, S. L., & Ashkanasy, N. M. (2017). Bridging the gap between green behavioral intentions and employee green behavior: The role of green psychological climate. *Journal of Organizational Behavior, 38*(7), 996–1015. https://doi.org/10.1002/job.2178

Paillé, P., & Meija-Morelos, J. H. (2019). Organisational support is not always enough to encourage employee environmental performance. The moderating role of exchange ideology. *Journal of Cleaner Production, 220*, 1061–1070. https://doi.org/10.1016/j.jclepro.2019.02.192

Paillé, P., Valéau, P., & Carballo-Penela, A. (2023). Green rewards for optimizing employee environmental performance: Examining the role of perceived organizational support for the environment and internal environmental orientation. *Journal of Environmental Planning and Management, 66*(14), 2810–2831. https://doi.org/10.1080/09640568.2022.2092723

Peng, J., Chen, X., Zou, Y., & Nie, Q. (2021). Environmentally specific transformational leadership and team pro-environmental behaviors: The roles of pro-environmental goal clarity, pro-environmental harmonious passion, and power distance. *Human Relations, 74*(11), 1864–1888. https://doi.org/10.1177/0018726720942306

Pham, N. T., Jabbour, C. J. C., Vo-Thanh, T., Huynh, T. L. D., & Santos, C. (2023a). Greening hotels: Does motivating hotel employees promote in-role green performance? The role of culture. *Journal of Sustainable Tourism, 31*(4), 951–970. https://doi.org/10.1080/09669582.2020.1863972

Pham, N. T., Lu, J., Ogbonnaya, C., Tuan, T. H., Degbey, W. Y., & Laker, B. (2024a). A more ethical workplace? How and why perceived socially responsible human resource management makes a difference. *Journal of Management Studies*. https://doi.org/10.1111/joms.13161

Pham, N. T., Thuy, V. T. N., Tuan, T. H., & Nguyen, Q. H. (2024b). Improving the workforce's competitive advantage in the remote work context: A moderated-mediation analysis of perceived organizational support, employee retention, and transformational leadership. *Journal of Competitiveness, 16*(3). https://doi.org/10.7441/joc.2024.03.07

Pham, N. T., Tuan, T. H., Le, T. D., Nguyen, P. N. D., Usman, M., & Ferreira, G. T. C. (2023b). Socially responsible human resources management and employee retention: The roles of shared value, relationship satisfaction, and servant leadership. *Journal of Cleaner Production, 414*, 137704. https://doi.org/10.1016/j.jclepro.2023.137704

Pham, N. T., Tučková, Z., & Jabbour, C. J. C. (2019). Greening the hospitality industry: How do green human resource management practices influence organizational citizenship behavior in hotels? A mixed-methods study. *Tourism Management, 72*, 386–399. https://doi.org/10.1016/j.tourman.2018.12.008

Rhoades, L., & Eisenberger, R. (2002). Perceived organizational support: A review of the literature. *Journal of Applied Psychology, 87*(4), 698. https://doi.org/10.1037/0021-9010.87.4.698

Robertson, J. L., & Barling, J. (2013). Greening organizations through leaders' influence on employees' pro-environmental behaviors. *Journal of Organizational Behavior, 34*(2), 176–194. https://doi.org/10.1002/job.1820

Saifulina, N., & Carballo-Penela, A. (2017). Promoting sustainable development at an organizational level: An analysis of the drivers of workplace environmentally friendly behaviour of employees. *Sustainable Development, 25*(4), 299–310. https://doi.org/10.1002/sd.1654

Scherbaum, C. A., Popovich, P. M., & Finlinson, S. (2008). Exploring individual-level factors related to employee energy-conservation behaviors at work 1. *Journal of Applied Social Psychology, 38*(3), 818–835. https://doi.org/10.1111/j.1559-1816.2007.00328.x

Triandis, H. C. (2001). Individualism-collectivism and personality. *Journal of Personality, 69*(6), 907–924. https://doi.org/10.1111/1467-6494.696169

Triandis, H. C., Leung, K., Villareal, M. J., & Clack, F. I. (1985). Allocentric versus idiocentric tendencies: Convergent and discriminant validation. *Journal of Research in Personality, 19*(4), 395–415. https://doi.org/10.1016/0092-6566(85)90008-X

Vallerand, R. J., Salvy, S. J., Mageau, G. A., Elliot, A. J., Denis, P. L., Grouzet, F. M., & Blanchard, C. (2007). On the role of passion in performance. *Journal of Personality, 75*(3), 505–534. https://doi.org/10.1111/j.1467-6494.2007.00447.x

Yoo, B., & Donthu, N. (2002). The effects of marketing education and individual cultural values on marketing ethics of students. *Journal of Marketing Education, 24*(2), 92–103. https://doi.org/10.1177/0273475302242002

Yusliza, M.-Y., Othman, N. Z., & Jabbour, C. J. C. (2017). Deciphering the implementation of green human resource management in an emerging economy. *Journal of Management Development, 36*(10), 1230–1246. https://doi.org/10.1108/JMD-01-2017-0027

Zacher, H., Rudolph, C. W., & Katz, I. M. (2023). Employee green behavior as the core of environmentally sustainable organizations. *Annual Review of Organizational Psychology and Organizational Behavior, 10*(1), 465–494. https://doi.org/10.1146/annurev-orgpsych-120920-050421

Zhang, J., & Huang, R. (2019). Employees' pro-environmental behaviours (PEBs) at international hotel chains (IHCs) in China: The mediating role of environmental concerns (ECs).

Journal of Hospitality and Tourism Management, 39, 129–136. https://doi.org/10.1016/j.jhtm.2019.03.007

Zhang, S., Ren, S., & Tang, G. (2024). From passive to active: The positive spillover of required employee green behavior on green advocacy. *Journal of Business Ethics, 192*(1), 57–76. https://doi.org/10.1007/s10551-023-05494-x

Zhang, Y., Wu, J., & Fan, Y. (2022). The effect of perceived organizational support toward the environment on team green innovative behavior: Evidence from Chinese green factories. *Emerging Markets Finance and Trade, 58*(8), 2326–2341. https://doi.org/10.1080/1540496X.2021.1977121

Frontline Employees' Green Behavior: The Role of Green Training, Green Perceived Organizational Support, and Green Knowledge Sharing

M. Nazmul Islam and Mohammad Rabiul Basher Rubel

Abstract Environmental sustainability in company operations has garnered attention recently. Environmental management companies may benefit from being the first to apply new tactics and boost their green image. Therefore, global banking management considers green banking the most important environmental management tool. Hence, Bangladesh's banking industry is developing green banking practices in response to the government's urge. Accordingly, employee green behavior (EGB) helps banking organizations maximize their green potential. Thus, promoting EGB has been difficult for practitioners. Hence, this study offered a new paradigm and examined how green training (GT) and green perceived organizational support (GPOS) affect EGB using social exchange theory (SET). The study also examined how green knowledge sharing (GKS) mediates the correlations. This study surveyed 326 frontline employees of a full-time Bangladeshi bank using a questionnaire. Partial least-squares structural equation modeling (PLS-SEM-4.1.1) was used to investigate the assumptions. The findings of the study reveal that GT and GPOS favorably affect EGB. The outcomes also showed that GKS mediates the proposed relationships. The findings allow Bangladesh's banking industry researchers and practitioners to promote green behavior among frontline employees.

1 Introduction

In today's business environment, organizations are placing greater emphasis on sustainable growth and corporate environmental sustainability (Lim, 2022). Organizations that prioritize environmental management have the potential to get first-mover advantages. This means they may implement unique differentiation methods, improve their reputation for being environmentally friendly, and gain a competitive edge over others in the market (Wu et al., 2018; Zhou & Zheng, 2024).

M. N. Islam · M. R. B. Rubel (✉)
BRAC Business School, BRAC University, Dhaka, Bangladesh
e-mail: islam.nazmul@bracu.ac.bd; rabiul.basher@bracu.ac.bd

M. Y. Yusliza et al. (eds.), *Employee Green Behavior*,
https://doi.org/10.1007/978-981-95-1422-9_9

Moreover, organizations that are deeply committed to implementing change to take environmentally friendly initiatives typically have shaped employee behavior (Islam, 2024a, b) and improved employee performance (Hameed et al., 2020). To be successful in sustainability initiatives, employees' pro-environmental behaviors can play crucial roles in the organization. Employee green behavior (EGB), a pro-environmental behavior, has been receiving increased attention and implementation in the workplace to achieve organizational sustainable goals (Mirahsani et al., 2024). According to Leung and Rosenthal (2019), EGB refers to the acts and behaviors that employees do to promote environmental sustainability in the organization. Therefore, organizations must encourage employees to engage in EGB to capitalize on the potential benefits of green initiatives within the organization (Zibarras & Coan, 2015; Kim et al., 2019; Fawehinmi et al., 2024). Moreover, frontline employees can play an essential role (Islam & Idris, 2025; Rubel et al., 2025) in becoming successful in environmental initiatives by demonstrating EGB. However, ensuring frontline EGB is a complex task for the organization. As a result, the question arises: What factors enhance frontline EBG in the organization?

Hence, the prediction of EGB is an evolving area (Zhang et al., 2021), and the examination of their precursors is crucial for enlightening future organizational decision-makers (Ahmad et al., 2021). Existing literature witnessed that both individual- and organizational-level predictors enhance EGB (Khalid et al., 2022). Environmental leadership (Ashraf et al., 2024), ethical leadership (Ahmed et al., 2020), green transformational leadership (Chen & Wu, 2022), green leadership (Abbas & Khan, 2023), inclusive leadership (Surucu, 2024), and green human resource management (HRM) (Ababneh, 2021) have all been shown to influence EGB. In addition to the leadership styles, researchers have studied various other factors, such as social entrepreneurs (Ahmed et al., 2020), green culture (Mirahsani et al., 2024), and corporate environmental strategy (Faraz et al., 2024), that can also influence EGB. Despite the numerous studies conducted, contemporary literature highlighted that GT and GPOS play important roles in nurturing EGB (Amrutha & Geetha, 2021; Aboramadan et al., 2022). However, there has been a lack of focus on the association between GT, GPOS, and frontline EGB.

GT enhances problem-solving skills in employees to address environmental challenges and integrate business policies with environmental protection strategies, promoting sustainability (Usman et al., 2023). GT nurtures EGB which in the long run helps the organization to achieve its environmental goals (Amrutha & Geetha, 2021). Also, GPOS refers to employees' perception of an organization's acknowledgment of its sustainability efforts (Lamm et al., 2015). According to Aboramadan et al. (2022), GPOS provides a suitable environment in the organization to enhance EGB. However, the impact of GT and GPOS is not simple and not straightforward. Previous literature supported different mediational effects in the relationship between GT and employee behavioral outcomes (Amrutha & Geetha, 2021; Pinzone et al., 2019) and GPOS and employee behavioral outcomes (Zhang et al., 2022), whereas existing literature witnessed that GKS is one of the key factors in enhancing EGB (Ahmed et al., 2023; Khan et al., 2022). According to Abbas and Khan (2023), GKS is the exchange of environmentally supportive knowledge among organizational members to improve their environmental performance. GKS plays a mediational role in enhancing EGB (Khan et al., 2022; Rubel et al., 2021).

Accordingly, through the lens of SET, the present study proposed a unique hypothesized model (Fig. 1) to examine the relationship between GT, GPOS, and EGB and the mediational role of GKS in the relationship between GT, GPOS, and EGB. SET claims that social behavior is a consequence of a reciprocal process of exchange (Blau, 1964). Therefore, when employees receive GT and GPOS in the workplace, as well as if the organization can ensure GKS in the organization, employees in exchange return this support to the organization by demonstrating EGB. Moreover, the proposed model of the present study was examined in the context of the banking sector of Bangladesh.

Bangladesh's banking sector has been going through a transition phase (Islam et al., 2022). The Bangladesh government's call for a green banking movement has influenced the banking industry to adopt green practices in their operations and lending policies. The industry has high green content in both in-house and green lending policies, promoting sustainable management (Bangladesh Bank, 2011; Rubel et al., 2021). According to Kulsum and Huda (2018), banking management considers green banking to be the most important aspect of environmental management. But managing change is a complex task (Islam, 2023). Therefore, the behavior that provides green services can have substantial impact on the environmental performance of banking organizations. Considering the influence of green service behavior on environmental performance, we examine the perceptions of frontline employees regarding the effects of GT and GPOS on EGB in the context of Bangladesh's banking sector. Furthermore, the mediating role of GKS on the association between GT and EGB, as well as GPOS and EGB, is investigated.

The study is organized in the following manner. Initially, the theoretical foundation is presented, subsequently leading to the formulation of the research

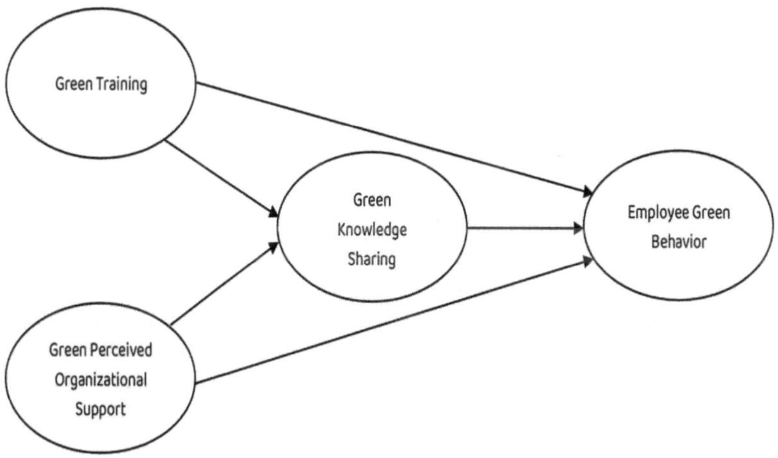

Fig. 1 Hypothesized model (Source: Author's work)

hypotheses. Next, the research methodology is presented, and then, the outcomes of the hypothesis testing are reported. Subsequently, the discussion is presented, exploring both the theoretical and practical implications of the study. Lastly, the limitations of the study and some areas for further research are discussed.

2 Underlying Theory

2.1 *Social Exchange Theory*

The present research used SET as a theoretical framework. One of SET's key concepts is that trust and mutual understanding are the foundations of any successful relationship, and reciprocity is the driving force that keeps relationships moving. When one person does something beneficial for another, the receiving party responds by changing their behavior for the better (Cropanzano et al., 2017). The principle of reciprocity fosters relationships that benefit both parties (Eisenberger et al., 1986). According to Lambe et al. (2001), when these kinds of reciprocal exchanges happen repeatedly, they lead to the development of exchange norms, which in turn boost trust and decrease uncertainty. Relationships between organizations and their employees revolve around reciprocal actions (Rubel et al., 2025). Organizations that invest in their employees and initiate growth-oriented initiatives will, in turn, reap benefits from them. In the current study, we draw upon SET to propose a hypothesized model (Fig. 1) and argue that providing GT and GPOS can foster GKS, which subsequently influences GKS and EGB. More specifically, when an organization ensures GT and GPOS for the employees, it can enhance reciprocal relationships between the organization and employees, which ensure GKS behavior and enhance EGB based on reciprocal relationships.

3 Literature Review and Hypothesis Development

3.1 *Employee Green Behavior*

EGB is defined as the actions that employees exhibit that have a beneficial impact on the environment (Unsworth et al., 2021). Researchers have characterized EGB as a form of pro-environmental behavior that is specific to the workplace (Ones et al., 2018; Norton et al., 2015). Stern (2000) describes EGB as an intentional behavior that contributes to the mitigation of human actions which make adverse effects on the environment. It could include efforts like saving water and energy, making better use of resources, cutting down on waste, and reusing (Norton et al., 2015). Furthermore, Ones and Dilchert (2013) highlighted that EGB encompasses not only actions that promote environmental sustainability but also actions that hinder or damage it. Norton et al. (2015) classify EGB into two categories: task green behavior and voluntary green behavior. Norton et al. (2015) characterized task EGB as the implementation of environmentally friendly actions within the limitations of organizational requirements, integral to job responsibilities. Voluntary EGB is defined as green behaviors that exceed organizational expectations and involve personal initiatives.

Similarly, Dumont et al. (2017) classify EGB into two categories: "in-role green behavior" and "extra-role green behavior." Task green behavior resembles an in-role behavior, while voluntary green behavior resembles an extra-role green behavior. For instance, adhering to organization norms and regulatory protocols for the appropriate disposal of hazardous waste items and taking steps to prevent air, water, and soil contamination are considered in-role EGB. In contrast, extra-role EGB, outside of one's official duties, includes actions such as switching off lights when they are not in use, reusing discarded papers for droughting memoranda, and opting for stairs instead of elevators. Researchers refer to "EGB" in the existing literature as "employee pro-environmental behavior" (Paillé & Paillé, 2020), "voluntary pro-environmental behavior" (Pinzone et al., 2019), and "employee ecological behavior" (Okumus et al., 2019).

3.2 Green Training and Employee Green Behavior

GT is known as an approach to enhancing employees' problem-solving skills to address environmental challenges and integrate business policies with environmental protection strategies, eventually achieving sustainability (Aragão & Jabbour, 2017; Usman et al., 2023). According to Renwick et al. (2013), GT is considered one of the most crucial components of green HRM. Evidence widely recognizes GT as a significant predictor of EGB (Pham et al., 2020). The lens of SET further explains the relationship between GT and EGB, as it raises awareness of energy conservation and fosters EGB as an exchange of GT (Chaudhary, 2020; Stefanelli et al., 2020; Usman et al., 2023). GT provides employees with formal awareness and comprehension of informal environmental-friendly workplace practices, resulting in their voluntary adoption of such practices. Hence, this study proposes the following hypothesis:

Hypothesis 1 GT has a positive effect on EGB.

3.3 Green Perceived Organizational Support and Employee Green Behavior

GPOS refers to the distinct value employees attribute to the extent to which an organization acknowledges their efforts toward sustainability (Lamm et al., 2015). For instance, when employees receive rewards for their pro-environmental actions, they perceive the organization's appreciation for their environmental protection efforts. This reinforces the perception of GPOS (Pinzone et al., 2019). Existing literature showed that GPOS enhances EGB and non-EGB (Karatepe et al., 2022); green creativity (Hameed et al., 2022); and EGB (Aboramadan et al., 2022). In an organizational setting, employees exhibit green behavior when their organizations offer them green support. Furthermore, the lens of SET offers a comprehensive analysis of the connection between GPOS and EGB.

According to SET (Blau, 1964), behavior among people arises from a process of reciprocal exchange. As a result, organizations provide adequate support for employees to become more environmentally conscious and encourage them to participate in green activities within the organization. As a result, employees exhibit environmental consciousness or EGB within the organization. Hence, this study proposes:

Hypothesis 2 GPOS has a positive effect on EGB.

3.4 Green Knowledge Sharing as Mediator

GKS refers to the act of exchanging knowledge about environmentally friendly practices among members of an organization to enhance the organization's environmental performance (Abbas & Khan, 2023; Lin & Chen, 2017). More specifically, employee GKS states as the interchange and dissemination of knowledge that enhances the environmental knowledge of organizations, co-workers, and colleagues, thereby ensuring improved environmental performance. Concerned and environmentally friendly organizations prioritize the growth of workforces' skills, understanding, and capabilities to promote GKS through GT. According to Rubel et al. (2021), organizations can arrange GT through green HRM practices to boost GKS among employees. When organizations provide GT or training to make employees knowledgeable about environmental issues, it enhances GKS within the organization. Using the concept of reciprocity, this study elucidates how the provision of green training fosters reciprocal relationships between the organization and its employees. This, in turn, leads to employees becoming knowledgeable about green practices, thereby enhancing GKS within the organization. Hence, this study proposes:

Hypothesis 3 GT has a positive effect on GKS.

Research on knowledge sharing has traditionally examined organizational support, a crucial type of management support (Wang & Noe, 2010). Similarly, organizations can foster GPOS by developing a culture that fosters environmental knowledge within the organization, thereby enhancing GKS within the organization. Moreover, SET can explain the association among GPOS and GKS. According to SET, when workforces think that the organization where they are working values their green or environmentally friendly efforts and treats them well, such as by providing good working conditions, fair compensation, and opportunities for advancement, they will feel obligated to return the favor by showing concern for the success of the organization and actively working toward its objectives (Rhoades & Eisenberger, 2002). More specifically, an organization enhances green knowledge sharing when it supports its employees and encourages them to become more ecologically mindful and knowledgeable. Hence, the present study proposes:

Hypothesis 4 GPOS has a positive effect on GKS.

Knowledge sharing enables individuals and teams within an organization to interact and consult with each other to create, share, and utilize knowledge. It entails the organization of knowledge regarding individual awareness within the team. Knowledge sharing is an indicator of employee performance. Kwahk and Park (2016) demonstrate that the alignment between knowledge sharing and real knowledge has a substantial impact on employee performance. Alike, Rubel et al. (2021) revealed that GKS is one of the important predecessors of EGB (Rubel et al., 2021). Moreover, sharing knowledge among co-workers enhances EGB. The research rationalizes the association among GKS and EGB, using the SET lens as a framework. Since knowledge sharing entails sharing information and support to others to attain organizational objectives, it involves a reciprocal component where employees demonstrate green behavior to collaborate toward the green organizational image, eventually leading to the advancement of EGB. Accordingly, this study proposes:

Hypothesis 5 GKS has a positive effect on EGB.

We discussed the connections among GT and EGB (H1), GT and GKS (H3), and GKS and EGB (H5) in the literature review and hypothesis development section above. It makes sense to suggest that GT has a positive impact on EGB, which ultimately influences EGB. More specifically, the degree of GKS may differ in an organizational context, and GT in an organization can increase the level of GKS that enhances EGB. In other words, the previously mentioned relationships between GT, and EGB; GT, and GKS; and GKS, and EGB suggest that GKS reacts as a potential mediator in the association among GT and EGB. GT independently influences EGB, but when combined with GKS, its impact on EGB reduces, indicating the mediating role of GKS. Previously, Rubel et al. (2021) observed the role of GKS in their study. To the best of our knowledge, no one has previously explored the role of GKS in the relationship between GT and EGB. Moreover, this study uses SET to explain the mediating effect of GKS in the connection among GT and EGB. SET defines exchange as the free choice of individuals, motivated by the anticipated and actual benefits obtained from others. Accordingly, when organizations facilitate GT for their employees, which enables GKS practices among them, it ultimately nurtures EGB. Hence, this study proposes:

Hypothesis 6 GKS positively mediates the relationship between GT and EGB.

Similarly, the relationship among GPOS and EGB (H2), GPOS and GKS (H4), and GKS and EGB (H5) is rational to propose that GKS acts as a potential mediator in the association among GPOS and EGB. More specifically, when an organization provides adequate support to its employees in creating an environmentally friendly organization and cultivating an environment of GKS, it ultimately enhances EGB. To put it another way, GPOS becomes more effective to promote EGB, when GPOS amplifies GKS within the organization. These logics indicate that GKS plays a mediating role in the relationship between GPOS and EGB. The mediational role of GKS can be further explained using SET. According to SET, when one does a favor for another, it results in a future return as reciprocity. Therefore, in an organizational context, when organizations provide sufficient support for their employees and encourage the GKS, employees reciprocate by demonstrating EGB within the organization. Hence, this study proposes:

Hypothesis 7 GKS positively mediates the relationship between GPOS and EGB.

4 Methodology

Using questionnaires, data were collected from FLEs working in different banks across Bangladesh in January and February 2024. In the initial stage of data collection, we formally contacted the HR divisions of all 40 private commercial banks in Bangladesh. This letter provided a concise overview of the study's objectives, methods for selecting participants, a cover letter, and the questionnaire. Only 20 of 40 private commercial banks were willing to take part in the research. With the support of the HR departments at these banks, we were able to disseminate a total of 500 questionnaires, with 25 sets going to each bank. This chapter utilized the judgment sampling technique, a nonprobability sampling method. Due to the absence of a list of FLEs, probability sampling was not feasible in this instance. In the context of nonprobability sampling, researchers utilize judgmental sampling to deliberately select samples that meet specific criteria.

We employed a two-wave survey methodology with a 6-week gap between waves to mitigate the potential influence of common method variance (CMV). During the initial phase (Time 1), employees supplied demographic information and responded to inquiries concerning GT and GPOS. After 6 weeks, during the second phase (Time 2), the participants were once again requested to fill out online questionnaires that specifically addressed the mediating variable (GKS) and the dependent variable (EGB) of this study. To increase the rate at which participants responded during the second phase, we carried out regular follow-ups with them.

Of the 500 questionnaires that were issued, 387 replies were obtained during the initial phase (Time 1), yielding a response rate of 66%. During the second phase (Time 2), a set of questionnaires was distributed to 387 participants, resulting in the collection of 362 responses. After removing incomplete or unengaged responses, the final sample comprised 342 participants, representing a final response rate of 61%. To assess the nonresponse bias, we conducted independent sample t tests to compare the responses from the first and last quartiles. Our analysis revealed no statistically significant changes in the study variables. Therefore, we determined that nonresponse bias was not a significant concern in this investigation.

5 Measures

All items of the variables were adapted from the earlier accepted scale. Three items of GT were adapted from the study of Pinzone et al. (2019). Another independent variable GPOS was adapted from the scale of Hameed et al. (2022). Furthermore, GKS, the mediating variable, was measured by five items from the scale of Rubel et al. (2021). Lastly, the dependent variable EGB was assessed by five items from the recognized scale of Hameed et al. (2022). The reliability coefficients for these altered scales varied between 0.734 and 0.926.

6 Data Analysis

We conducted a two-step analysis of the measurement and structural models employing Smart-PLS (4.0.9), adhering to the guidelines provided by Hair and Alamer (2022). Furthermore, the Statistical Package for the Social Sciences (SPSS Version 22) was used to enter data and conduct the descriptive analysis.

7 Common Method Variance (CMV)

To mitigate CMV, we have employed a variety of methods, as suggested by Podsakoff et al. (2012). Initially, we provided employees with a guarantee of absolute anonymity to reduce the influence of social desirability bias and encourage candid responses. Participants were assured throughout data collection that there were no correct or incorrect answers, which effectively mitigated evaluation fear. In addition, we included various scaling approaches in the questionnaire to further reduce the impact of CMV. Both the independent and mediating variables were assessed using a 5-point Likert scale, but the dependent variable was rated using a 7-point Likert scale. To quantitatively evaluate the magnitude of CMV, we performed Harman's single-factor test. The findings indicated that the first component accounted for about 28.91% of the variability, which falls much below the 50% benchmark. This suggests that CMV was not a noteworthy issue in our dataset.

8 Results

8.1 Demographic Profile

Nearly three-quarters of those who took the survey were married. The following breakdown of qualifications was found in terms of education: 58.3% had an MBA, 38.7% had a bachelor's degree in business, and 13% had a master's degree in another subject. The age bracket within 31–35 years old made up 48% of the participants, while 23.6% fell into the 41–45 years old category. Nearly half of the employees had been with their current employer for more than 2 years but less than 5 years, while 38% had been there for 5–10 years and only 25% for more than 10 years. Notably, 68.4% were with more than 5 years of experience.

8.2 Measurement Model

The measurement model was evaluated by the computation of item loadings, average variance extracted (AVE), and composite reliability (CR). According to the standards outlined by Hair and Alamer (2022), it is recommended that all loading values ought to be equal to or greater than 0.708, AVE must be equal to or more than 0.50, and CR should be equal to or greater than 0.70. The findings displayed in Table 1 have substantiated the convergent validity of the constructs examined in this study. In addition, the research revealed the Rho_A values to assess the appropriateness of the measurement model. Henseler et al. (2016) found that the reliability coefficient Rho_A generally ranges between CR and CA. All Rho_A values complied with the necessary criteria and were deemed acceptable.

The model's discriminant validity was established by applying the HTMT criteria presented in Table 2. Given that the square root of the average variance extracted (AVE) for each construct was higher than the associated correlation coefficients, there was no evidence of a multicollinearity problem. The findings shown in Table 2 corroborate the known discriminant validity of the constructs under examination, as indicated by the HTMT scores varying between −1 and 1.

8.3 Structural Model

To test the appropriateness of structural models, various measures were used including coefficients of determination (R^2), path coefficients (β), cross-validated redundancy (Q^2), and effect size (f^2) (Hair & Alamer, 2022). The results indicated that GT,

Table 1 Measurement model

Constructs	Items	Loading	CA	RHO_A	CR	AVE	VIF
EGB	EGB 1	0.881	0.862	0.882	0.900	0.645	2.89
	EGB 2	0.76					1.60
	EGB 3	0.720					1.60
	EGB 4	0.779					1.93
	EGB 5	0.863					2.92
GKS	GKS 1	0.831	0.828	0.836	0.876	0.585	3.13
	GKS 2	0.775					3.06
	GKS 3	0.737					2.08
	GKS 4	0.745					1.50
	GKS 5	0.732					1.42
GPOS	GPOS 1	0.845	0.832	0.850	0.888	0.666	1.88
	GPOS 2	0.858					2.03
	GPOS 3	0.830					1.98
	GPOS 4	0.723					1.49
GT	GT 1	0.728	0.701	0.730	0.826	0.613	1.27
	GT 2	0.759					1.37
	GT 3	0.858					1.40

Source: Author's own work

Table 2 Discriminant validity

HTMT	**EGB**	**GKS**	**GPOS**	**GT**
EGB				
GKS	0.435			
GPOS	0.476	0.654		
GT	0.37	0.422	0.548	
Fornell/Larcker	**EGB**	**GKS**	**GPOS**	**GT**
EGB	**0.803**			
GKS	0.391	**0.765**		
GPOS	0.426	0.579	**0.816**	
GT	0.291	0.359	0.424	**0.783**

Note: Green POS Green perceived organizational support, *GKS* Green knowledge sharing, *EGB* Employee green behavior
Source: Author's own work

GPOS, and GKS together explained 41.9% of the variance of EGB, whereas both GT and GPOS together described 35.1% of the variance of GKS. Based on Cohen's (1988) criterion, the R^2 values for both GKS and EGB were determined to be substantial. The results of the structural model (Fig. 2) showed that both GT and GPOS had significant positive influences on EGB, and the values were ($\beta = 0.107, p < 0.01$) and ($\beta = 0.265$, $p < 0.01$). Furthermore, the path from GT to GKS ($\beta = 0.137$, $p < 0.01$) and GPOS to GKS ($\beta = 0.520, p < 0.01$) was also shown to be positive and significant. Additionally, GKS had a significant impact on EGB ($\beta = 0.284, p < 0.01$). Therefore, all investigated direct paths were found to be statistically significant, as hypothesized and indicated in Table 3.

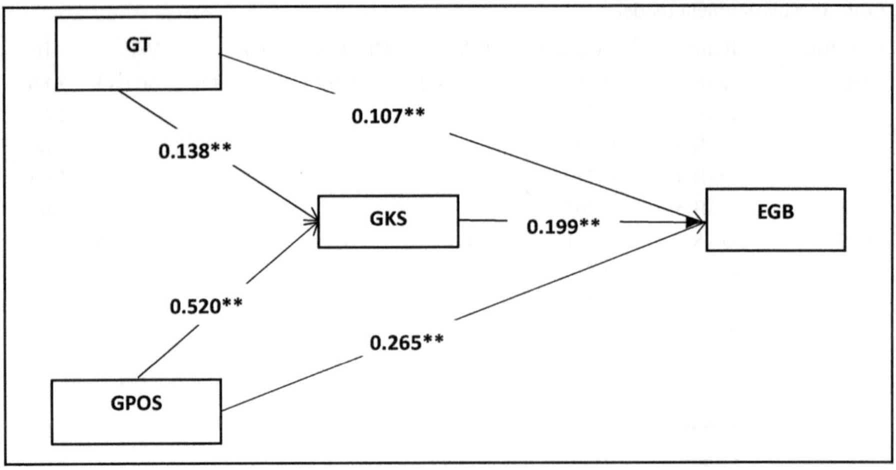

Note: GPOS= Green perceived organizational support, GKS = Green knowledge sharing, EGB = Employee green behavior.

Fig. 2 Structural model (Source: Author's own work). *Note: GPOS* Green perceived organizational support, *GKS* Green knowledge sharing, *EGB* Employee green behavior

Table 3 Structural model

Direct hypotheses							
Hypotheses	**Std. beta**	**Std. error**	***t*-value**	***p*-value**	***f2***	**VIF**	**Decision**
GT > EGB	0.107	0.057	1.88*	0.03	0.012	1.25	S
GPOS > EGB	0.265	0.065	4.07**	0.00	0.055	1.64	S
GT > GKS	0.138	0.053	2.6**	0.005	0.024	1.21	S
GPOS > GKS	0.520	0.049	10.6**	0.00	0.342	1.22	S
GKS > EGB	0.199	0.055	3.64**	0.00	0.033	1.54	S
Significant at $^{**}p < 0.01$(based on one-tailed); S = Supported							
Mediation analysis							
Direct path	**Std. beta**	**Std. error**	***t*-value**	***p*-value**	**95% LL**	**95% UL**	**Decision**
GT > GKS > EGB	0.028	0.015	1.84*	0.033	0.009	0.056	S
GPOS > GKS > EGB	0.104	0.028	3.67**	0	0.059	0.152	S
Significant at $^{**}p < 0.01$(based on two-tailed); S = Supported							

Source: Author's own work

In addition, the present research explored the role of GKS as a mediator between GT, GPOS, and EGB, considering the recommendation of Preacher and Hayes (2008). The results demonstrated a substantial mediating influence of GKS on the relationship between GT and EGB ($\beta = 0.123$, $p < 0.01$), as well as between GPOS and EGB ($\beta = 0.123$, $p < 0.01$). Therefore, all indirect hypotheses were determined to be statistically significant and were accepted considering the obtained results.

9 Discussion

The main goal of this study was to investigate how GT and GPOS have influences on EGB with the mediating effect of GKS. The results show that both GT and GPOS have positive significant influences on employee GKS as well as EGB. GKS and EGB are positively related. Hence, the findings support the hypothesized relationship. The study finds that EGB of FLEs in the banking industry of Bangladesh could be strongly predicted by GPOS and GT. A previous study finds the substantial correlations among GT, GPOS, and green service behavior (Amrutha & Geetha, 2021). This might be because the responders really care about the environment, and their actions mirror the green duties that are taught in GT and supported by GPOS. Rubel et al. (2021) verified that GT has a substantial impact on the green service behavior of FLEs in banking organizations. Employees are more likely to participate in organizational green services when they receive suitable environmental assistance (Aboramadan et al., 2022). It is worth noting that the current research confirms that bank FLEs are more likely to carry out their assigned and volunteer green obligations comprising EGB when they receive overall organizational GT and GPOS.

Furthermore, the results indicate a direct correlation between employees' GKS inside the organization and their engagement in an environmentally friendly behavior, i.e., EGB. This is consistent with previous research conducted by Rubel et al. (2021), which found a significant and favorable correlation between GKS and green service behaviors in the banking industry of Bangladesh. In addition, the present study revealed a noteworthy beneficial mediation impact of GKS on the relationship between GT, GPOS, and EGB.

10 Implications

Using SET as a foundation, this study adds to the existing literature on the topic of GT, GPOS, and the effect they have on EGB in the banking sector of Bangladesh. It further solidifies our belief that firms may enhance their EGB by offering valuable training and suitable assistance on green issues. The second contribution of this study is to fill a knowledge vacuum by using SET to explain the connection between GPOS, GT, and environmentally conscious actions taken by employees in the service industry of an emerging economy, Bangladesh. The third important takeaway is the study's addition to the GKS research; specifically, it provides empirical proof of the mediating role of GKS in the relationship between GPOS, GT, and EGB. As a result, GKS is emphasized as a crucial gateway to an environmentally conscious behavior. Thus, according to the current research framework, it clarifies that when a company provides all the necessary support to its employees, it may lead to an increase in their ability to share information and green conduct on the job. Thus, the company should put more effort into GT and employee-GPOS, which can encourage green information sharing and promote EGB and, in turn, improve the organization's overall green performance.

Important consequences for managers are also offered by the current study. For instance, the study's results suggest that businesses should make sure their employees get GT and all the GPOS they need to become more environmentally conscious at work. Accordingly, the goal of this study is to bring attention among managers to the importance of GT and GPOS in the workplace. Furthermore, to foster a green organizational culture where employees feel comfortable sharing what they know about sustainability and how they can improve EGB, organizations should try to offer time-oriented GT and provide GPOS. Creating a shared perspective among employees for the ecological well-being of the organization requires managers to ensure that their organizations are implementing GT and all essential, green-oriented services. Managers are responsible for establishing a green vision, providing staff with GT and other GPOS and nurturing GKS so that EGB can be put into action. On top of that, they should be committed to the green goals of the organization and devise necessary support mechanisms in the forms of GT and GPOS in creating GKS culture to promote EGB.

11 Limitations and Future Research Directions

While this research does add to the literature by being the first to examine GT, GPOS, GKS, and EGB in banking organizations, further research into these topics could yield different results if applied to banking organizations not practicing green. Future researchers may look at how GT and GPOS affect EGB through GKS in different industries. In addition, the study sampled those who work as front-line positions. Thus, different outcomes may be observed if the identical models have been evaluated on different management levels. To fill gaps in our understanding, future research might look at the ways in which various employee groups directly impact a company's sustainable performance via providing environmentally conscious efforts. Additionally, future studies may think about multidimensional green behavior variables and assess them, as our study only looked at one dimension. The potential moderating and mediating roles of green knowledge and the GKS culture in which organizations function might be investigated in future studies.

12 Conclusion

To encourage employees to achieve the organization's green policies through their own actions, businesses should create suitable GT and mandate GPOS. Such training and support attend numerous rationales. GT provides workers with the knowledge, ethics, and abilities that are essential for achieving green management objectives. GPOS aligns the objectives of the organization's environmental management strategy with the arrangements of various forms of green supports for the staff. Moreover, they raise the levels of environmental consciousness among workers. Additionally, GT and GPOS could enhance GKS culture via mentoring, coaching and facilitation of related projects, schemes, or strategies, and further promote EGB. Such blending of GT, GPOS, and GKS could be more effective in influencing EGB than the effect of organizational, green-oriented assistance, or GT alone. Therefore, businesses should promote GT and GPOS and nurture GKS culture to enhance the environment-friendly behaviors of employees.

References

Ababneh, O. M. A. (2021). How do green HRM practices affect employees' green behaviors? The role of employee engagement and personality attributes. *Journal of Environmental Planning and Management, 64*(7), 1204–1226.

Abbas, J., & Khan, S. M. (2023). Green knowledge management and organizational green culture: An interaction for organizational green innovation and green performance. *Journal of Knowledge Management, 27*(7), 1852–1870.

Aboramadan, M., Kundi, Y. M., & Becker, A. (2022). Green human resource management in nonprofit organizations: Effects on employee green behavior and the role of perceived green organizational support. *Personnel Review, 51*(7), 1788–1806.

Ahmad, S., Islam, T., Sadiq, M., & Kaleem, A. (2021). Promoting green behavior through ethical leadership: A model of green human resource management and environmental knowledge. *Leadership & Organization Development Journal, 42*(4), 531–547.

Ahmed, M., Zehou, S., Raza, S. A., Qureshi, M. A., & Yousufi, S. Q. (2020). Impact of CSR and environmental triggers on employee green behavior: The mediating effect of employee wellbeing. *Corporate Social Responsibility and Environmental Management, 27*(5), 2225–2239.

Ahmed, I., Islam, T., & Umar, A. (2023). Bridging organisational and individual green actions through green knowledge sharing & individual values. *Knowledge Management Research & Practice, 21*(6), 1071–1083.

Amrutha, V. N., & Geetha, S. N. (2021). Linking organizational green training and voluntary workplace green behavior: Mediating role of green supporting climate and employees' green satisfaction. *Journal of Cleaner Production, 290*, 125876.

Aragão, C. G., & Jabbour, C. J. C. (2017). Green training for sustainable procurement? Insights from the Brazilian public sector. *Industrial and Commercial Training, 49*(1), 48–54.

Ashraf, S., Ali, S. Z., Khan, T. I., Azam, K., & Afridi, S. A. (2024). Fostering sustainable tourism in Pakistan: Exploring the influence of environmental leadership on employees' green behavior. *Business Strategy & Development, 7*(1), e328.

Bangladesh Bank. (2011). Policy guidelines for green banking, Circular, No. 2, Banking Regulation and Policy Department, Bangladesh Bank. Retrieved from www.bangladesh-bank.org.

Blau, P. M. (1964). Justice in social exchange. *Sociological Inquiry, 34*(2), 193.

Chaudhary, R. (2020). Green human resource management and employee green behavior: An empirical analysis. *Corporate Social Responsibility and Environmental Management, 27*(2), 630–641.

Chen, T., & Wu, Z. (2022). How to facilitate employees' green behavior? The joint role of green human resource management practice and green transformational leadership. *Frontiers in Psychology, 13*, 906869.

Cohen, J. (1988). *Statistical power analysis for the behavioral sciences* (2nd ed.). Lawrence Erlbaum Associates.

Cropanzano, R., Anthony, E. L., Daniels, S. R., & Hall, A. V. (2017). Social exchange theory: A critical review with theoretical remedies. *Academy of Management Annals, 11*(1), 479–516.

Dumont, J., Shen, J., & Deng, X. (2017). Effects of green HRM practices on employee workplace green behavior: The role of psychological green climate and employee green values. *Human Resource Management, 56*(4), 613–627.

Eisenberger, R., Huntington, R., Hutchison, S., & Sowa, D. (1986). Perceived organizational support. *Journal ofApplied Psychology, 71*(3), 500.

Faraz, N. A., Ahmed, F., & Xiong, Z. (2024). How firms leverage corporate environmental strategy to nurture green behavior: Role of multi-level environmentally responsible leadership. *Corporate Social Responsibility and Environmental Management, 31*(1), 243–259.

Fawehinmi, O., Yusliza, M. Y., Tanveer, M. I., & Abdullahi, M. S. (2024). Influence of green human resource management on employee green behavior: The sequential mediating effect of perceived behavioral control and attitude toward corporate environmental policy. *Corporate Social Responsibility and Environmental Management, 31*(3), 2514–2536.

Hair, J., & Alamer, A. (2022). Partial least squares structural equation modeling (PLS-SEM) in second language and education research: Guidelines using an applied example. *Research Methods in Applied Linguistics, 1*(3), 100027.

Hameed, Z., Khan, I. U., Islam, T., Sheikh, Z., & Naeem, R. M. (2020). Do green HRM practices influence employees' environmental performance? *International Journal of Manpower, 41*(7), 1061–1079.

Hameed, Z., Naeem, R. M., Hassan, M., Naeem, M., Nazim, M., & Maqbool, A. (2022). How GHRM is related to green creativity? A moderated mediation model of green transformational leadership and green perceived organizational support. *International Journal of Manpower, 43*(3), 595–613.

Henseler, J., Ringle, C. M., & Sarstedt, M. (2016). Testing measurement invariance of composites using partial least squares. *International Marketing Review, 33*(3), 405–431.

Islam, M. N. (2023). Managing organizational change in responding to global crises. *Global Business and Organizational Excellence, 42*(3), 42–57.

Islam, M. N. (2024a). Mind the change championing behaviour of employees while leaders work on engaging employees with the change process. *International Journal of Learning and Intellectual Capital, 21*(6), 625–645.

Islam, M. N. (2024b). Leaders' part in boosting change encouraging attitude of employee: The sequential mediation model of dealing change in the organisation. *South Asian Journal of Human Resources Management, 11*(2), 203–227.

Islam, M. N., & Idris, A. (2025). Unleashing employee creativity: The role of knowledge-based HRM, thriving at work and leader's effectiveness. *South Asian Journal of Business Studies*. Epub ahead of print.

Islam, M. N., Furuoka, F., & Idris, A. (2022). Transformational leadership and employee championing behavior during organizational change: The mediating effect of work engagement. *South Asian Journal of Business Studies, 11*(1), 1–19.

Karatepe, O. M., Hsieh, H., & Aboramadan, M. (2022). The effects of green human resource management and perceived organizational support for the environment on green and non-green hotel employee outcomes. *International Journal of Hospitality Management, 103*, 103202.

Khalid, B., Shahzad, K., Shafi, M. Q., & Paille, P. (2022). Predicting required and voluntary employee green behavior using the theory of planned behavior. *Corporate Social Responsibility and Environmental Management, 29*(5), 1300–1314.

Khan, K., Shams, M. S., Khan, Q., Akbar, S., & Niazi, M. M. (2022). Relationship among green human resource management, green knowledge sharing, green commitment, and green behavior: A moderated mediation model. *Frontiers in Psychology, 13*, 924492.

Kim, Y. J., Kim, W. G., Choi, H. M., & Phetvaroon, K. (2019). The effect of green human resource management on hotel employees' eco-friendly behavior and environmental performance. *International Journal of Hospitality Management, 76*, 83–93.

Kulsum, R., & Huda, S. S. (2018). Re-thinking about the green banking model in the context of Bangladesh. *The Journal of Developing Areas, 52*(2), 197–214.

Kwahk, K. Y., & Park, D. H. (2016). The effects of network sharing on knowledge-sharing activities and job performance in enterprise social media environments. *Computers in Human Behavior, 55*, 826–839.

Lamm, E., Tosti-Kharas, J., & King, C. E. (2015). Empowering employee sustainability: Perceived organizational support toward the environment. *Journal of Business Ethics, 128*, 207–220.

Leung, Y. W., & Rosenthal, S. (2019). Explicating perceived sustainability-related climate: A situational motivator of pro-environmental behavior. *Sustainability, 11*(1), 231.

Lim, W. M. (2022). The sustainability pyramid: A hierarchical approach to greater sustainability and the United Nations sustainable development goals with implications for marketing theory, practice, and public policy. *Australasian Marketing Journal, 30*(2), 142–150.

Lin, Y. H., & Chen, Y. S. (2017). Determinants of green competitive advantage: The roles of green knowledge sharing, green dynamic capabilities, and green service innovation. *Quality & Quantity, 51*, 1663–1685.

Lambe, C. J., Wittmann, C. M., & Spekman, R. E. (2001). Social exchange theory and research on business-to-business relational exchange. *Journal of Business-to-Business Marketing, 8*(3), 1–36.

Mirahsani, N., Azizan, O., Shahriari, M., Rexhepi, G., & Najmi, A. (2024). Green culture toward employee green behavior; the mediation roles of perceived support and green identity. *Environment, Development and Sustainability, 26*(6), 16149–16172.

Norton, T. A., Parker, S. L., Zacher, H., & Ashkanasy, N. M. (2015). Employee green behavior: A theoretical framework, multilevel review, and future research agenda. *Organization & Environment, 28*(1), 103–125.

Okumus, F., Köseoglu, M. A., Chan, E., Hon, A., & Avci, U. (2019). How do hotel employees' environmental attitudes and intentions to implement green practices relate to their ecological behavior? *Journal of Hospitality and Tourism Management, 39*, 193–200.

Ones, D. S., & Dilchert, S. (2013). Measuring, understanding, and influencing employee green behaviors. In *Green organizations* (pp. 115–148). Routledge.

Ones, D. S., Wiernik, B. M., Dilchert, S., & Klein, R. M. (2018). Multiple domains and categories of employee green behaviours: More than conservation. In *Research handbook on employee pro-environmental behaviour* (pp. 13–38). Edward Elgar Publishing.

Paillé, P., & Paillé, P. (2020). Employees and pro-environmental behaviors: Obstacles, constraints, and barriers. In: *Greening the workplace: Theories, methods, and research* (pp. 91–112).

Pham, N. T., Vo-Thanh, T., Shahbaz, M., Huynh, T. L. D., & Usman, M. (2020). Managing environmental challenges: Training as a solution to improve employee green performance. *Journal of Environmental Management, 269*, 110781.

Pinzone, M., Guerci, M., Lettieri, E., & Huisingh, D. (2019). Effects of 'green' training on pro-environmental behaviors and job satisfaction: Evidence from the Italian healthcare sector. *Journal of Cleaner Production, 226*, 221–232.

Podsakoff, P. M., MacKenzie, S. B., & Podsakoff, N. P. (2012). Sources of method bias in social science research and recommendations on how to control it. *Annual Review of Psychology, 63*(1), 539–569.

Preacher, K. J., & Hayes, A. F. (2008). Asymptotic and resampling strategies for assessing and comparing indirect effects in multiple mediator models. *Behavior Research Methods, 40*(3), 879–891.

Renwick, D. W., Redman, T., & Maguire, S. (2013). Green human resource management: A review and research agenda. *International Journal of Management Reviews, 15*(1), 1–14.

Rhoades, L., & Eisenberger, R. (2002). Perceived organizational support: A review of the literature. *Journal of Applied Psychology, 87*(4), 698.

Rubel, M. R. B., Kee, D. M. H., & Rimi, N. N. (2021). The influence of green HRM practices on green service behaviors: The mediating effect of green knowledge sharing. *Employee Relations: The International Journal, 43*(5), 996–1015.

Rubel, R. B., Islam, M. N., Islam, M. A., & Rimi, N. N. (2025). Unleashing employees ethical voice behavior: A mediated moderation model of ethical leadership, person-job fit, person-organization fit and co-workers' ethical support. *International Journal of Ethics and Systems*. Epub ahead of print.

Stefanelli, N. O., Teixeira, A. A., Caldeira De Oliveira, J. H., Antonio Ferreira, M., & Sehnem, S. (2020). Environmental training: A systematic review of the state of the art of the theme. *Benchmarking: An International Journal, 27*(7), 2048–2076.

Stern, P. C. (2000). New environmental theories: Toward a coherent theory of environmentally significant behavior. *Journal of Social Issues, 56*(3), 407–424.

Surucu, L. (2024). The influence of green inclusive leadership on green creativity: A moderated mediation model. *Journal of Hospitality Marketing & Management, 33*, 1–24.

Unsworth, K. L., Davis, M. C., Russell, S. V., & Bretter, C. (2021). Employee green behaviour: How organizations can help the environment. *Current Opinion in Psychology, 42*, 1–6.

Usman, M., Rofcanin, Y., Ali, M., Ogbonnaya, C., & Babalola, M. T. (2023). Toward a more sustainable environment: Understanding why and when green training promotes employees' eco-friendly behaviors outside of work. *Human Resource Management, 62*(3), 355–371.

Wang, S., & Noe, R. A. (2010). Knowledge sharing: A review and directions for future research. *Human Resource Management Review, 20*(2), 115–131.

Wu, H. C., Wei, C. F., Tseng, L. Y., & Cheng, C. C. (2018). What drives green brand switching behavior? *Marketing Intelligence and Planning, 36*(6), 694–708.

Zhang, Z., Zhang, Y., & Jia, M. (2021). Does a sense of calling facilitate sustainability? Research on the influence of calling on employee green behavior. *Business Strategy and the Environment, 30*(7), 3145–3159.

Zhang, Y., Wu, J., & Fan, Y. (2022). The effect of perceived organizational support toward the environment on team green innovative behavior: Evidence from Chinese green factories. *Emerging Markets Finance and Trade, 58*(8), 2326–2341.

Zhou, Q., & Zheng, X. (2024). Socially responsible human resource management and employee green behavior at work: The role of learning goal orientation and moral identity. *The International Journal of Human Resource Management, 35*(1), 1–35.

Zibarras, L. D., & Coan, P. (2015). HRM practices used to promote pro-environmental behavior: A UK survey. *The International Journal of Human Resource Management, 26*(16), 2121–2142.

Impact of Green Employee Behavior on Employee Well-being: Evidence from UAE Hospitality Sector

Alaa Amin Abdalla, M. Y. Yusliza, and Khalid Farooq

Abstract The aim of the present chapter is to establish the correlation between the employee green behavior (EGB) and employee well-being in the UAE hospitality industry. Based on data obtained from the administrative staff and using SPSS and AMOS for analysis, the study postulates four dimensions of EGB—green learning, individual practices, influencing others, and organizational voice. The findings indicate that EGB contributes toward self-esteem, which in turn promotes the well-being of employees. This chapter contributes to the EGB literature by documenting a four-dimensional scaling system of measuring EGB behavior in the hotel sector of the GCCs, which is still less attended by researchers. Additionally, it provides policymakers an appropriate model of increasing EGB levels as part of green human resource management strategies where benefits accrue to the employees and the organization.

1 Introduction

It is very recent that studies looking at individual traits such as well-being and research pro-social behavior have gained traction (Williams & Eberhardt, 2020; Johnson et al., 2021). In the literature, employee green behavior (EGB), or a type of pro-social behavior, also ranks as one of the critical micro-activities purposeful for addressing the environmental and sustainable developmental challenges in entities (Singh & Smith, 2022; Yusliza et al., 2021). Managing waste, reducing the use of

A. A. Abdalla (✉)
College of Business, Abu Dhabi University, Abu Dhabi, United Arab Emirates
e-mail: alaa.abdalla@adu.ac.ae

M. Y. Yusliza · K. Farooq
Faculty of Business, Economics and Social Development, Universiti Malaysia Terengganu, Kuala Nerus, Terengganu, Malaysia
e-mail: yusliza@umt.edu.my; khalid.farooq@umt.edu.my

M. Y. Yusliza et al. (eds.), *Employee Green Behavior*,
https://doi.org/10.1007/978-981-95-1422-9_10

resources, and seeking environmental knowledge are only some of the activities that are part of Hwang's and Zhao's EGB model (Huang et al., 2021; Kim et al., 2022). EGB contributes significantly to organizational sustainability by improving environmental performance and reinforcing a collective sense of cohesion among staff at the workplace (Zhang et al., 2022; Thompson & Lee, 2023).

It is the networks of organizations and employees who wish to operate in a green environment and, as well, who work toward making green behavior a rewardable essence (Smith et al., 2021). Through various measures of environmental protection pressures, companies begin to put green behavior into their work processes (Williams et al., 2022). There are a number of studies that emphasize aspects necessary for EGB to occur, but research on the EGB impact on the employees is still very limited (Kim et al., 2022).

EGB, as a form of pro-environmental conduct, enhances eco-efficiency at the organizational level and improves the firm's environmental performance at the organizational level. At a lower scale, it not only achieves job or environmental objectives but also improves job satisfaction and physical and mental health in addition to career growth. Well-being is of great importance for individuals and organizations, as those who experience great well-being tend to be highly motivated and are more productive in performing their tasks (Cameron & Yates, 2022). According to EGB and social exchange theory, the more an employee contributes to the environment, the more fulfilled and effective the employee feels (Miller et al., 2021).

The self-esteem viewpoint, as a measure of well-being, is expected to be relevant in EGB and employee well-being (EWB) association (Yang et al., 2022). Employees who understand that the green concern is appreciated by the organization demonstrate high levels of well-being, and as such, POS for environmental initiatives is necessary (Zhou & Lin, 2023). This is essential considering that in the future, concern for the environment will be a major issue, and it is essential to comprehend how EGB enhances EWB.

The UAE hospitality sector faces unique challenges, including high energy consumption, ecological impacts, and employee stress linked to demanding work environments, making the study of EGB and employee well-being (EWB) especially relevant (Abdalla, 2024). By addressing these dual challenges, this chapter provides critical insights into improving sustainability and employee satisfaction in the UAE hospitality industry (Abdalla, 2024).

In general, extensive research exists on EGB (e.g., Huang et al., 2021; Kim et al., 2022); however, many aspects regarding the interaction of EGB with EWB in the particular socio-economic and cultural context of the UAE are yet to be investigated. Most of the previous studies have narrowed their emphasis on environmental performance as well as organizational results and boundaries, neglecting the employees' benefits of EGB. With the region being reliant on capital- and energy-intensive economies, it calls for new ways of doing business that incorporate sustainability into human resource practices. In situating the results of the research in the context of the UAE, the study emphasizes the significance of green human resource management (HRM) practices and offers specific recommendations to the government and structures of business that seek to improve the health of employees while fostering labor productivity in accordance with environmental goals.

This chapter focuses on the link between EGB and EWB, discussing the processes and time frames when EGB has an effect on EWB. It describes and confirms four components of EGB: green learning, individual practice, influencing others, and organizational voices. By increasing the awareness of the dimensions of EGB and their relations with outcome variables, the research places emphasis on the benefits of green behavior for the employees who practice it. The results are particularly beneficial for the managers, giving theoretical implications on the ways of encouraging EGB at the workplace and improving the well-being of workers.

2 Literature Review

2.1 Theoretical Foundation

According to Blau, 1964, Social Exchange Theory (SET) assumes that any social interaction between two individuals is the result of certain reward exchange relations, whereby an individual renders a service to another person and expects some return services. Here, we extend the model of SET in relation to EGB by arguing that when the employer makes the resources available for developing and pushing green behaviors, it is possible through positive exchanges to achieve a higher state of well-being (EWB) in the employed force.

The studies show that by recognizing and rewarding employees who engage in green practices like green voice and green organizational learning, organizations are able to secure the actual perception of their support among their employees. This perception encourages them, in particular, to raise their self-esteem, which results in higher levels of engagement in other pro-environmental behaviors and thus benefits them even more. This is consistent with the idea of SET that people in social exchanges gain trust, commitment, and satisfaction in relation (Cropanzano & Mitchell, 2005).

As noted, this study also addresses the relationships between EGB and EWB through the lens of self-esteem which is part of intrinsic motivators. Specifically, in the context of this chapter, self-esteem is conceptualized as a mediating variable that fills a gap in self-schema model literature on trust and emotions that are assumed to be integral in employer–employee relationships, yet less attention is given to longer-term established relationships. Thus, the study advances the concept of SET by proposing additional constructs and showing their significance within the model of stakeholder motives and actions in the area of the environmental dimension of business practices.

2.2 *Theoretical Basis and Research Hypothesis*

2.2.1 Employee Green Behavior

EGB embedded in workplace practices is more specific than individual green behavior which involves taking steps to reduce or protect the environment (Robertson & Barling, 2013). EGB encompasses these pro-environmental behaviors that are directly related to the ecological concerns and the routine functions of an organization, thus enabling employees and employers to assess their overall workplace green behaviors (Abdalla, 2024). This, as a result, improves the efficiency of green programs, thus optimizing the growth of the society and the economy.

The positive effects of EGB in organizations have attracted considerable attention. For instance, studies in the field of construction revealed that employees' green attitude promotes waste reduction and saves an organization's resources (Chen et al., 2023). Such unplanned green acts, like sharing environmental knowledge and issues and proposing management-related recommendations, have contributed positively to the environmental performance of the organizations (Chen et al., 2023). Other researchers also pointed out the contributions of EGBs in developing environmentally friendly products and building an organizational culture of environmental citizenship that improves/remedies the environmental performance of an organization in general (Smith et al., 2021).

In terms of personal development, the practice of EGB has been associated with career advancement and increased job satisfaction. EGB contributes to meeting not only the demands of the organization but also the needs and expectations of the individual in the form of personal objectives and rewards (Williams & Eberhardt, 2020). Existing scholarship has proven that the performance of green behaviors toward work is positively related to the progress of employees' careers. In a comprehensive way, EGB can be understood as a complex phenomenon whose antecedents include personal factors, including emotions, self-efficacy, and job satisfaction, as well as contextual factors, including CSR, leadership patterns, and green culture in organizations (Kim et al., 2022). Even though the advantages are conspicuous, there are still controversies over the measurement of EGB, as various researchers came up with several alternative scales that provide a multitude of viewpoints concerning individual practice, organizational perspective, or the role of green behaviors (Williams et al., 2021).

EGB encourages both individual and organizational green performance, but a gap exists between the employees' knowledge and their practice. There is a need to reinforce employee green training so that the full positive effects of EGB can be realized (Zhou & Lin, 2023). Further, employees who have taken green training and have environmental knowledge are likely to have been trained and have worked toward ensuring the employees adopt pro-environmental practices. It, therefore, follows that this study develops a four-pronged approach to EGB through green learning, individual practice, influencing, and organizational voices.

Hypothesis 1 EGB can be characterized into one of four dimensional categories: green learning, individual practice, influencing others, and organizational voices.

2.2.2 Employee Green Behavior and Employee Well-being

"If we can define 'well-being' as pursuing pleasure and pain as an abstract principle," then we can argue that it is a more encompassing and fundamental construct than that of "health" (Smith, 2022). EWB can be regarded as the level of satisfaction with physical activity and psychological stress of employees based at their workplace and can also be considered a measure of the mental health of the people in organizations (Johnson & Lee, 2023a, b). Various studies show that numerous factors, such as moral values (Green et al., 2021), psychological capital (Brown & Taylor, 2022), and willingness to help, can be salient to EWB. The positive impact of altruistic behavior on social interactions, integration, and sense of life strengthens subjective well-being—as Wang and Zhang highlight (2024).

The involvement of pro-social behavior is receiving wider attention from researchers due to its benefits on the self-improvement of those who implement the behavior, the beneficiaries of the behavior, and the experiences of the two parties (Abdalla, 2025). Such behaviors include those that relate to self-others, groups, or societies in general (Lee et al., 2023). Green behavior, which denotes actions that assist in organizational and environmental health (Johnson & Brown, 2022), has also been revealed to be rewarding to the performer. For example, volunteering assists in improving psychological stress coping mechanisms and alleviating depression symptoms. Higher levels of mental health are associated with volunteers, thus making them more likely to report increased happiness and increasing their levels of life satisfaction (Smith & Jones, 2022a, b).

Well-being is not simply being happy; it is also a process of an individual obtaining his or her goals willingly (Harris, 2023). When individual goals are in line with personal or employment values and are self-directed instead of imposed, employees feel more contented and well (Lee, 2024). Employees carrying out green behavior, who are personally oriented and not pressured externally, experience more fulfillment and purpose (Brown & Lee, 2024). EGB, is driven by environmental concerns, and EGB further supports employee self-image and helps them in more subtle areas such as their careers, physical, and mental health. EGB promotes developing organizations in a sustainable manner as well as assisting in the protection of the environment, and within this context, it has been shown to support employees' sense of meaning and satisfaction, enhancing their well-being (Green et al., 2022).

Recent studies have also looked into the green behavior of employees and its impact on their psychology. For instance, Smith et al. (2023a, b) gathered and performed a systematic meta-analysis on conceptual studies and research conducted under this topic and reported positive results, including elements of well-being and increased job satisfaction and motivation. Moreover, Johnson and Lee (2023a, b), in their case study on the hospitality sector, reported that workers that undertook active engagement in green practices experienced less stress and greater satisfaction in their job. In the same manner, Kim et al. (2023) in the construction sector claimed that green practices affect the purpose and burnout level of the employees positively. These recent studies provide an explanation as to why the employees engage in more environmental practices and how engaging in environmental practices can improve employees' mental health and job satisfaction in the different sectors.

Through reviews of the current literature, this chapter suggests that EGB has a favorable effect on EWB. EGB is a voluntary activity that encourages organizational and environmental sustainability, making employees appreciated and respected, thus elevating positive feelings and self-contentment. On the other hand, disregard of organizational green behavior and wasteful practices might lead to reproaches, loss of trust, and decline of positive psychological experiences and EWBs, while it may also lead to negative emotions (Smith & Taylor, 2023). In this regard, as a pro-environmental behavior, EGB aids participants in counteracting work-related stresses and challenges, which has an overall positive contribution to EWB (Johnson & Zhang, 2024). Therefore, the researcher puts forward the following hypothesis:

Hypothesis 2 EGB has a positive effect on EWB.

2.3 *The Mediating Role of Self-esteem*

Leary and Baumeister (2023) defined self-esteem as a stable trait of personality that involves the individual's self-evaluation and regards it as a favorable one. It includes the self-appraisals of a member's competencies and worth to an organization as a member (Smith & Green, 2022). According to Coopersmith (2024), self-esteem has four basic components: importance, ability, virtue, and strength. Such individuals do possess a favorable image of themselves, have belief in their many abilities, and do not easily succumb to setbacks, hence their confidence when facing failures (Brown et al., 2023). As such, self-esteem has been associated with subjective well-being (Miller & Johnson, 2023). Such a claim is further supported as self-esteem is suggested to mitigate the effects other determinants may have on well-being.

Actions such as volunteering cultivate feelings of skill, engagement, and effectiveness. Instrumental support that is offered elicits feelings of indebtedness and further prompts the recipient of support to undertake actions that could assist the giver (Green et al., 2022; Wang & Zhang, 2024). With regard to EGB, it closely relates to the self-image of an individual in the workplace setting. Through such recognition, people's self-worth and self-esteem are elevated. As affirmed by Grube and Piliavin (2022), active participation in selfless actions over an extended duration tended to increase the level of self-comprehensiveness as well as self-regard. On the other hand, Lee (2024) state that employees who are involved in green activities gain respect and appreciation in return, which have a positive impact on their self-image. Therefore, EGB may promote self-worth by meeting psychological demands for significance, capability, moral values, and power (Smith & Davis, 2023). This brings forth the following hypothesis:

Hypothesis 3 EGB has a positive effect on self-esteem.

According to Johnson and Brown (2023), self-esteem and positive emotions are positively correlated. Self-esteem can contribute positively to positive functioning, which is consistent with the self-determination theory that internal need satisfaction is important for positive functioning (Harris, 2023). This theory emphasizes that the satisfaction of internal needs—autonomy, competence, and belonging—is of greater importance than the external factors such as money. The term "need for autonomy" can be defined as the need to have control over one's own behavior and choice; "need for competence" refers to having the ability and power to perform tasks in a satisfactory manner; and the "need for belonging" means being cared for, respected, and connected to others.

On the positive side, when self-esteem is high for employees, they have a working sense of meaning in the work that they do, they believe that they can complete the tasks assigned to them, they feel self-governed, and they perceive they have a stake in decisions taken in regard to the organizations. This set of perceptions satisfies their basic psychological needs, hence resulting in an increase in EWB. In contrast, low self-esteem brings about a work orientation where work has less meaning, belief in ability is low, preferred autonomy is low, and perceived impact on organizational strategy is low, and all these lead to lower functioning (Smith et al., 2023a, b). Hence the researcher states the following hypotheses:

Hypothesis 4 Self-esteem has a positive effect on EWB.

Hypothesis 5 EGB has a positive effect on EWB through self-esteem.

Based on the literature reviewed as well as the outlined hypotheses in the study, the research model is depicted in Fig. 1.

This figure illustrates the conceptual framework linking EGB, self-esteem, and EWB. The model highlights the four dimensions of EGB—green learning,

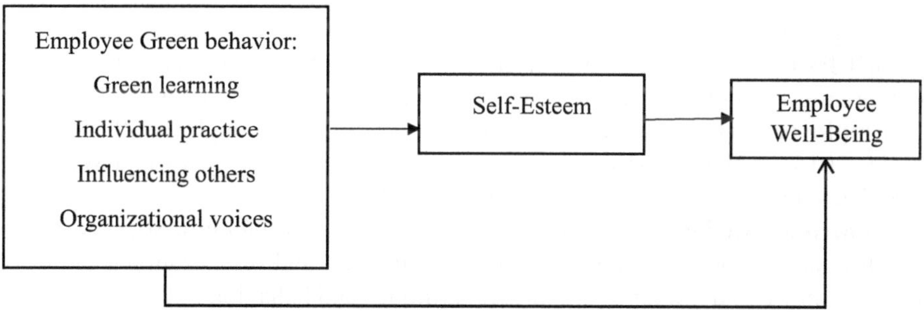

Fig. 1 Research model (Source: Author's own work)

individual practice, influencing others, and organizational voice—and their direct and mediated effects on EWB through self-esteem.

2.4 Sample and Procedure

The data were gathered from 430 respondents, employees of 10 hotels located in different emirates of the UAE. Data collection was carried out between February and June 2024. The participating hotels were distributed as follows: 3 hotels in Dubai, 2 in Abu Dhabi, 2 in Sharjah, 1 in Ras Al Khaimah, and 2 in Fujairah. These hotels ranged from one in luxury to mid-range and budget accommodations.

In the first place, the researcher has made contact with a designated person for each hotel, introduced the objectives, structure, and course of the survey and assured total confidentiality of the responses in order to win their confidence. Thanks to the human resource managers of the hotels, the researcher distributed a letter explaining that all employees received cover letters from us, which defined the areas of EGB to be studied along with its purpose. Since it was deemed necessary for the HR departments to assist in the completion of the questionnaires, they were asked to do so. In total, 360 questionnaires were valid out of 430 distributed.

2.5 Measures

The scales used in this research study, apart from the EGB scales developed in this study, were drawn from previously validated scales. All measures were rated on a scale ranging from 1 to 5, where 1 represented strongly disagree and 5 represented strongly agree.

Employee Green Behavior (EGB): In this study, the EGB scale was developed, which consists of four subdimensions. Three items were used to measure green learning; individual practice also comprised four items; influencing others was assessed using three items; organizational voices included items such as "I suggest environmentally-friendly practices to managers to improve the company's environmental performance." The Cronbach alpha coefficient for this scale was 0.909.

Self-Esteem: Self-esteem was determined from the Rosenberg self-esteem scale, comprising 10 items. The Cronbach alpha coefficient for this scale was 0.938.

Employee Well-Being (EWB): EWB was measured through the Zheng et al. scale (2015), which consists of 18 items and includes three types of EWB. The Cronbach alpha coefficient for this scale was 0.941.

Controlled Variables: This investigation took into consideration socially demographic factors such as gender, age, education level, and organizational seniority because they could influence employees' attitudes and behavior.

3 Data Results and Analysis

3.1 Exploratory Factor Analysis

The data on EFA were sourced from 360 valid questionnaires administered among the employees of 10 hotels, located in different emirates of the UAE. The responses were obtained during a structured survey conducted between February and June, 2024. The sample incorporated representation from the different scales of hotels ranging from luxury, middle, and lower. The EFA was carried out in SPSS 25.0 through the principal component analysis (PCA) method, with the aim of determining the factors or constructs underlying EGB. Using Bartlette's Statistical Test of Sphericity ($p < 0.001$) was found adequate, and the Kaiser–Meyer–Olkin (KMO) measure of sampling adequacy gave a figure of 0.80, which is good for factor analysis. Four factors were extracted based on eigenvalues of more than 1, and these accounted for 78.1% of the total variance.

The researcher performed EFA, followed by the reliability and validity of the 360 questionnaires. The first Kaiser–Meyer–Olkin (KMO) measure of sampling adequacy and Bartlett's test of sphericity were conducted using SPSS version 25.0. The KMO value was 0.80, indicating a meritorious level of sampling adequacy. Additionally, Bartlett's test of sphericity was significant ($p < 0.001$), confirming that the data were suitable for exploratory factor analysis (EFA).

Next, the principal component method was used to conduct the extraction and rotation of factors, the criteria of orthogonal covariance maximum, and Kaiser methods (eigenvalue >1 rule), and the factor loading value not less than 0.5 was adopted for EFA. Ultimately, 78.111% explains the total variance obtained after 13 multidimensional questions were formulated and the four-dimensional structure was proposed. A coefficient of 0.8 and above is the safest for internal reliability estimates, and thus, all factors demonstrated a good EFA of greater than 0.8. This suggests that the EFA also shows that the green learning dimension in organizations exists in parallel with the other three practitioners practicing horizontally as they influence other people, organizational voice, etc. In order to verify the EFA results (see Table 1) and conduct related hypothesis tests, the researcher changed the survey subjects and expanded the sample data to do further empirical research.

3.2 Model Testing

The researcher first carried out a sequence of eight CFAs, using AMOS 25.0 software, so as to test the discriminant validity of the four subdimensions of EGB, and then proceeded to the Anderson and Gerbing two-step technique for SEM analysis. It was first necessary to specify and estimate a CFA model (see Anderson & Gerbing, 1988). This model comprised four subdimensions of EGB that were allowed to correlate freely with one another. It met the criteria for a very good fit of the data relative to the model ($\chi^2 = 116.837$, df = 58, χ^2/df = 1.956, IFI = 0.984, TLI = 0.971,

Table 1 Four-dimensional exploratory factor analysis results of EGB

Questionnaire item	Green learning	Individual practice	Influencing others	Organizational voices
1. I stay informed of my company's environmental initiatives	0.716			
2. I actively participate in environmental protection-related training provided by the company	0.989			
3. I take the initiative to learn environmental protection knowledge to improve environmental protection capabilities	0.886			
4. I print double-sided whenever possible		0.784		
5. I use personal water cups instead of disposable paper cups in the office		0.700		
6. I complete the tasks assigned by the company in an environmentally friendly way		0.888		
7. I perform the duties specified in the job description in an environmentally friendly manner		0.820		
8. I spontaneously give my time to help my colleagues take the environment into account in everything they do at work			0.884	
9. I encourage my colleagues to adopt more environmentally conscious behavior			0.889	
10. I encourage my colleagues to express their ideas and opinions on environmental issues			0.903	
11. I make suggestions about environmentally friendly practices to managers to increase the company's environmental performance				0.848
12. I try to draw management's attention to potentially environmentally unfriendly activities				0.855
13. I inform the management of potentially environmentally irresponsible policies and practices				0.855
Interpretable variation (%) (78.111% overall)	18.311	21.482	20.567	18.655
Cronbach alpha coefficient (total 0.852)	0.845	0.841	0.924	0.872

Source: Author's own work

CFI = 0.980, GFI = 0.955, RMSEA = 0.044). In addition, the fit of the four-dimensional model was better than that of other models; therefore, the four-dimensional structure of EGB was confirmed. EGB's four-dimensional structure was thus suitable for data collection, and its appropriate design structure was good for testing the hypotheses in the study. Hypothesis 1 was therefore confirmed. Table 2 provides the fit statistics for each of the models used in the analysis.

Next, the researcher tried to apply the model as a mediation effect (three-factor measurement model), whereby all indicators were loaded on their latent variables "EGB," "self-esteem," and "EWB." An acceptable model fit was demonstrated (χ^2 = 391.061, df = 107, χ^2/df = 3.486, IFI = 0.951, TLI = 0.932, CFI = 0.939, GFI = 0.921, RMSEA = 0.071), and all standardized factor loadings were significant at $p < 0.001$. As a result, the researcher tested the mediation mechanism.

3.3 Descriptive Statistics and Correlations

Table 3 shows the means, standard deviations, latent intercorrelations, and Cronbach alpha coefficients on the studied variables. All scales had an internal reliability that was acceptable to the norms set by the internal consistency for the study. Self-esteem was also found to correlate positively with EWB, and these correlations show that EGB is positively correlated with EWB, and self-esteem also reveals a significant correlation with EWB (Table 4).

Table 2 Confirmatory factor analysis results of the four-dimensional model for EGB

Models	χ^2	DF	χ^2/df	RMSEA	GFI	CFI	TLI	IFI
Single-factor model	688.868	64	10.412	0.145	0.782	0.806	0.745	0.801
Two-factor model	367.542	61	5.862	0.103	0.905	0.901	0.875	0.895
Three-factor model	556.342	63	8.817	0.142	0.847	0.841	0.806	0.855
Four-factor model	116.837	58	1.956	0.044	0.955	0.980	0.971	0.984

Source: Author's own work

Table 3 Descriptive statistics and intercorrelations

Variable	M	SD	1	2	3	4	5	6	7
Gender	1.522	0.561							
Age	1.545	0.661	−0.216**						
Education	1.889	0.585	0.006	0.104*					
Organizational tenure	2.564	1.412	−0.267**	0.781	0.157**				
EGB	4.307	0.571	0.007	0.024	−0.185**	−0.023	(0.907)		
Self-esteem	4.143	0.577	0.047	0.065	−0.132**	−0.003	0.282**	(0.935)	
EWB	3.835	0.587	−0.017	0.055	−0.241**	0.015	0.305**	0.367**	(0.941)

Notes: $^{*}p < 0.05$. $^{**}p < 0.01$; $N = 360$; scale reliabilities (coefficient alpha) are on the diagonal. All the scales range from 1 to 5

Source: Author's own work

Table 4 Mediating role of self-esteem

Variable type		Employee well-being			Self-esteem		
		Model 1	Model 2	Model 3	Model 4	Model 5	Model 6
Control variable	Gender	0.004	0.003	−0.015	−0.013	0.051	0.52
	Age	0.084	0.065	−0.147*	0.124	−0.164*	−0.194**
	Education	−0.251**	−0.205***	−0.203***	−0.175***	−0.143**	−0.092*
	Organizational tenure	−0.011	0.009	−0.073	−0.046	0.161*	0.181*
Independent variable	EGB		0.273***		0.186***		0.278***
Mediating variable	Self-esteem			**0.355*****	**0.307*****		
F		7.856	13.815	20.786	20.881	3.754	10.441
R^2		0.068***	0.133***	0.195***	0.224***	0.037*	0.109***
ΔR^2		0.007	0.009	0.008	0.011	0.010	0.009

Notes: $^{*}p < 0.05$. $^{**}p < 0.01$. $^{***}p < 0.001$; $N = 360$
Source: Author's own work

The test of mediation effect: After incorporating the mediator variable self-esteem into the relationship, self-esteem, and EWB, the researcher established that the influencing coefficient of EGB on self-esteem and EWB (model 4, $\beta = 0.186$, $p < 0.001$) got smaller. Self-esteem however still had a positive impact on EWB which is significant (model 4, $\beta = 0.307$, $p < 0.001$). The results of the analysis portray that self-esteem set in a mediation dimension in a relationship that existed among EGB and EWB. Therefore, Hypothesis 5 was proven.

4 Discussion and Implications

This chapter proposes a framework of EGB that consists of four dimensions: green learning, practicing, influencing others, and organizational voices. As the empirical results show, this four-dimensional framework shows good internal reliability and discriminant validity. Past studies emphasized measuring EGB through practice, impact, and others' voices only. However, for effective practice of EGBs, popular and suitable green behaviors need to be first identified and selected. Green learning behaviors provide essential bases for the effective and efficient performance of green behaviors in practice science (Clark, 2023; Li et al., 2023a, b). This behavior not only demonstrates the commitment of the employees toward the ecological conservation but also provides encouragement for their routine activities along with subtle green behaviors within the organization, namely practice, impact, and voice. This study is the first to bring in green learning behavior in the measurement of EGB. As regards well-being, this chapter analyzed the dependent variables of EGB and formulated a theoretical framework that included EGB, self-esteem, and EWB. The findings show that undertaking EGB positively impacts one's EWB, thus supporting our research hypothesis. Also, self-esteem is understood to have a positive impact on EWB which is consistent with what has been observed before (Smith & Green, 2022). In addition, self-esteem acts as a mediator between EGB and EWB. All these findings provide theoretical insights for organizations wanting to promote green behavior in their employees and improve their well-being.

In the UAE hospitality industry, green learning is defined by staff education on conserving energy and limiting waste, while individual practice covers actions such as practicing plastic neutrality and utilizing linen reuse programs. Other inclusiveness regarding influence is observing fellow employees encouraging their peers to recycle or practice eco-friendly behaviors, while organizational voice is suggestions to adopt paper-free operations or complain about inefficient use of resources. These dimensions show how GEB responds to the sustainability problems in the cited industry as represented.

In the hospitality industry of the UAE, appropriate HRM policies and strategies are necessary for fostering GEB. Luxury hotels have the option of targeting complex sustainability training as well as eco-luxury certification while providing benefits such as being recognized publicly as green champions. Budget accommodations and other hostels may leverage their cheap approaches by embarking on waste-saving campaigns as well as certificate- and reward-based simple recognition. Such strategies are consistent with organizational capabilities while facilitating GEB throughout the organization.

This UAE-focused study offers insights adaptable to regions with similar environmental challenges, such as other GCC countries, by addressing localized sustainability needs. Adjustments such as the incorporation of community-based green approaches are likely to be necessary in regions characterized by different cultural priorities. In turn, longitudinal studies could show the long-term effects of GEB in EWB, and cross-sector comparison could be useful in the reference for implementing greater EWB.

4.1 Theoretical Implication

As a preliminary step, this study seeks to conceptualize EGB using four dimensions grounded on the initial sustainability of EGB. The analysis conducted in this work shows that EGB causes an increase in EWB. EWB is defined as an employee's improvement perception in their life, work, and psychological emotions (Zheng et al., 2021). Such employees, when present, are more performance-oriented, experience a low rate of job burnout, and kindly advocate for eco-friendly approaches in the organization. Accordingly, EWB bears vital significance in relation to the survival and the core value of the organizational growth and sustainability (Kim & Park, 2022). The practice of green behavior in the workplace brings about different organizational effects and influences, while EWB is influenced by emotional interaction with other people in the organization (Liu et al., 2024). The fact that EGB was investigated from the EWB point of view in this case study is important because it enriches already existing literature on pro-social behavior and EWB, as well as helps to emphasize the under-researched area of EGB in a theoretical manner. This addition paves the way for more attention to this area of literature.

Second, the empirical findings of the study support that the self-esteem of individuals is a mediating factor in the link between EGB and EWB. This research tries to explain such relationships by addressing the internal structures and psychodynamics that assist in understanding the positive facets of EGB on EWB. Such findings are more consistent with the self-determination theory, which posits that self-esteem, in this case, is a strong self-regulatory mechanism that helps channel the advantages of EGB to EWB. There was an early theoretical development, where self-determination theory was used to predict EWB (Deci & Ryan, 2000), and further empirical investigations are drawn on within-country variations in well-being (Ryan & Deci, 2020; Van den Broeck et al., 2022).

4.2 Practical Implications

This study's results emphasize the importance of developing EGB order to support EWB in the hospitality sector of the UAE. It proposes mechanisms to reinforce GEB as an embedded culture within organizations' operations through green learning, environmental behaviors, influencing people, and organizational voice. Such employees will be encouraged to undertake responsibility and get involved in training, saving energy, and working with other departments, which will not only help greener targets but also improve their occupational and psychological well-being.

One important consideration is the impact of green learning on the well-being of the employees. Empowering employees with the knowledge on how to be environmentally responsible and involving them in a green program increases the sense of empowerment and belonging. This dimension of GEB enhances self-esteem and other self-expressions and brings internal satisfaction, both of which are vital for the well-being.

One more practical implication concerns the individual green practices aimed at waste prevention, resource conservation, and sustainable habits in the workplace. Employees who make such attempts do suffer less stress and have more job satisfaction because they help in the bigger picture of sustainability. These actions also foster a sense of achievement and pride, making them better off.

Furthermore, climbing up the ladder toward a workplace voice fostering organizational engagement can also be expressed by the approach for inviting employees to express their thoughts on environmental issues. The green policies endorsed by certain employees are the ones who are engaging, as they are respected, and their initiation is appreciated. This not only boosts their morale but also encourages the whole in-group toward the motive of collective sustainability.

Furthermore, it is the duty of the managers to endorse GEB by encouraging the employees and creating an avenue for joint efforts on green purposes. By making such pro-appropriate action routines and rewarding them, there is a cycle that is created where such behavior is chronic and employees feel good doing the right things. This reinforcement not only strengthens GEB practices but also constitutes a great deal toward improved mental health and satisfaction at the workplace.

These are the main facets of GEB that hospitality companies in the UAE should focus on in order to optimize sustainability issues and satisfaction among employees. This is a great deal for HR managers and policymakers who aim to achieve the targets set by the organization while focusing on employees' interactions.

4.3 Methodological Limitations

One key limitation of the current study is the reliance on cross-sectional data, which does not allow for the observation of changes in EGB and EWB over time. Future research could adopt longitudinal designs to track causal relationships. Additionally, the use of self-reported measures introduces the potential for social desirability bias, as participants might overstate their engagement in green practices to align with perceived social norms. To address this, future studies could incorporate observational or peer-reported measures. Another limitation is the focus on the UAE hospitality sector, which limits the generalizability of findings to other sectors or cultural contexts. Comparative studies across different industries and countries could enrich the understanding of EGB and EWB dynamics.

5 Conclusion

This chapter proposes that EGB encompasses green learning as a key dimension of EGB, in this case, in addition to individual practice of EGB, influencing others to practice EGB as well as organizational voices advocating for EGB adoption.

Several studies have identified EGB as a significant predictor of both social and organizational outcomes, such as EWB. Moreover, EGB has been found to exert a mediating effect, particularly through mechanisms such as self-esteem.

This research contributes to EGB studies through expanding the scope of behavioral measures and incorporation of the outcome variables of EGB. The redesigned EGB enabled employees to gain EWB, which enhanced more EGB, including those of the employees who have not engaged in EGB before. Policymakers should use these findings to promote EGB in their workplaces and help foster positive behavior in the organizations while providing feedback to employees that improves their self-esteem and general well-being.

References

Abdalla, A. A. (2024). Managing for sustainability: The impact of green human resource management practices on employee green behavior and green innovation. In *Reference module in social sciences*. https://doi.org/10.1016/B978-0-443-13701-3.00423-0

Abdalla, A. A. (2025). Boosting employee eco-friendly behaviors and innovation through green HR practices. In *Modern trends and future innovation in human resource management* (pp. 73–104). IGI Global. https://doi.org/10.4018/979-8-3693-6402-4.ch003

Anderson, J. C., & Gerbing, D. W. (1988). Structural equation modeling in practice: A review and recommended two-step approach. *Psychological Bulletin, 103*(3), 411–423. https://doi.org/10.1037/0033-2909.103.3.411

Blau, P. M. (1964). *Exchange and power in social life*. Wiley.

Brown, M., & Lee, J. (2024). Fulfillment through employee Green behavior: A self-driven approach. *Journal of Environmental Management, 58*(3), 215–231. https://doi.org/10.1016/j.jenvman.2024.01235

Brown, J., & Taylor, S. (2022). Psychological capital and employee well-being: A systematic review. *Journal of Management Psychology, 37*(2), 120–138. https://doi.org/10.1108/JMP-05-2021-0279

Brown, T., Smith, L., & Lee, A. (2023). Resilience and self-esteem: Confidence in the face of setbacks. *Journal of Personality Development, 32*(4), 78–95. https://doi.org/10.1037/pd2023.00034

Cameron, L., & Yates, P. (2022). Employee well-being and productivity: A critical review. *Journal of Workplace Studies, 19*(4), 315–330.

Chen, L., Zhang, X., & Wang, Y. (2023). The impact of employee Green attitude on waste reduction and resource savings in the construction industry. *Journal of Construction Management and Sustainability, 12*(1), 54–70. https://doi.org/10.1016/j.conman.2023.104576

Clark, R. (2023). Green learning behaviors: The foundation for effective employee Green behavior practices. *Journal of Environmental Psychology, 27*(1), 115–130. https://doi.org/10.1016/j.jenvp.2023.00167

Coopersmith, S. (2024). The four basic components of self-esteem: Importance, ability, virtue, and strength. *Journal of Psychological Studies, 45*(1), 10–25. https://doi.org/10.1016/j.psyst.2024.00103

Cropanzano, R., & Mitchell, M. S. (2005). Social exchange theory: An interdisciplinary review. *Journal of Management, 31*(6), 874–900. https://doi.org/10.1177/0149206305279602

Deci, E. L., & Ryan, R. M. (2000). The "what" and "why" of goal pursuits: Human needs and the self-determination of behavior. *Psychological Inquiry, 11*(4), 227–268. https://doi.org/10.1207/S15327965PLI1104_01

Green, P., Smith, R., & Turner, L. (2021). The role of moral values in enhancing employee well-being. *Journal of Business Ethics, 169*(3), 535–550. https://doi.org/10.1007/s10551-021-04739-1

Green, P., Wang, Y., & Davis, L. (2022). Sustainable development and employee well-being through Green behavior. *Journal of Environmental Psychology, 19*(4), 283–300. https://doi.org/10.1016/j.jenvp.2022.00345

Grube, J., & Piliavin, J. (2022). The impact of long-term participation in selfless actions on self-comprehensiveness and self-regard. *Journal of Social and Personality Development, 48*(2), 112–130. https://doi.org/10.1080/00224545.2022.1183458

Harris, P. (2023). Well-being and goal attainment: A process beyond happiness. *Journal of Human Development and Psychology, 38*(1), 45–60. https://doi.org/10.1007/jhdp.2023.0101045

Huang, R., Zhao, T., & Wang, S. (2021). Modeling employee Green behavior: Integrating environmental knowledge and sustainable practices. *Journal of Environmental Management, 289*, 112461. https://doi.org/10.1016/j.jenvman.2021.112461

Johnson, H., & Brown, A. (2022). Green behavior in organizations: Actions for organizational and environmental health. *Sustainable Business Review, 14*(2), 89–105. https://doi.org/10.1016/j.sbr.2022.04.005

Johnson, P., & Brown, M. (2023). The correlation between self-esteem and positive emotions: Insights for well-being. *Journal of Positive Psychology, 19*(3), 250–265. https://doi.org/10.1080/17439760.2023.001089

Johnson, M., & Lee, A. (2023a). Employee well-being: Satisfaction with physical activity and psychological stress in the workplace. *Journal of Occupational Health Psychology, 28*(1), 25–38. https://doi.org/10.1037/ocp0000456

Johnson, H., & Lee, S. (2023b). Green behaviors and employee well-being in the hospitality industry: A sector-specific study. *Journal of Hospitality Management, 43*(3), 175–190. https://doi.org/10.1016/j.jhm.2023.02.004

Johnson, M., & Zhang, L. (2024). Counteracting work-related stress: The role of employee Green behavior in enhancing well-being. *Journal of Environmental Psychology, 19*(2), 145–160. https://doi.org/10.1016/j.jenvp.2024.00235

Johnson, M., Davis, L., & Patel, R. (2021). The role of individual traits in driving pro-social behavior in organizations. *Journal of Human Resource Development Quarterly, 32*(2), 150–168. https://doi.org/10.1002/hrdq.21456

Kim, S., & Park, H. (2022). Employee well-being and organizational sustainability: The role of Green behavior at the workplace. *Journal of Environmental Management, 37*(4), 245–260. https://doi.org/10.1016/j.jenvman.2022.12348

Kim, H., Park, J., & Lee, Y. (2022). Green behavior at work: Reducing resource usage and managing waste. *Business and the Environment, 47*(3), 182–194. https://doi.org/10.1002/bate.22234
Kim, E., Zhao, J., & Zhang, Y. (2023). Green behaviors and employee purpose in the construction sector. *Sustainability and Work, 12*(2), 102–120. https://doi.org/10.1080/23555229.2023.2011007
Leary, M. R., & Baumeister, R. F. (2023). Self-esteem as a stable personality trait and its role in self-evaluation. *Journal of Personality and Social Psychology, 75*(2), 123–142. https://doi.org/10.1037/pspi.2023.0056
Lee, S. (2024). Aligning personal and employment values for greater well-being. *Journal of Organizational Psychology, 49*(1), 90–106. https://doi.org/10.1108/JOP.2024.01.093
Lee, K., Kim, J., & Choi, M. (2023). Understanding pro-social behavior: Connections to self and society. *Journal of Community Psychology, 51*(4), 337–352. https://doi.org/10.1002/jcop.22867
Li, W., Abdalla, A. A., Mohammad, T., Khassawneh, O., & Parveen, M. (2023a). Towards examining the link between green HRM practices and employee green in-role behavior: Spiritual leadership as a moderator. *Psychology Research and Behavior Management, 16*, 383–396. https://doi.org/10.2147/PRBM.S396114
Li, S., Chen, Y., & Zhao, L. (2023b). Green learning and its role in enhancing employee Green behaviors in practice science. *Journal of Sustainable Organizational Performance, 16*(2), 90–105. https://doi.org/10.1016/j.jsop.2023.00135
Liu, Q., Wang, L., & Chen, M. (2024). Emotional interactions in the workplace: Impact on employee well-being and organizational growth. *Journal of Applied Psychology, 49*(1), 58–75. https://doi.org/10.1037/apl0000400
Miller, D., & Johnson, A. (2023). The relationship between self-esteem and subjective well-being. *Journal of Positive Psychology, 19*(1), 57–72. https://doi.org/10.1080/17439760.2023.00107
Miller, J., Thompson, A., & White, K. (2021). Employee green behavior and social exchange theory: Linking environmental contribution to job fulfillment and effectiveness. *Journal of Organizational Behavior, 42*(7), 1124–1140. https://doi.org/10.1002/job.2579
Robertson, J. L., & Barling, J. (2013). Greening organizations through leaders' influence on employees' pro-environmental behaviors. *Journal of Organizational Behavior, 34*(2), 176–194. https://doi.org/10.1002/job.1820
Ryan, R. M., & Deci, E. L. (2020). The theoretical developments and future of self-determination theory. *Journal of Human Motivation, 50*(1), 135–155. https://doi.org/10.1016/j.jhm.2020.00119
Singh, P., & Smith, A. (2022). Employee green behavior as a mechanism to address environmental and sustainability challenges. *Sustainability Management Journal, 9*(4), 556–573. https://doi.org/10.3390/susman9050678
Smith, J. (2022). The concept of well-being: A fundamental construct beyond health. *Journal of Psychological Inquiry, 29*(2), 145–158. https://doi.org/10.1007/jpi-2022-145158
Smith, J., & Green, P. (2022). Self-appraisals and organizational worth: The role of self-esteem in employee competency perception. *Journal of Organizational Behavior, 38*(3), 212–228. https://doi.org/10.1002/job.2502
Smith, J., & Jones, L. (2022a). The mental health benefits of volunteering: Links to happiness and life satisfaction. *Journal of Positive Psychology, 17*(2), 123–139. https://doi.org/10.1080/17439760.2022.2025814
Smith, R., & Jones, T. (2022b). Green behavior in HR management: A new paradigm. *Journal of Human Resources, 59*(4), 701–719. https://doi.org/10.1080/00221628.2022.2084531
Smith, J., & Taylor, R. (2023). Consequences of ignoring organizational Green behavior: Impacts on trust and employee well-being. *Journal of Organizational Behavior, 42*(6), 915–932. https://doi.org/10.1002/job.2506
Smith, A., Johnson, P., & Roberts, L. (2021). Building rewardable Green behavior in organizational networks. *Journal of Environmental Sustainability, 13*(2), 122–135. https://doi.org/10.1002/jes.21456
Smith, J., Green, P., & Taylor, R. (2023a). The impact of low self-esteem on work orientation and organizational functioning. *Journal of Organizational Psychology, 45*(2), 134–150. https://doi.org/10.1002/job.2023.00456

Smith, A., Johnson, P., & Roberts, L. (2023b). The psychological impact of green behaviors in organizations: A meta-analysis. *Journal of Environmental Psychology, 54*(1), 45–67. https://doi.org/10.1016/j.jenvp.2023.01.001

Thompson, G., & Lee, K. (2023). Green behavior and social cohesion at work: A new framework. *Workplace Ecology Review, 15*(1), 105–124. https://doi.org/10.1093/workeco/rev011

Van den Broeck, A., Ferris, D. L., & Chang, C. H. (2022). Within-country variations in Well-being: Insights from self-determination theory. *Journal of International Well-Being Studies, 32*(3), 250–267. https://doi.org/10.1016/j.intwellstudies.2022.10231

Wang, Y., & Zhang, X. (2024). The role of altruistic behavior in enhancing subjective well-being: Social interactions and life integration. *Journal of Happiness Studies, 24*(1), 55–72. https://doi.org/10.1007/s10902-022-00546-1

Williams, P., & Eberhardt, R. (2020). Exploring the influence of well-being on pro-social behavior in the workplace. *Journal of Organizational Behavior, 41*(3), 234–252. https://doi.org/10.1002/job.2448

Williams, P., Brown, R., & Davis, K. (2021). Controversies in measuring employee Green behavior: A review of alternative scales and perspectives. *Journal of Environmental Behavior Research, 14*(3), 215–230. https://doi.org/10.1007/jebr-214230

Williams, D., Taylor, M., & Green, S. (2022). Incorporating Green behavior into organizational processes under environmental protection pressures. *Environmental Management Journal, 45*(3), 301–318. https://doi.org/10.1007/emj045301

Yang, S., Li, W., & Chen, D. (2022). Self-esteem as a measure of well-being in the employee green behavior and well-being relationship. *Journal of Environmental Psychology, 80*, 101759. https://doi.org/10.1016/j.jenvp.2022.101759

Yusliza, M. Y., Fawehinmi, O., & Mohd-Yusoff, Y. (2021). Exploring the interplay of green human resource management, employee green behavior, and personal moral norms. *SAGE Open, 11*(1), 2158244020982292. https://doi.org/10.1177/2158244020982292

Zhang, L., Wu, Y., & Chen, X. (2022). Social cohesion and environmental performance: The role of employee green behavior. *Journal of Sustainable Business Research, 6*(2), 240–258. https://doi.org/10.1007/sb22-240

Zheng, X., Zhu, W., Zhao, H., & Zhang, C. (2015). Employee well-being in organizations: Theoretical model, scale development, and cross-cultural validation. *Journal of Organizational Behavior, 36*(5), 621–644. https://doi.org/10.1002/job.1990

Zheng, H., Liu, Y., & Zhang, J. (2021). Defining employee well-being: The role of work, life, and psychological emotions. *Journal of Organizational Behavior, 42*(6), 789–804. https://doi.org/10.1002/job.2534

Zhou, L., & Lin, J. (2023). Perceived organizational support for environmental initiatives and its impact on employee well-being. *Journal of Sustainability and Human Resource Management, 11*(1), 45–62. https://doi.org/10.1007/suhrm-11014562

Green Human Resource Management and Employee Green Behaviors: Proposing a Conceptual Framework

Kashif Javed, Mehran Nejati, Janice Redmond, and Azadeh Shafaei

Abstract Climate change has become one of the world's biggest global challenges. This has created a sense of urgency for pro-environmental practices within organizations by taking climate actions. These climate actions are often informed by organizational policies and implemented by the human resource department to foster employee green behaviors and determine the outcomes achieved. This chapter proposes a conceptual framework to explore the relationships between green human resource management (HRM) and employee green behaviors and articulates how organizational citizenship behaviors for the environment can have an impact on the relationship between green HRM practices and employee green behaviors. In addition, the framework suggests some boundary conditions that influence the proposed relationships by examining the potential moderating role of ethical and inclusive leadership as well as corporate social responsibility (CSR).

K. Javed (✉)
Centre for Sustainability and Governance Research (CSGR), School of Business and Law, Edith Cowan University, Joondalup, Australia
e-mail: k.javed@ecu.edu.au

M. Nejati
Centre for People, Place & Planet (CPPP), School of Business and Law, Edith Cowan University, Joondalup, Australia
e-mail: m.nejati@ecu.edu.au

J. Redmond
School of Business and Law, Edith Cowan University, Joondalup, Australia
e-mail: j.redmond@ecu.edu.au

A. Shafaei
Mental Awareness, Respect and Safety (MARS) Centre, School of Business and Law, Edith Cowan University, Joondalup, Australia
e-mail: a.shafaei@ecu.edu.au

M. Y. Yusliza et al. (eds.), *Employee Green Behavior*,
https://doi.org/10.1007/978-981-95-1422-9_11

1 Introduction

Human society is determined to address the looming environmental crisis resulting from global warming, which is evident from environmental targets set by various countries and regions as well as the 2030 agenda for sustainable development as a global response (Tu et al., 2023). With the increasing concern to address issues associated with rapid climate change and environmental damage, businesses are constantly being called on to be more responsible for solving these problems (Ren et al., 2021).

In response to increasing environmental issues affecting organizations, green management has evolved as a strategic tool for the corporate world, and green human resource management (HRM) has become instrumental in the areas of environmental training, management, and leadership development in businesses (Sarwar & Mustafa, 2023). Green HRM is emerging as a key strategy for organizations to address sustainability challenges at both the individual and organizational levels (Shafaei et al., 2020). Industry plays an important role in ensuring environmental sustainability, and organizations can attain business sustainability by adopting green HRM as a strategy to develop employee green behaviors (EGB) (Yong et al., 2020). EGBs are defined as "scalable actions and behaviors that employees engage in or bring about that are linked with, and contribute to environmental sustainability" (Ones & Dilchert, 2012).

Green HRM is a diverse concept, which defines the HRM's role in environmental management (Renwick et al., 2013). Green HRM includes HR strategies and practices focused on an organization's financial, social, and environmental goals. Simultaneously, it supports the objective of minimizing the negative impacts on the environment and sheds light on the role of different levels of management in implementing an organization's green agenda (Kramar, 2014).

Scholarly literature on green HRM shows that leaders play an integral role in implementing any green initiative, as leaders influence employees' green knowledge, skills, abilities, and opportunities (Baldassari et al., 2023). Leader behavior has been found to be a key determinant in the success of any green initiative (Mansoor et al., 2022). Therefore, leaders shape an organization's sustainability commitment and approach. Research has established that leadership plays an important role in implementing green HRM practices, promoting sustainability and ensuring environmental performance (Memon et al., 2022).

While green HRM has been considered to achieve organizational sustainability goals, it is still unclear how it influences EGBs; hence, there is a need to study the mechanism for this relationship (Yuan et al., 2024). Although the link between green HRM practices and pro-environmental performance has long been established, there is a dearth of research on the relationship between green HRM practices and individual sustainable behaviors and performance (Darvishmotevali & Altinay, 2022). This chapter addresses this gap by exploring a conceptual framework that focuses on green HRM.

The leadership style of supervisors can significantly influence the achievement of an organization's green agenda and the shaping of employee behavior (Ahmad et al.,

2022). Consequently, there are calls in the literature to examine a combination of leadership styles to enhance the implementation of an organization's green agenda (Mansoor et al., 2022). Given that unethical leadership behavior has been associated with certain contemporary corporate scandals, there are calls for organizations to transition from traditional leadership styles to those that can foster ethical, moral, and responsible business practices (Domínguez-Escrig & Mallén-Broch, 2023). Considering the significant role that leaders play, it is imperative to investigate whether ethical leadership EL can influence employee pro-environmental behaviors and enhance follower trust in organizational commitment (Ali & Hassan, 2023).

In addition to the importance of ethical leadership in advancing an organization's sustainability goals, organizations need leaders who can adapt their behaviors to the circumstances of the group being led (Haslam et al., 2024). In the context of advocating EGBs, leaders who exhibit ethical conduct are also expected to demonstrate inclusivity and equitable treatment of their employees (Liu et al., 2017). Inclusive leaders facilitate their followers in bringing their uniqueness (abilities and perspectives) toward the achievement of organizational goals and developing belongingness with the team and organization (Randel et al., 2018). Given the attributes of inclusive leadership, it is important to determine whether inclusive leaders can influence EGBs and improve trust in followers, which in turn can build organizational commitment. Organizations can encourage managers to adopt an inclusive leadership style to strengthen the relationship between green HRM and employee innovative behavior, ultimately resulting in the enhanced achievement of organizational sustainable goals and encouraging pro-environmental behaviors in employees (Shafaei & Nejati, 2023).

Research on green HRM has been the focus of attention of many scholars. Recently, researchers have attempted to identify the antecedents of green HRM and the mediating factors that influence the relationship between green HRM and outcomes (Ren et al., 2022). Attention is required to study the mediation mechanisms between green HRM practices and EGBs (Garavan et al., 2023). There is a certain vacuum in studies that focus on the mediating role of organizational citizenship behavior in the environment (OCBE) between green HRM practices and EGBs. OCBE has been the focus of many recent studies. However, it has mostly been studied as an outcome variable (Islam et al., 2021; Ismail et al., 2022; Khan & Khan, 2022; Lu et al., 2022; Muisyo et al., 2022) but not a mediating variable.

Nonetheless, the mediating role of OCBE has gained interest in recent years. Chang et al. (2019) found that OCBE mediates the relationship between green organizational identity, green shared vision, and green product development. Organizational green initiatives influence OCBE which results in employees increasing the display of green behaviors; this then encourages their colleagues to also do more (Yue et al., 2023). There is still a gap in the literature that needs to be addressed to empirically investigate the mediating role of OCBE.

In their recent literature review, Nadeem and Singh (2025) identified that between 2013 and 2024, researchers of green HRM have heavily focused on studying green HRM in the context of the Hospitality and Tourism industry, mostly applying the lens of ability–motivation–opportunity (AMO) theory and using a limited set of mediators and moderators. Therefore, there is a need to investigate less-explored relationships with fewer identified boundary conditions. The framework proposed in this chapter investigates the relationships between green HRM and EGBs with the mediating role of OCBE and the moderating role of Corporate Social Responsibility (CSR), ethical leadership, and inclusive leadership. This study aimed to answer the following questions to explore under-researched propositions and identify future research areas.

RQ1: Do green HRM practices impact EGBs?

RQ2: Does OCBE mediate the relationship between green HRM and EGBs?

RQ3: Does CSR influence the relationship among green HRM, EGBs, and OCBE?

RQ4: Do ethical and inclusive leadership styles influence the relationship between green HRM, EGBs, and OCBE?

A significant aspect of the proposed research is theorizing the potential moderating influence of CSR, along with ethical and inclusive leadership styles, on the proposed framework. This endeavor aims to address the literature gap by responding to calls for investigation into the boundary conditions of these relationships (Mansoor et al., 2022; Masud et al., 2023; Sachdeva and Singh 2024) and helps advance theorizing the green HRM research.

2 Key Concepts and Issues

2.1 Theoretical Underpinnings

A systematic literature review by Chowdhury et al. (2023) revealed that 24 different theories have underlined green HRM research. However, 5 of these 24 theories have received extensive interest: ability–motivation–opportunity (AMO), resource-based view (RBV), social identity theory (SIT), social exchange theory (SET), and stakeholder theory (SHT).

AMO theory by Appelbaum et al. (2000) has been widely used in green HRM research to explain the impact of green recruitment and green training on developing ability, green rewards, and green performance appraisals in green motivation and flexible work environments in explaining performance results (Abualigah et al., 2023; Aggarwal & Agarwala, 2023; Muisyo et al., 2022; Parida et al., 2021; Rizvi & Garg, 2021).

RBV by Barney (1991) has also been extensively used by green HRM researchers in underpinning strategic HRM in enhancing organizational performance to achieve competitive advantage (Agrawal & Pradhan, 2023; Chen & Yan, 2022; Hameed et al., 2022; Mansoor et al., 2022).

Another widely used theory in green HRM research is SIT by Ashforth and Mael (1989), which has been used by researchers in explaining an individual's urge to become part of a group or being identified as members of a unit to create a positive identification for themselves (Ahmad et al., 2022; Al-Swidi et al., 2021).

Blau (1964) proposed SET, which has been widely used to explain the transactional or reciprocal nature of interactions between employees and an organization (Abualigah et al., 2023; Ismail et al., 2022; Moin et al., 2021).

SHT was initially proposed by Freeman (1984) and has been used by green HRM researchers to explain the relationships between different stakeholder groups and how they affect or are affected by the organization (Chowdhury et al., 2023; Khan et al., 2023).

The conceptual model proposed in the current chapter is grounded in SET because it helps explain the relationships between different variables in this research. SET (Blau, 2017) provides a foundation for understanding the reciprocal relationship between green HRM and EGBs through OCBE. The following section reviews SET in more detail.

2.1.1 Social Exchange Theory

Social exchange relationships are everyday routines for everyone as an individual or part of the corporate world (Paillé et al., 2023). Blau (2017) suggested that individuals' actions are motivated by the expected returns from others, and this assumption formed the basis for SET when it was initially proposed by Blau (1964). SET proposes that employees' behaviors and attitudes depend on their level of organizational trust and support (Chen & Yan, 2022). Social exchange mechanisms and principles are well-established; however, researchers still need to explore negotiated exchange mechanisms (Paillé et al., 2023). In this study, we hypothesize that employees perceive green HRM as discretionary, which elicits reciprocal EBGs. Organizations that demonstrate commitment to sustainability and implement green HRM are perceived by their employees as being more committed to overall sustainability goals (Malik et al., 2021). In turn, this perception can engender a propensity among employees to become committed to organizational green objectives and respond by exhibiting more EGBs (Ali Ababneh et al., 2021).

This chapter also explores the mechanism by which green HRM fosters EGBs through the mediating role of OCBE. According to the principles of reciprocity, when organizations support their employees through trust and inclusion, they respond by becoming more involved and championing organizational sustainability advocacy and demonstrating EGBs (Fasih et al., 2025). Furthermore, the boundary conditions of transactional behavior in response to green HRM are investigated by assessing the moderating roles of leadership (ethical leadership and inclusive leadership) and CSR. This aligns with the notion that a leader's behavior influences followers' responses to organizational practices (Baldassari et al., 2023; Mansoor et al., 2022; Memon et al., 2022) and CSR leads to pro-environmental behaviors (Su & Swanson, 2019).

2.2 *Conceptual Framework*

Drawing on SET, the following figure presents the proposed conceptual framework which will be elaborated upon in this chapter. The framework conjectures a link between green HRM and EGBs through the mediating role of OCBE. Moreover, it proposes three boundary conditions that can affect the proposed relationships.

2.3 Green Human Resource Management

Scholars in the field of green HRM have recommended that it be a multidimensional concept and should be measured by individual dimensions, including recruitment and selection, training, performance appraisal, pay and rewards, environmental vision, and teamwork (Jabbour & Santos, 2008; Renwick et al., 2013). Based on the literature review, Tang et al. (2018) posited that the five most significant dimensions of green HRM are recruitment and selection, training and development, performance management, pay and reward, and employee involvement.

2.3.1 Green Recruitment and Selection

Green recruitment and selection has been considered an important part of green HRM by many researchers (Ahmad, 2015; Jackson et al., 2011). Green recruitment and selection help organizations in choosing candidates who have a commitment toward environmental issues and try to reduce the negative impact of their actions on the environment (Guerci et al., 2016; Jabbour & Santos, 2008), and it can be subdivided into three components, namely, prospective employee's green awareness, green employer branding, and green recruitment and selection criteria (Tang et al., 2018).

2.3.2 Green Training and Development

Green training and development activities focus on developing and improving employees' understanding of environmental issues and motivating them to adopt environmentally friendly practices to conserve energy and reduce waste (Jabbour et al., 2015). Green training and development includes awareness enhancement, knowledge management, and climate building, and it should be extended to all employees across the organization, irrespective of whether they work in any environmental department (Tang et al., 2018). Frontline employees are closest to waste, and if trained well in environmental issues, they can identify and reduce waste significantly (Renwick et al., 2013).

2.3.3 Green Performance Management

Green performance management is the process of assigning activities to employees and evaluating their performance with reference to organizational environmental goals (Jabbour & Santos, 2008). Environmental responsibilities, reduction of carbon footprint, environmental incidents, and communication of green policies and concerns are criteria for formulating green performance indicators (Tang et al., 2018). Employee rewards and compensation were based on performance appraisals.

Therefore, it is the most important element of green performance management for managers and employees (Ahmad, 2015).

2.3.4 Green Pay and Reward

Green pay and rewards, which are sometimes referred to as green compensation, aim to attract, retain, and motivate employees to work toward achieving organizational green/environmental goals by offering employees financial (cash) and nonfinancial (awards, recognition, and praise) rewards (Jabbour & Santos, 2008). Some researchers argue that nonfinancial rewards may be more useful in motivating employees to improve their green performance (Jabbour & Santos, 2008). However, most researchers suggest that a combination of financial and nonfinancial rewards serves as a better strategy for firms in enhancing and motivating EGBs (Jabbour & Santos, 2008; Renwick et al., 2013).

2.3.5 Green Involvement

Green involvement is the process of ensuring employees' participation in environmental management activities to achieve organizational environmental goals (Rizvi & Garg, 2021). Employees' participation in environmental management activities can help stimulate the urge to identify environmental issues and contribute to minimizing negative environmental impacts (Renwick et al., 2013). Green involvement can be measured on the basis of five aspects of "clear green vision, a greening learning climate, various communication channels, offering green practices, and encouraging green involvement" (Renwick et al., 2013; Tang et al., 2018).

2.4 Green HRM and Employee Green Behaviors

Understanding environmental issues and aspects of business activities leads to social and economic benefits at both the individual and organizational levels (Ahmad, 2015; Karatepe et al., 2022). Green HRM is key to encouraging EGBs (Cahyadi et al., 2022). Green HRM has the potential to improve employees' skills, abilities, and green behaviors (Guerci et al., 2016). Many contemporary scholars have studied the relationship between green HRM and EGBs (Dumont et al., 2017; Ercantan & Eyupoglu, 2022; Fawehinmi et al., 2020; Gill et al., 2021; Karmoker et al., 2021; Le Ha & Uyen, 2021). Green HRM engenders a sense of meaningfulness that fosters employee values, thereby reinforcing their green behaviors (Tuan, 2022). Therefore, based on the literature and rooted in SET, we contend that green HRM is seen as a discretionary practice and, in turn, employees feel obligated to reciprocate by displaying more EGBs.

Proposition 1 Green HRM has a positive effect on EGBs.

2.5 Mediating Role of Organizational Citizenship Behavior for the Environment

The contemporary literature on green HRM suggests that green HRM does not directly inculcate EGBs, but various mediation mechanisms are required to foster EGBs (Boxall et al., 2016). Many researchers have called for research on mediating mechanisms to explain the relationship between green HRM practices and more environmental outcomes (Anwar et al., 2020; Chaudhary, 2018; Luu, 2019; Saeed et al., 2019).

OCBE is an extra-role behavior of employees, whereby they display voluntary pro-environmental behaviors to support the organization's green objectives and goals (Anwar et al., 2020; Boiral & Paillé, 2012). OCBE is considered to be a key element in enhancing an organization's environmental performance (Robertson & Barling, 2017). Green HRM is often considered to be the best approach for promoting OCBE (Pham et al., 2020). Paillé et al. (2014) suggested that OCBE should be used as a mediator between HRM and organizational performance and found evidence on the mediating role of OCBE in the link between strategic HRM and environmental performance.

Elshaer et al. (2022) categorized OCBE behaviors into three categories: eco-helping, eco-civic engagement, and eco-initiatives. Eco-helping behaviors use shared green knowledge and support each other in addressing environmental concerns and demonstrating more EGBs. Eco-civic engagement allows employees more freedom to actively participate in organizational green initiatives through their own choice. Eco-initiatives are organizational green initiatives based on employees' suggestions (including green HRM practices).

Therefore, drawing on SET, it can be proposed that green HRM develops a sense of responsibility and provides a focused objective toward protecting the environment, and employees feel obligated to display OCBE, which subsequently motivates them to respond by demonstrating EGBs and displaying them at work.

Proposition 2 OCBE mediates the positive relationship between green HRM and EGBs.

2.6 Moderating Role of Ethical Leadership

In the last few decades, the corporate world has witnessed several corporate scandals that have led researchers to study concepts such as ethical leadership and green HRM. Ethical leadership is based on concepts such as ethics, integrity, and moral obligations, which also form the basis of green HRM (Ahmad et al., 2022). Ethical leadership of CEOs and top management is fundamental to the effectiveness of green HRM initiatives and practices (Ren et al., 2021). Ethical leadership is considered to influence EGBs positively (Ahmad et al., 2021; Chreif & Farmanesh, 2022). Ethical leadership is also considered to positively influence the OCBE of employees (Islam et al., 2021).

Applying the lens of SET, ethical leaders demonstrate commitment toward organizational objectives, green HRM, the environment, the community, employees, and customers, which induces a reciprocal response from employees to display green and OCBE behaviors. Employees are more likely to demonstrate OCBE when they see role models in their ethical leaders enacting green HRM practices (Islam et al., 2021). Ethical leaders can accentuate this relationship because employees are conjectured to be more committed to organizational goals when they perceive their leaders to be ethical (Philipp & Lopez, 2013).

Proposition 3 Ethical leadership moderates the relationship between green HRM, OCBE, and EGBs.

2.7 Moderating Role of Inclusive Leadership

When leaders are available, approachable, and more collaborative, employees are more likely to invest their time and energy in their work and toward organizational goals (Decuypere & Schaufeli, 2020). Inclusive leadership complements green HRM (Shafaei & Nejati, 2023) and focuses on employees' needs to develop a more motivated workforce. Therefore, it positively impacts green HRM practices (Fang et al., 2021). Inclusive leadership encourages OCBE and EGBs (Thabet et al., 2023) by ensuring equality in recognizing the efforts of each team member and appreciating diversity at the same time; hence, employees are more likely to show green behaviors if they are encouraged to perform routine work (Quan et al., 2022).

Inclusive leadership shares numerous characteristics with other leadership styles, such as transformational leadership, empowering leadership, servant leadership, leader–member exchange, and authentic leadership. However, inclusive leadership is distinctive in its acceptance by members as they are, allowing them to bring their uniqueness to work and feel welcomed. In return, members take pride in themselves as part of the team and accept and advocate the team's agenda (Randel et al., 2018). Underpinning SET, we contend that inclusive leadership induces a reciprocal reaction from employees to demonstrate EGBs and OCBE.

Proposition 4 Inclusive leadership moderates the relationship between green HRM, OCBE, and EGBs.

2.8 Moderating Role of Corporate Social Responsibility

CSR is a very old concept in the business world and suggests that organizations have a responsibility toward society and should look beyond maximizing profit (Carroll & Shabana, 2010). CSR is considered an important tool for generating social responsibility and encouraging organizations to incorporate environmental and social concerns into their operations, ultimately leading to pro-environmental behaviors (Su & Swanson, 2019). The aim of green HRM is to ensure sustainable results and economic, social, and environmental factors; therefore, it can be considered a component of organizational CSR programs (Ahmad, 2015). Most contemporary studies on green HRM and CSR have included CSR as an exogenous or endogenous concept (Freitas et al., 2020; Malik et al., 2021; Zhao et al., 2021) or as a mediator (Ziyadeh et al. 2024). Freitas et al. (2020) have called for research to further explore the relationship between green HRM and CSR.

The instrumental view of CSR–HRM suggests that the aim of both CSR and HRM activities is to improve organizational performance (Úbeda-García et al., 2021). Underpinned in SET, CSR activities can support the implementation of green HRM and influence employees to reciprocate EGBs and OCBE. The literature suggests that CSR moderates the employee–organization relationship as well as employees' behavioral and attitudinal outcomes. Additionally, ethical CSR initiatives have a more moderating impact on stakeholders and the organizations (Chen et al., 2018). However, few researchers have focused on green HRM, CSR, and OCBE together (Malik et al., 2021). Based on the literature on CSR–HRM and SET dimensions, we propose that CSR induces reciprocal EGBs via OCBE.

Proposition 5 CSR moderates the relationship between green HRM, OCBE, and EGBs.

3 Future Direction

Future researchers may endeavor to investigate additional mediating and moderating variables to elucidate the relationship between green HRM and EGBs. Moreover, they may consider conducting similar research across multiple sectors, such as manufacturing and services, concurrently and employ mixed methods to address the challenges of endogeneity. This approach aligns with the recommendations of Nadeem and Singh (2025).

In addition, we encourage future researchers to explore multilevel outcome variables and study the impact of this model on individual and organizational outcomes, such as sustainable organizational performance. Moreover, future researchers could employ alternative research designs, such as randomized experiments, to address endogeneity concerns.

4 Discussion of Implications

The findings of this research will advance the research based on SET by testing the principles of the theory on exchange between organizational commitment toward green HRM which encourages OCBE among employees who subsequently respond by adopting more EGBs. The findings of this research will also support the corporate sector in identifying and implementing the green HRM and other important factors in achieving sustainable outcomes, which will contribute toward the United Nations' Sustainable Development Goal 13.

5 Conclusion

This chapter proposes a conceptual framework that suggests a potential mechanism through which green HRM can foster EGBs. Moreover, it proposes three possible boundary conditions that can affect the proposed relationships. This chapter emphasizes the importance of empirically testing a less-explored dimension in green HRM research, namely green recruitment and selection, green training and development, green performance management, green pay and rewards, and green involvement, to examine the relationship between green HRM and EGBs and to investigate the mediating role of OCBE under the boundary conditions of ethical leadership, inclusive leadership, and CSR.

In consideration of the proposed framework illustrated in Fig. 1, we encourage researchers to empirically evaluate the model and identify additional variables that may mediate or influence the implementation of green HRM practices by organizations and their impact on EGBs to address corporate and global sustainability.

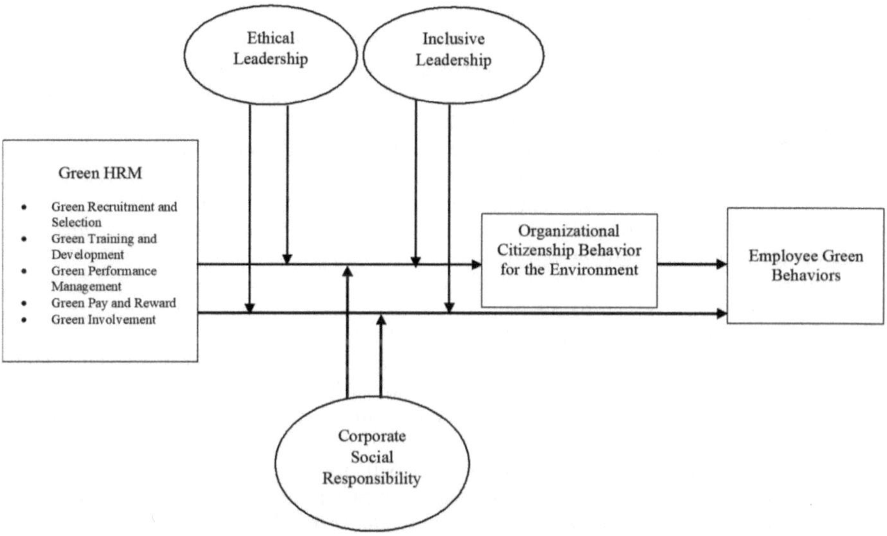

Fig. 1 Conceptual framework (Source: Author's own work)

References

Abualigah, A., Koburtay, T., Bourini, I., Badar, K., & Gerged, A. M. (2023). Towards sustainable development in the hospitality sector: Does green human resource management stimulate green creativity? A moderated mediation model. *Business Strategy and the Environment, 32*(6), 3217–3232.

Aggarwal, P., & Agarwala, T. (2023). Relationship of environmental strategy and performance: Role of green human resource management. *International Journal of Business Environment, 14*(1), 37–66. https://doi.org/10.1504/IJBE.2022.10048217

Agrawal, S., & Pradhan, S. (2023). Employee green behavior in hotels: The role of green human resource management, green transformational leadership and value congruence. *Consumer Behavior in Tourism and Hospitality, 18*(2), 241–255. https://doi.org/10.1108/CBTH-11-2022-0191

Ahmad, S. (2015). Green human resource management: Policies and practices. *Cogent Business & Management, 2*(1), 1030817.

Ahmad, S., Islam, T., Sadiq, M., & Kaleem, A. (2021). Promoting green behavior through ethical leadership: A model of green human resource management and environmental knowledge [Leadership and green behavior]. *Leadership & Organization Development Journal, 42*(4), 531–547. https://doi.org/10.1108/LODJ-01-2020-0024

Ahmad, I., Ullah, K., & Khan, A. (2022). The impact of green HRM on green creativity: Mediating role of pro-environmental behaviors and moderating role of ethical leadership style. *The International Journal of Human Resource Management, 33*(19), 3789–3821. https://doi.org/10.1080/09585192.2021.1931938

Ali Ababneh, O. M., Awwad, A. S., & Abu-Haija, A. (2021). The association between green human resources practices and employee engagement with environmental initiatives in hotels: The moderation effect of perceived transformational leadership. *Journal of Human Resources in Hospitality & Tourism, 20*(3), 390–416.

Ali, M., & Hassan, M. (2023). Green management practices and trust for green behavioral intentions and mediation of ethical leadership. An attribution theory perspective in tourism. *International Journal of Contemporary Hospitality Management, 35*(9), 3193–3215. https://doi.org/10.1108/IJCHM-04-2022-0506

Al-Swidi, A. K., Gelaidan, H. M., & Saleh, R. M. (2021). The joint impact of green human resource management, leadership and organizational culture on employees' green behaviour and organisational environmental performance. *Journal of Cleaner Production, 316*, 128112.

Anwar, N., Nik Mahmood, N. H., Yusliza, M. Y., Ramayah, T., Noor Faezah, J., & Khalid, W. (2020). Green human resource management for organisational citizenship behaviour towards the environment and environmental performance on a university campus. *Journal of Cleaner Production, 256*, 120401. https://doi.org/10.1016/j.jclepro.2020.120401

Appelbaum, E., Bailey, T., Berg, P., & Kalleberg, A. L. (2000). *Manufacturing advantage: Why high-performance work systems pay off.*

Ashforth, B. E., & Mael, F. (1989). Social identity theory and the organization. *Academy of Management Review, 14*(1), 20–39.

Baldassari, P., Eberhard, S., Jiang, Y., Muller-Camen, M., Obereder, L., Schiffinger, M., & Thiele, R. (2023). Looking up and fitting in: Team leaders' and members' behaviors and attitudes toward the environment in an MNC. *Human Resource Management, 62*(3), 267–282.

Barney, J. (1991). Firm resources and sustained competitive advantage. *Journal of Management, 17*(1), 99–120.

Blau, P. (2017). *Exchange and power in social life*. Routledge.

Boiral, O., & Paillé, P. (2012). Organizational citizenship behaviour for the environment: Measurement and validation. *Journal of Business Ethics, 109*, 431–445.

Boxall, P., Guthrie, J. P., & Paauwe, J. (2016). Editorial introduction: Progressing our understanding of the mediating variables linking HRM, employee well-being and organisational performance. *Human Resource Management Journal, 26*(2), 103.

Blau, P. M. (1964). Exchange & power in social life. New York, NY: Wiley.

Cahyadi, A., Natalisa, D., Poór, J., Perizade, B., & Szabó, K. (2022). Predicting the relationship between green transformational leadership, green human resource management practices, and employees' green behavior. *Administrative Sciences, 13*(1), 5.

Carroll, A. B., & Shabana, K. M. (2010). The business case for corporate social responsibility: A review of concepts, research and practice. *International Journal of Management Reviews, 12*(1), 85–105. https://doi.org/10.1111/j.1468-2370.2009.00275.x

Chang, T.-W., Chen, F.-F., Luan, H.-D., & Chen, Y.-S. (2019). Effect of green organizational identity, green shared vision, and organizational citizenship behavior for the environment on green product development performance. *Sustainability, 11*(3), 617.

Chaudhary, R. (2018). Can green human resource management attract young talent? An empirical analysis. *Evidence-based HRM: A Global Forum for Empirical Scholarship, 6.*

Chen, Y. S., & Yan, X. (2022). The small and medium enterprises' green human resource management and green transformational leadership: A sustainable moderated-mediation practice. *Corporate Social Responsibility and Environmental Management, 29*(5), 1341–1356.

Chen, Z. F., Hong, C., & Occa, A. (2018). How different CSR dimensions impact organization-employee relationships: The moderating role of CSR-culture fit. *Corporate Communications: An International Journal, 24*(1), 63–78.

Chowdhury, S. R., Mendy, J., & Rahman, M. (2023). A systematic literature review of GHRM: Organizational sustainable performance reimagined using a new holistic framework. *Sustainability, 15*(9), 7513. https://doi.org/10.3390/su15097513

Chreif, M., & Farmanesh, P. (2022). Applying green human resource practices toward sustainable workplace: A moderated mediation analysis. *Sustainability, 14*(15), 9250. https://doi.org/10.3390/su14159250

Darvishmotevali, M., & Altinay, L. (2022). Green HRM, environmental awareness and green behaviors: The moderating role of servant leadership. *Tourism Management, 88*, 104401. https://doi.org/10.1016/j.tourman.2021.104401

Decuypere, A., & Schaufeli, W. (2020). Leadership and work engagement: Exploring explanatory mechanisms. *German Journal of Human Resource Management, 34*(1), 69–95. https://doi.org/10.1177/2397002219892197

Domínguez-Escrig, E., & Mallén-Broch, F. F. (2023). Leadership for sustainability: Fostering organizational learning to achieve radical innovations. *European Journal of Innovation Management, 26*(2), 309–330. https://doi.org/10.1108/EJIM-03-2021-0151

Dumont, J., Shen, J., & Deng, X. (2017). Effects of green HRM practices on employee workplace green behavior: The role of psychological green climate and employee green values. *Human Resource Management, 56*(4), 613–627.

Elshaer, I. A., Abdelrahman, M. A., Azazz, A. M., Alrawad, M., & Fayyad, S. (2022). Environmental transformational leadership and green innovation in the hotel industry: Two moderated mediation analyses. *International Journal of Environmental Research and Public Health, 19*(24), 16800.

Ercantan, O., & Eyupoglu, S. (2022). How do green human resource management practices encourage employees to engage in green behavior? Perceptions of university students as prospective employees. *Sustainability, 14*(3), 1718.

Fang, Y.-C., Ren, Y.-H., Chen, J.-Y., Chin, T., Yuan, Q., & Lin, C.-L. (2021). Inclusive leadership and career sustainability: Mediating roles of supervisor developmental feedback and thriving at work. *Frontiers in Psychology, 12*, 671663. https://doi.org/10.3389/fpsyg.2021.671663

Fasih, S. T., Zia, M. Q., Ellahi, R. M., & Mahmood, A. (2025). The green soft and hard talent management and social sustainability: A moderated serial mediation model. *Journal of the Knowledge Economy*, 1–31.

Fawehinmi, O., Yusliza, M., Wan Kasim, W. Z., Mohamad, Z., & Sofian Abdul Halim, M. A. (2020). Exploring the interplay of green human resource management, employee green behavior, and personal moral norms. *SAGE Open, 10*(4), 2158244020982292.

Freitas, W. R. S., Caldeira-Oliveira, J. H., Teixeira, A. A., Stefanelli, N. O., & Teixeira, T. B. (2020). Green human resource management and corporate social responsibility: Evidence from Brazilian firms. *Benchmarking, 27*(4), 1551–1569. https://doi.org/10.1108/BIJ-12-2019-0543
Freeman, R. E. (1984). Strategic management: A stakeholder approach. Pitman Publishing.
Garavan, T., Ullah, I., O'Brien, F., Darcy, C., Wisetsri, W., Afshan, G., & Mughal, Y. H. (2023). Employee perceptions of individual green HRM practices and voluntary green work behaviour: A signalling theory perspective. *Asia Pacific Journal of Human Resources, 61*(1), 32–56.
Gill, A., Ahmad, B., & Kazmi, S. (2021). The effect of green human resource management on environmental performance: The mediating role of employee eco-friendly behavior. *Management Science Letters, 11*(6), 1725–1736.
Guerci, M., Longoni, A., & Luzzini, D. (2016). Translating stakeholder pressures into environmental performance—The mediating role of green HRM practices. *The International Journal of Human Resource Management, 27*(2), 262–289. https://doi.org/10.1080/09585192.2015.1065431
Hameed, Z., Naeem, R. M., Hassan, M., Naeem, M., Nazim, M., & Maqbool, A. (2022). How GHRM is related to green creativity? A moderated mediation model of green transformational leadership and green perceived organizational support. *International Journal of Manpower, 43*(3), 595–613.
Haslam, S. A., Alvesson, M., & Reicher, S. D. (2024). Zombie leadership: Dead ideas that still walk among us. *The Leadership Quarterly, 35*, 101770.
Islam, T., Hussain, D., Ahmed, I., & Sadiq, M. (2021). Ethical leadership and environment specific discretionary behaviour: The mediating role of green human resource management and moderating role of individual green values. *Canadian Journal of Administrative Sciences, 38*(4), 442–459. https://doi.org/10.1002/cjas.1637
Ismail, H., El Irani, M., & Kertechian, K. S. (2022). Green HRM and nongreen outcomes: The mediating role of visionary leadership in Asia. *International Journal of Manpower, 43*(3), 660–681. https://doi.org/10.1108/IJM-04-2020-0162
Jabbour, C. J. C., & Santos, F. C. A. (2008). Relationships between human resource dimensions and environmental management in companies: Proposal of a model. *Journal of Cleaner Production, 16*(1), 51–58.
Jabbour, C. J. C., Jugend, D., de Sousa Jabbour, A. B. L., Gunasekaran, A., & Latan, H. (2015). Green product development and performance of Brazilian firms: Measuring the role of human and technical aspects. *Journal of Cleaner Production, 87*, 442–451.
Jackson, S. E., Renwick, D. W., Jabbour, C. J., & Muller-Camen, M. (2011). State-of-the-art and future directions for green human resource management: Introduction to the special issue. *German Journal of Human Resource Management, 25*(2), 99–116.
Karatepe, O. M., Hsieh, H., & Aboramadan, M. (2022). The effects of green human resource management and perceived organizational support for the environment on green and non-green hotel employee outcomes. *International Journal of Hospitality Management, 103*, 103202. https://doi.org/10.1016/j.ijhm.2022.103202
Karmoker, K., Kona, F. A., Oyshi, A. H., & Yasmin, K. S. (2021). Effects of green human resource management on employee green behavior: Moderating role of employee environmental knowledge. *International Journal of Sustainable Development & World Policy, 10*, 64–80.
Khan, A. N., & Khan, N. A. (2022). The nexuses between transformational leadership and employee green organisational citizenship behaviour: Role of environmental attitude and green dedication. *Business Strategy and the Environment, 31*(3), 921–933.
Khan, M., Usman, M., Shafique, I., Ogbonnaya, C., & Roodbari, H. (2024). Can HR managers as ethical leaders cure the menace of precarious work? Important roles of sustainable HRM and HR manager political skill. *The International Journal of Human Resource Management, 35*, 1824–1850. https://doi.org/10.1080/09585192.2023.2241821
Kramar, R. (2014). Beyond strategic human resource management: Is sustainable human resource management the next approach? *The International Journal of Human Resource Management, 25*(8), 1069–1089.

Le Ha, N. T., & Uyen, P. T. B. (2021). The effect of green human resource management on employee green behaviors in a developing country, Vietnam. *International Journal of Information, Business and Management, 13*(1), 107–124.

Liu, W., Song, Z., Li, X., & Liao, Z. (2017). Why and when leaders' affective states influence employee upward voice. *Academy of Management Journal, 60*(1), 238.

Lu, H., Xu, W., Cai, S., Yang, F., & Chen, Q. (2022). Does top management team responsible leadership help employees go green? The role of green human resource management and environmental felt-responsibility. *Corporate Social Responsibility and Environmental Management, 29*(4), 843–859. https://doi.org/10.1002/csr.2239

Luu, T. T. (2019). Green human resource practices and organizational citizenship behavior for the environment: The roles of collective green crafting and environmentally specific servant leadership. *Journal of Sustainable Tourism, 27*, 1167.

Malik, S. Y., Mughal, Y. H., Azam, T., Cao, Y., Wan, Z., Zhu, H., & Thurasamy, R. (2021). Corporate social responsibility, green human resources management, and sustainable performance: Is organizational citizenship behavior towards environment the missing link? *Sustainability, 13*(3), 1044. https://doi.org/10.3390/su13031044

Mansoor, A., Farrukh, M., Jahan, S., Lee, J. K., & Abd Wahab, S. (2022). Promoting green performance through green human resource practices and green servant leadership. *Asia Pacific Journal of Human Resources, 60*(4), 900–918.

Masud, M. M., Jafrin, N., Mohammad Saif, A. N., & Al-Mamun, A. (2023). The moderating effect of corporate social responsibility between green human resource management and organizations' environmental performance. *Journal of Environmental Planning and Management, 66*(12), 2424–2446.

Memon, S. B., Rasli, A., Dahri, A. S., & Hermilinda Abas, I. (2022). Importance of top management commitment to organizational citizenship behaviour towards the environment, green training and environmental performance in Pakistani industries. *Sustainability, 14*(17), 11059. https://doi.org/10.3390/su141711059

Moin, M. F., Omar, M. K., Wei, F., Rasheed, M. I., & Hameed, Z. (2021). Green HRM and psychological safety: How transformational leadership drives follower's job satisfaction. *Current Issues in Tourism, 24*(16), 2269–2277.

Muisyo, P. K., Qin, S., Ho, T. H., Julius, M. M., & Barisoava Andriamandresy, T. (2022). Implications of GHRM on organisational citizenship behaviour: The mediating role of enablers of green culture. *International Journal of Manpower, 43*(3), 719–741. https://doi.org/10.1108/IJM-05-2020-0245

Nadeem, R., & Singh, R. (2025). Charting the future of green HRM practices: Insights from theories, context, characteristics and methodologies (TCCM) framework and analytical hierarchy process (AHP) analysis. *Journal of Management Development, 44*, 154.

Ones, D. S., & Dilchert, S. (2012). Environmental sustainability at work: A call to action. *Industrial and Organizational Psychology, 5*(4), 444–466.

Paillé, P., Chen, Y., Boiral, O., & Jin, J. (2014). The impact of human resource management on environmental performance: An employee-level study. *Journal of Business Ethics, 121*, 451–466.

Paillé, P., Sanchez-Famoso, V., Valéau, P., Ren, S., & Mejia-Morelos, J. H. (2023). Green HRM through social exchange revisited: When negotiated exchanges shape cooperation. *The International Journal of Human Resource Management, 34*(17), 3277–3307.

Parida, S., Ananthram, S., Chan, C., & Brown, K. (2021). Green office buildings and sustainability: Does green human resource management elicit green behaviors? *Journal of Cleaner Production, 329*, 129764. https://doi.org/10.1016/j.jclepro.2021.129764

Pham, N. T., Vo Thanh, T., Tučková, Z., & Thuy, V. T. N. (2020). The role of green human resource management in driving hotel's environmental performance: Interaction and mediation analysis. *International Journal of Hospitality Management, 88*, 102392. https://doi.org/10.1016/j.ijhm.2019.102392

Philipp, B. L. U., & Lopez, P. D. J. (2013). The moderating role of ethical leadership: Investigating relationships among employee psychological contracts, commitment, and citizenship behavior. *Journal of Leadership & Organizational Studies, 20*(3), 304–315. https://doi.org/10.1177/1548051813483837

Quan, D., Tian, L., & Qiu, W. (2022). The study on the influence of green inclusive leadership on employee green behaviour. *Journal of Environmental and Public Health, 2022*, 5292184. https://doi.org/10.1155/2022/5292184

Randel, A. E., Galvin, B. M., Shore, L. M., Ehrhart, K. H., Chung, B. G., Dean, M. A., & Kedharnath, U. (2018). Inclusive leadership: Realizing positive outcomes through belongingness and being valued for uniqueness. *Human Resource Management Review, 28*(2), 190–203.

Ren, S., Tang, G., & Jackson, S. E. (2021). Effects of green HRM and CEO ethical leadership on organizations' environmental performance. *International Journal of Manpower, 42*(6), 961–983.

Ren, S., Jiang, K., & Tang, G. (2022). Leveraging green HRM for firm performance: The joint effects of CEO environmental belief and external pollution severity and the mediating role of employee environmental commitment. *Human Resource Management, 61*(1), 75–90. https://doi.org/10.1002/hrm.22079

Renwick, D. W., Redman, T., & Maguire, S. (2013). Green human resource management: A review and research agenda. *International Journal of Management Reviews, 15*(1), 1–14.

Rizvi, Y. S., & Garg, R. (2021). The simultaneous effect of green ability-motivation-opportunity and transformational leadership in environment management: The mediating role of green culture. *Benchmarking, 28*(3), 830–856. https://doi.org/10.1108/BIJ-08-2020-0400

Robertson, J. L., & Barling, J. (2017). Toward a new measure of organizational environmental citizenship behavior. *Journal of Business Research, 75*, 57–66.

Sachdeva, C., & Singh, T. (2024). Green transformational leadership and pro-environmental behaviour: Unravelling the underlying mechanism in the context of hotel industry. *International Journal of Organizational Analysis, 32*(2), 255–271.

Saeed, B. B., Afsar, B., Hafeez, S., Khan, I., Tahir, M., & Afridi, M. A. (2019). Promoting employee's proenvironmental behavior through green human resource management practices. *Corporate Social Responsibility and Environmental Management, 26*(2), 424–438.

Sarwar, A., & Mustafa, A. (2023). Analysing the impact of green intellectual capital on environmental performance: The mediating role of green training and development. *Technology Analysis & Strategic Management, 36*, 1–14.

Shafaei, A., & Nejati, M. (2023). Green human resource management and employee innovative behaviour: Does inclusive leadership play a role? *Personnel Review, 53*, 266.

Shafaei, A., Nejati, M., & Yusoff, Y. M. (2020). Green human resource management: A two-study investigation of antecedents and outcomes. *International Journal of Manpower, 41*(7), 1041–1060.

Su, L., & Swanson, S. R. (2019). Perceived corporate social responsibility's impact on the well-being and supportive green behaviors of hotel employees: The mediating role of the employee-corporate relationship. *Tourism Management, 72*, 437–450.

Tang, G., Chen, Y., Jiang, Y., Paillé, P., & Jia, J. (2018). Green human resource management practices: Scale development and validity. *Asia Pacific Journal of Human Resources, 56*(1), 31–55.

Thabet, W. M., Badar, K., Aboramadan, M., & Abualigah, A. (2023). Does green inclusive leadership promote hospitality employees' pro-environmental behaviors? The mediating role of climate for green initiative. *The Service Industries Journal, 43*(1–2), 43–63. https://doi.org/10.1080/02642069.2022.2120982

Tu, Y., Li, Y., & Zuo, W. (2023). Arousing employee pro-environmental behavior: A synergy effect of environmentally specific transformational leadership and green human resource management. *Human Resource Management, 62*(2), 159–179. https://doi.org/10.1002/hrm.22138

Tuan, L. T. (2022). Promoting employee green behavior in the Chinese and Vietnamese hospitality contexts: The roles of green human resource management practices and responsible leadership. *International Journal of Hospitality Management, 105*, 103253. https://doi.org/10.1016/j.ijhm.2022.103253

Úbeda-García, M., Claver-Cortés, E., Marco-Lajara, B., & Zaragoza-Sáez, P. (2021). Corporate social responsibility and firm performance in the hotel industry. The mediating role of green human resource management and environmental outcomes. *Journal of Business Research, 123*, 57–69.

Yong, J. Y., Yusliza, M. Y., Ramayah, T., Chiappetta Jabbour, C. J., Sehnem, S., & Mani, V. (2020). Pathways towards sustainability in manufacturing organizations: Empirical evidence on the role of green human resource management. *Business Strategy and the Environment, 29*(1), 212–228. https://doi.org/10.1002/bse.2359

Yuan, Y., Ren, S., Tang, G., Ji, H., Cooke, F. L., & Wang, Z. (2024). How green human resource management affects employee voluntary workplace green behaviour: An integrated model. *Human Resource Management Journal, 34*(1), 91–121.

Yue, G., Wei, H., Khan, N. U., Saufi, R. A., Yaziz, M. F. A., & Bazkiaei, H. A. (2023). Does the environmental management system predict TBL performance of manufacturers? The role of green HRM practices and OCBE as serial mediators. *Sustainability, 15*(3), 2436. https://doi.org/10.3390/su15032436

Zhao, F., Kusi, M., Chen, Y., Hu, W., Ahmed, F., & Sukamani, D. (2021). Influencing mechanism of green human resource management and corporate social responsibility on organizational sustainable performance. *Sustainability, 13*(16), 8875. https://doi.org/10.3390/su13168875

Ziyadeh, M. W., Othman, M., & Zaid, A. A. (2024). Effects of green human resource management on organisational sustainability: The mediating role of corporate social responsibility and organisational citizenship behaviour. *International Journal of Organizational Analysis, 32*(2), 357–372.

Green Leadership in Action: How Green Human Resource Management and Green Engagement Foster Green Behavior in Chinese Enterprises

Feige You, Daisy Mui Hung Kee, Pengyu Chen, and Gary Peng Liang Tan

Abstract This study investigates the mechanisms through which green transformational leadership (GTL) promotes employee green behavior (EGB) within Chinese private enterprises. Specifically, we investigate the mediating influence of green engagement (GE) and the moderating impact of green human resource management (HRM). While existing literature has explored green HRM's effect on environmental EGB, our research addresses a notable gap by being the first to comprehensively analyze the integrated pathways connecting GTL to GB via both green HRM practices and GE. Utilizing SmartPLS for data analysis, we examined responses from 266 employees in the Chinese private sector. Our findings confirm that GTL had a positive and significant effect on GE, which in turn positively influenced EGB. Additionally, GE was found to mediate the relationship between GTL and EGB. The interaction between GHRM and GTL also significantly predicted GE, indicating a moderating effect. However, the interaction between GHRM and GE did not significantly affect EGB. These insights not only enrich the theoretical understanding of green leadership and employee behavior but also provide actionable recommendations for organizations striving to cultivate a robust culture of environmental responsibility.

F. You (✉) · D. M. H. Kee · P. Chen
School of Management, Universiti Sains Malaysia, George Town, Penang, Malaysia
e-mail: feige.you@outlook.com; daisy@usm.my; pengyuc69@gmail.com

G. P. L. Tan
School of Business and Administration, Wawasan Open University,
George Town, Penang, Malaysia
e-mail: pltan@wou.edu.my

M. Y. Yusliza et al. (eds.), *Employee Green Behavior*,
https://doi.org/10.1007/978-981-95-1422-9_12

1 Introduction

In response to escalating global environmental concerns and the urgency of combating climate change, businesses worldwide are under mounting pressure to adopt sustainable practices (Saleh & Al-Swidi, 2019). These practices comply with environmental regulations while proactively seeking innovative ways to minimize ecological footprints and enhance environmental sustainability (Dumont et al., 2017). For enterprises, integrating green practices is more than a regulatory obligation; it is a strategic necessity that strengthens long-term resilience and competitive advantage (Taamneh et al., 2025; World Health Organization, 2021). In this context, promoting environmental employee green behavior (EGB) is important for organizational success and reputation. Given that China stands at the forefront of global economic advancements and remains committed to the Sustainable Development Goals (SDGs), understanding the green initiatives of its private sector is essential (You & Kee, 2024). The extent to which these initiatives promote sustainability will significantly contribute to China's broader environmental efforts. Understanding EGB within the context of private enterprises in China provides insights for leaders, policymakers, and stakeholders dedicated to advancing environmental responsibility (You & Kee, 2024).

Private enterprises are a foundational pillar of China's economic landscape, demonstrating substantial contributions across key economic indicators (Ministry of Ecology and Environment, the People's Republic of China, 2020). In 2024, these firms were responsible for over 50% of the nation's tax revenue, underscoring their crucial role in fiscal stability. Furthermore, they accounted for more than 60% of China's Gross Domestic Product (GDP), highlighting their significant impact on economic output and growth. Their expansive presence also plays a vital role in the labor market, generating over 80% of urban employment (Frimpong, 2025). These figures collectively emphasize the indispensable nature of the private sector in driving China's economic vitality and job creation, establishing it as a primary engine for national development (China Public Private Partnerships Center, 2020). However, as China's private sector continues its remarkable economic expansion, these enterprises are simultaneously encountering intensified examination concerning their environmental footprint and corporate social responsibility (China Public Private Partnerships Center, 2020). Therefore, understanding and promoting EGB within these enterprises become crucial for sustainable development and societal well-being (Fawehinmi et al., 2022).

Despite burgeoning research on EGB predominantly in Europe and North America, studies focusing on Asia, especially China, remain scarce (Francoeur & Paillé, 2022; Taamneh et al., 2025). This study seeks to address this gap by examining the relationships between GTL, green HRM, GE, and EGB within private enterprises in China. Specifically, this research aims to answer the following questions:

- How GTL influences GE in Chinese private enterprises?
- How GE influences EGB in Chinese private enterprises?
- How GTL can influence EGB through GE in Chinese private enterprises?

- How does green HRM moderate the relationship between GTL and GE in Chinese private enterprises?
- How does green HRM moderate the relationship between GE and EGB in Chinese private enterprises?

Unlike previous studies primarily exploring green HRM's impact on EGB, our study contributes uniquely by bridging the gap in the existing literature. This study is the first to explore the mechanisms connecting GTL to EGB via green HRM practices and GE. By elucidating these mechanisms, our research deepens theoretical comprehension and offers valuable guidance for organizations seeking to foster a culture of environmental accountability.

2 Literature Review

2.1 Social Cognitive Theory

Our research framework draws on the Social Cognitive Theory (SCT), proposed by Bandura in 1986, as a foundational theoretical model for exploring environmental behaviors and organizational psychology. SCT posits that individuals acquire knowledge through observation, are influenced by their social environment, and that their behaviors are shaped by their self-efficacy, their belief in their ability to perform specific tasks (Bandura, 1986; Machida & Schaubroeck, 2011).

SCT is highly relevant to our framework as it explains how GTL can influence EGB through various mechanisms. Green transformational leaders serve as powerful role models (Farrukh et al., 2022), inspiring individuals to emulate their eco-friendly behaviors and values by communicating a compelling vision of sustainability and providing resources for eco-friendly initiatives (Chen & Chang, 2013). This aligns with SCT's emphasis on observational learning and the impact of influential figures.

Furthermore, SCT supports the mediating roles of green HRM and GE. Green HRM practices, by integrating environmental considerations into HR policies (Khan & Khan, 2022), create a supportive social environment that provides employees with the necessary knowledge, abilities, and tools for EGB (Patwa & Kee, 2024; Sabokro et al., 2021). This reflects SCT's emphasis on how the social environment and reinforcement influence behavior (Bandura, 2001).

Similarly, GE, defined as employees' emotional and cognitive dedication to sustainability, aligns with SCT by suggesting that individuals are more inclined to adopt behaviors that resonate with their values and beliefs, especially when they feel confident in their ability to contribute positively (Hartman & Barber, 2020). GTL fosters this sense of purpose and confidence, leading to greater engagement.

In essence, SCT provides a robust theoretical lens through which to understand how GTL can cultivate EGB within China's private sector, both directly and indirectly, by shaping the organizational environment through green HRM and fostering individual dedication through GE. This is crucial for promoting sustainable practices and achieving SDGs in a sector facing significant environmental challenges (Al-Swidi et al., 2021).

2.2 GTL and GE

When employees are actively involved and invested in their jobs, they naturally exhibit higher levels of motivation and effort (Amin et al., 2024). This study extends this concept to GE, defining it as the dedication and energy employees commit specifically to environmentally friendly tasks. This state of mind is characterized by positive feelings, satisfaction, and profound dedication toward one's work (Schaufeli, 2017), encompassing high morale, strong commitment, and complete absorption in their sustainable duties (Schaufeli, 2017). Ultimately, GE describes the extent to which employees dedicate their intellectual, emotional, and physical energy to their environmentally focused work (Amin et al., 2024).

Employees who are deeply committed to sustainable practices become active supporters of green initiatives, aligning their actions with the organization's overarching environmental goals (Amin et al., 2024). This commitment not only positively impacts individual employee performance but also drives broader organizational improvement. GTL plays a vital role in shaping both GE and corresponding behaviors by cultivating a deep commitment to environmental sustainability (Sun et al., 2022). This leadership approach involves leaders actively encouraging employees to pursue environmental objectives and inspiring them to surpass expected standards of eco-friendly behavior (Chen & Chang, 2013; Aboramadan, 2022; Salama et al., 2022; Gustiah & Nurhayati, 2022; Huang et al., 2021). An employee's enthusiasm for their work is frequently considered a crucial determinant of their overall productivity (Gustiah & Nurhayati, 2022). Notably, various studies have also established a clear link between transformational leadership and heightened work engagement (Suliman et al., 2023; Huang et al., 2021; Gustiah & Nurhayati, 2022). This suggests that GTL, with its emphasis on inspiration and environmental objectives, can effectively foster GE, leading to more sustainable organizational outcomes. Therefore, we propose the following hypothesis:

Hypothesis 1 GTL has a positive effect on GE.

2.3 GE and EGB

Engaged employees are more likely to build strong, positive connections with their organizations (Amin et al., 2024; Taamneh et al., 2025), which in turn contribute to favorable job-related outcomes (Saks, 2006). Similarly, this engagement not only drives employees to perform their job duties but also inspires them to take part in voluntary activities beyond their designated roles (Kahn, 1990). Prior studies (Aboramadan et al., 2020; Agarwal et al., 2012; Alfes et al., 2013; Haynie et al., 2016; Rich et al., 2010; Rodwell et al., 2017) have demonstrated that work engagement serves as a key predictor of job performance and extra-role behaviors, including innovation and organizational citizenship. Regarding the relationship between GE and EGB, employees with higher GE levels are more inclined to develop positive connections with their organizations. Under such circumstances, GE can lead to beneficial outcomes associated with EGB. It can motivate employees to embrace eco-friendly practices, actively participate in green initiatives, and encourage others to understand the significance of sustainable behaviors for the long-term success of both the organization and the community (Luu, 2017). From the preceding debate, we propose the following hypothesis:

Hypothesis 2 GE has a positive effect on EGB.

2.4 *The Mediator Role of GE Between GTL and EGB*

A variety of studies have highlighted the significant mediating role of work engagement (Aboramadan et al., 2019, 2020; Agarwal et al., 2012; Karatepe & Olugbade, 2016; Rich et al., 2010; Sulea et al., 2012; Sahan et al., 2025). As a motivating force that impacts performance results, work engagement is commonly recognized (Karatepe & Olugbade, 2016). Research (Chen & Chang, 2013; Khan & Khan, 2022) has investigated how GTL affects employees' environmentally conscious actions. Transformational leaders who prioritize sustainability have an impact on their teams by sharing their values, principles, and long-term objectives in this area (Chen & Chang, 2013). Al-Swidi et al. (2021) identified that GTL directly influences environmentally friendly EGB. Similarly, Visamitanan and Assarut (2021) reported that GTL has a positive and direct effect on environmental job engagement. These findings encourage the proposition of a clear and favorable relationship between GTL and eco-friendly actions. The outcomes underscore the critical role of leadership in fostering sustainable behavior among employees. By promoting a strong sense of enthusiasm, ambition, and vision, along with offering the necessary support, GTL motivates employees to actively contribute to their organization's environmental sustainability objectives (Irfan et al., 2022; Sobaih et al., 2022). Gustiah and Nurhayati (2022) proposed that GTL indirectly impacts employees' green performance through the mediating role of GE. Similarly, Suliman et al. (2023) demonstrated that GTL positively influences employees' environmental performance, with this effect being mediated by GE. Drawing from prior studies that have shown that GTL significantly enhances GE, and that GE strongly affects employees' environmental sustainability efforts, we suggest that GE could act as a mediator in the relationship between GTL and employees' EGB aimed at sustainability. In essence, GTL is believed to foster higher levels of GE, which subsequently promotes greater environmentally friendly behaviors among employees. Therefore, we propose the following hypothesis:

Hypothesis 3 GE positively mediates the relationship between GTL and EGB.

2.5 *Moderating Effect of Green HRM*

Leaders play a role in motivating employees, thereby influencing the stability and longevity of organizations (Tran et al., 2024). Over the past 20 years, transformational leadership has become a key focus in leadership studies (Chen & Chang, 2013; Hu & Kee, 2023; Kee et al., 2020; Wang & Shao, 2024), with research providing evidence of its connection to green HRM (Awan et al., 2023; Chen & Wu, 2022; Hameed et al., 2022). Chen and Chang (2013) describe GTL as leaders who motivate and inspire their teams to surpass environmental expectations and accomplish sustainability objectives. The impact of transformational leadership plays a crucial role in guiding an organization's strategic direction both in the present and in the future (Awan et al., 2023; Wang & Shao, 2024).

Research on green HRM practices began in the early 2000s (Muster & Schrader, 2011). Renwick et al. (2013, p. 1) define green HRM practices as "HRM aspects of environmental management." These practices include a variety of activities, systems, policies, strategies, programs, processes, and designs that align HRM with environmental sustainability concerns among employees (Tanova & Bayighomog, 2022; Darvishmotevali & Altinay, 2022; Arulrajah & Opatha, 2016; Yusoff et al., 2020; Mousa & Othman, 2020; Pham et al., 2020; Tang et al., 2018). The goal of green HRM is to raise environmental awareness among employees and improve their understanding of the environmental impact of their actions (Sahan et al., 2025; Gilal et al., 2019; Malik et al., 2020; Irani & Kilic, 2022). Several studies (Arshad, 2020; Cahyadi et al., 2022; Jia & Kilic, 2022; Singh et al., 2020; Sun et al., 2022) have found a connection between GTL and green HRM. Renwick et al. (2013) examined how green HRM helps encourage environmentally responsible behaviors among employees. Zhang et al. (2019) discovered that green HRM has a direct impact on both in-role and extra-role EGB in Chinese companies, which is consistent with the findings of Dumont et al. (2017) and Shen et al. (2019), who also identified a positive relationship between green HRM and environmentally friendly behaviors among employees. Recent studies (Sahan et al., 2025; Jabeen et al., 2024; Shahzad et al., 2023; Zhu et al., 2021) have reinforced the positive link between green HRM and EGB. Recent studies have emphasized the strategic role of green HRM in cultivating pro-environmental behaviors among employees, particularly when supported by a strong green work climate (Rubel et al., 2021a). Furthermore, green HRM practices such as green recruitment, training, and performance appraisal have been found to influence green service behaviors through mechanisms such as green knowledge sharing (Rubel et al., 2021b).

EGB involve actions that contribute to environmental well-being (Norton et al., 2015, 2017). Research indicates a positive correlation between green HRM and behaviors such as green task performance, empowerment for green initiatives, green job crafting, and organizational citizenship for environmental sustainability (Chaudhary, 2019; Dumont et al., 2017; Fawehinmi et al., 2020; Hameed et al., 2020; Renwick et al., 2013). According to Chen and Chang (2013), leaders' behaviors motivate employees to achieve environmental goals and inspire them to exceed the expected levels of environmental performance. Green leadership encourages individuals to prioritize organizational objectives over personal interests, guides them in various situations, offers support when needed, and ignites enthusiasm in employees to generate innovative environmental ideas (Begum et al., 2022). However, there is limited research in the existing literature on whether GTL can promote environmentally friendly behaviors among employees in private businesses in China. To address this gap, we broaden the concept of GTL and explore its impact on employees' green actions, with a focus on the dynamics between leaders and their subordinates.

Green HRM plays a critical role in a company's environmental strategy and in shaping employees' environmentally friendly behaviors (Taamneh et al., 2025; Alshaabani et al., 2021; Renwick et al., 2013; Singh et al., 2020; Jabeen et al., 2024). Research in HRM suggests that well-planned and intentional HRM strategies enhance employee commitment and the sense of purpose. Dumont et al. (2017) propose that green HRM initiatives help individuals meet organizational objectives and improve performance. Kim et al. (2019) and Muisyo and Qin (2021) discovered that green HRM fosters increased EGB and transforms organizational environmental actions. In the context of environmental management, green HRM has become a key concept. It integrates HRM with environmental management practices (Renwick et al., 2013; Anwar et al., 2020). Green HRM refers to HR activities aimed at enhancing environmental sustainability (Kramar, 2014). Hameed et al. (2022) assert that green HRM involves efforts to establish, implement, and maintain a framework that transitions an organization toward environmental sustainability. Zoogah (2011) defines green HRM as the implementation of policies and practices within an organization that promote sustainable resource use and prevent environmental harm. The elements of green HRM serve as tools for predicting environmental performance, meaning that incorporating green practices within an organization influences EGB (Tamimi et al., 2024). Green HRM is responsible for tasks such as designing green-focused jobs, setting rewards for eco-friendly behaviors, and defining environmental responsibilities within the workplace. It also involves embedding corporate environmental management within HRM practices (Renwick et al., 2013). The goal of green HRM is to build a strong commitment to sustainability and encourage employees to align with environmental goals (Fawehinmi et al., 2022). Therefore, we propose the following hypotheses:

Hypothesis 4 Green HRM strengthens the positive relationship between GTL and GE.

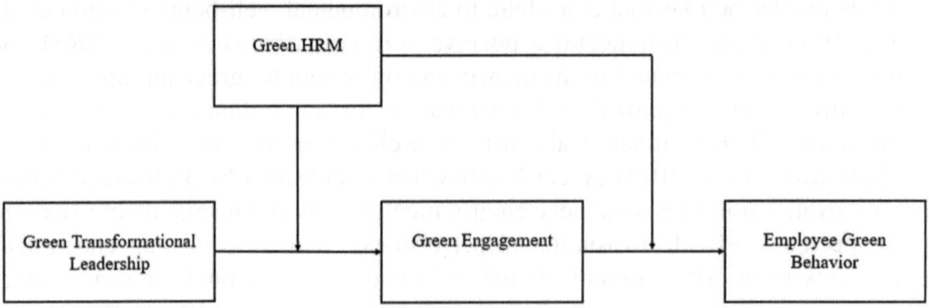

Fig. 1 Research framework (Source: Authors' work)

Hypothesis 5 Green HRM strengthens the positive relationship between GE and EGB.

Our research model is presented in Fig. 1.

3 Methodology

3.1 Sample and Procedures

Our research aims to investigate how GTL influences employees' environmental awareness in selected private companies in China. This research also examines the potential mediating roles of green HRM and GE in this relationship. Specifically, we focus on the top 100-ranked private enterprises in the region. The survey was initially developed in English and subsequently translated into Chinese for data collection from employees in the private sector in China. We employed a back translation method to ensure the accuracy of the translation. A total of 600 questionnaires were distributed to employees in private enterprises across China. Of these, 293 questionnaires were returned, with 27 being deemed invalid, resulting in 266 valid responses. The response rate was 44.3%. Data collection spanned 4 weeks, from March to April 2024. During this period, questionnaires were distributed to employees working in various industrial sectors within the private sector, mainly focusing on the top 100-ranked private enterprises in 2023 in China. The response rate obtained was higher than the average response rates commonly seen in the field of operations management (Synodinos, 2003). We followed recommendations from the literature to mitigate common method biases that can occur when the same data source is used to measure both the predictors and the criteria (Podsakoff et al., 2003).

Given that the original questionnaire was developed and validated in English, and this study was conducted within Chinese private enterprises, a rigorous translation process was essential to ensure clarity and comprehension among participants. To achieve both linguistic and conceptual equivalence, we employed the widely recognized translation and back-translation method, as proposed by Brislin (1970).

This robust approach involved an initial translation of the instrument from English into Mandarin Chinese. Subsequently, the Chinese version was back-translated into English. This iterative process allowed us to identify and resolve any inconsistencies, effectively addressing potential idiomatic and cultural nuances that might otherwise compromise the instrument's validity and reliability (Furlan et al., 2009; Berkanovic, 1980). The entire translation process was independently carried out by three experts, each highly proficient in both English and Mandarin.

Our study focuses on full-time employees with over 1 year of tenure within China's top 100 private enterprises. These enterprises were selected based on their 2023 ranking by the China Enterprises Association and their demonstrated commitment to green HRM practices. This strategic focus ensures our research delves into organizations that are leaders in the private sector and are actively engaged in green initiatives. We employed a purposive sampling technique to select our respondents, driven by specific criteria for both the organizations and the employees. The decision to target full-time employees stems from their typically deeper commitment and engagement within their organizations, making them more likely to be fully invested in initiatives related to green HRM, GTL, GE, and EGB. Their role in driving organizational goals also means their perceptions significantly influence the success of sustainability practices. By focusing on full-time staff, we anticipate collecting more reliable data, as these individuals are likely to possess a more profound understanding of their organization's sustainability efforts.

Furthermore, we mandated a minimum tenure of 1 year for respondents. This criterion is crucial because employees with longer tenure tend to have a deeper understanding of organizational culture, policies, and practices, including those related to green HRM and green initiatives. This sustained experience provides more reliable and balanced feedback, reducing the likelihood of initial biases. Ultimately, this rigorous sampling approach ensures that our data reflects meaningful insights into leadership styles and green HRM practices from a highly relevant and informed employee group. Table 1 presents the demographic profile of the participants involved in the study. More than half of the respondents are male (56.4%). The largest age group is 31–40 years (41.4%), followed by those aged 21–30 years (33.1%). A substantial majority are employees (66.2%). More than two-thirds of the participants (64.7%) have an organizational tenure of 5 years or less.

3.2 *Measures and Instrument Development*

In this research, a questionnaire was developed using established scales. The initial section contained demographic questions, including gender, age, job title, and length of service with the organization. Green transformational leadership used six items by Chen and Chang (2013) to understand EGB through green HRM and GE (e.g., "*The leader provides a clear environmental vision for the employees to follow*"). It was measured on a 7-point Likert scale from 1 (Strongly Disagree) to 7 (Strongly Agree) (α = 0.87). Green HRM was measured on a 6-item scale by

Table 1 Respondents' profile ($N = 266$)

Demographics		Frequency	Percentage (%)
Gender	Male	150	56.4
	Female	116	43.6
Age (years)	18≤20	4	1.5
	21–30	88	33.1
	31–40	110	41.4
	41–50	45	16.9
	51 and above	19	7.1
Position	Top manager	9	3.4
	Middle manager	20	7.5
	Lower manager	61	22.9
	Employee	176	66.2
Organizational tenure (years)	1–5 years	172	64.7
	6–10 years	61	22.9
	>10 years	33	12.4

Source: Authors' work

Dumont et al. (2017) using a 7-point Likert scale from 1 (Strongly Disagree) to 7 (Strongly Agree). A sample item is "*My company sets green goals for its employees*" ($\alpha = 0.88$). The green engagement was measured using six items of Suliman et al. (2023) using a 7-point Likert scale from 1 (Strongly Disagree) to 7 (Strongly Agree). A sample item is "*My environmental-related tasks inspire me*" ($\alpha = 0.92$). EGBs were measured using five items developed by De Roeck and Farooq (2018). Using 1 (never) and 7 (always), a sample item is "*I adequately complete assigned duties in environmentally friendly ways*" ($\alpha = 0.89$).

3.3 *Data Analysis*

The data were analyzed using SPSS version 22 and SmartPLS. Descriptive statistics, such as percentage, mean, standard deviation, and frequency, were utilized to summarize the participants' demographic information and confirm the study's content. Harman's single-factor test was performed to examine the potential presence of Common Method Variance (CMV). We employed Confirmatory Factor Analysis (CFA) and Cronbach's Alpha to assess the validity and reliability of the measurement instrument. Convergent validity was evaluated by calculating the average variance extracted (AVE) and composite reliability (CR). The study also employed the Fornell-Larcker criterion to check for discriminant validity. Structural equation modeling (SEM) and bootstrapping were used to test the hypotheses and explore the relationships between the variables.

4 Findings

4.1 Measurement Model

This research employed partial least-squares structural equation modeling (PLS-SEM) for data analysis due to its flexible distributional assumptions, its focus on exploration and prediction, the ability to handle complex models, and its straightforward model construction and interpretation. Structural equation modeling offers an efficient way to manage multiple dependence relationships involving latent variables with statistical effectiveness (Hair et al., 2017). Path modeling using partial least squares (PLS) is particularly advantageous over covariance-based methods when the primary aim of the study is causal-predictive analysis rather than theory testing (Hair et al., 2014). Based on searches in scientific databases like ISI Web of Science and Scopus, it is evident that there is a lack of prior research linking GTL to environmentally friendly behaviors. As a result, PLS-SEM was selected as the most suitable analytical method for this study.

The analysis commenced with an evaluation of the measurement model. Following the confirmation of the measurement theory and the model's validity and reliability, the structural model was analyzed to verify the proposed hypotheses. In SmartPLS, the assessment of the measurement model included examining item loadings, average variance extracted (AVE), composite reliability (CR), and the reliability of measurement constructs, as shown in Table 1. Convergent validity was established for all constructs, as indicated by item loadings above 0.70 and AVE values greater than 0.50. The reliability of the measurement models was verified through internal consistency, with Cronbach's alpha and CR values exceeding 0.70 for all constructs. Following Fornell and Larcker's (1981) criteria, discriminant validity was demonstrated when the AVE of a latent variable was higher than the squared correlations it shared with other variables in the model. This study confirmed the discriminant validity of the measurement model (see Table 2).

Discriminant validity was assessed using the Fornell–Larcker criterion (Fornell & Larcker, 1981), as presented in Table 3. The results confirm that the square root of the average variance extracted (AVE) for each construct (diagonal values: GB = 0.854, GE = 0.855, green HRM = 0.88, GTL = 0.8) is greater than its correlation with any other construct (off-diagonal values). This pattern indicates that all constructs, GB, GE, green HRM, and GTL, are empirically distinct and possess adequate discriminant validity. The results are shown in Table 3.

Table 2 Measurement model

Construct scale	Items	Loadings	AVE	Composite reliability	Cronbach's alpha
EGB	EGB1	0.851	0.729	0.931	0.907
	EGB2	0.872			
	EGB3	0.863			
	EGB4	0.826			
	EGB5	0.856			
GHRM	GHRM1	0.864	0.775	0.954	0.942
	GHRM2	0.883			
	GHRM3	0.890			
	GHRM4	0.886			
	GHRM5	0.888			
	GHRM6	0.870			
GTL	GTL1	0.864	0.639	0.910	0.875
	GTL2	0.384			
	GTL3	0.846			
	GTL4	0.843			
	GTL5	0.874			
	GTL6	0.867			
GE	GE1	0.852	0.732	0.942	0.927
	GE2	0.908			
	GE3	0.873			
	GE4	0.854			
	GE5	0.791			
	GE6	0.851			

Note. *CR* composite reliability, *AVE* average variance extracted, *EGB* employee green behavior, *Green HRM* green human resource management, *GE* green engagement, *GTL* green transformational leadership
Source: Authors' work

Table 3 Fornell–Larcker criterion

	GB	GE	Green HRM	GTL
GB	0.854			
GE	0.543	0.855		
Green HRM	0.524	0.541	0.880	
GTL	0.494	0.495	0.429	0.800

Note. *EGB* employee green behavior, *Green HRM* green human resource management, *GE* green engagement, *GTL* green transformational leadership
Source: Authors' work

4.2 Structural Model

The results shown in Table 4 indicate a significant positive correlation between GTL and GE ($\beta = 0.314$, $P < 0.001$), providing support for H1. Additionally, a positive and significant relationship was found between GE and EGB ($\beta = 0.365$, $P < 0.001$),

Table 4 Path analysis

H	Relationship	Standardized beta	Standard error	T value	P values	Results
H1	GTL- > GE	0.314	0.070	4.513	0.000***	Supported
H2	GE- > GB	0.365	0.064	5.705	0.000***	Supported
H3	GTL- > GE- > GB	0.115	0.032	3.630	0.000***	Supported
H4	GHRM × GTL- > GE	0.192	0.057	3.359	0.001***	Supported
H5	GHRM × GE- > GB	−0.015	0.039	0.393	0.694	Not supported

Note. *EGB* employee green behavior, *Green HRM* green human resource management, *GE* green engagement, *GTL* green transformational leadership
Source: Authors' work
* $P < 0.05$, ** $P < 0.01$, *** $P < 0.001$

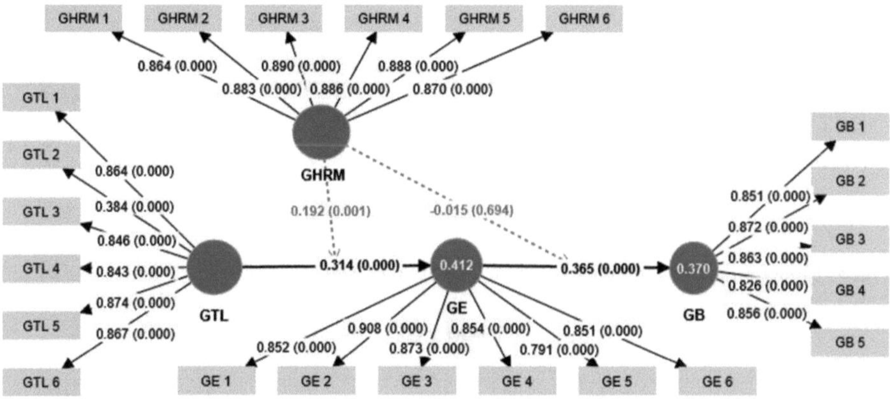

Fig. 2 Structural model (Source: Authors' work)

which further supports H2. Therefore, both H1 and H2 are supported. To test the mediation hypotheses, bootstrapping for the indirect effect was performed according to the guidelines of Preacher and Hayes (2004, 2008). The results in Table 4 indicate that GTL–WE–EGB ($\beta = 0.115$, $P < 0.001$) is significant, providing support for H3. The structural model is illustrated in Fig. 2. The model demonstrates significant positive relationships between GTL, green HRM, GE, and EGB, with strong factor loadings and significant p-values, indicating a good fit. Specifically, GE mediates the relationship between GTL and EGB, and green HRM significantly influences GE, highlighting its crucial role in fostering EGB within organizations.

As illustrated in Fig. 3, the graph demonstrates a significant moderation effect of green HRM on the relationship between GTL and GE. High levels of green HRM strengthen the positive impact of GTL on GE. Similarly, Fig. 4 demonstrates that green HRM significantly moderates the relationship between GE and EGB. Higher levels of green HRM strengthen the positive impact of GE on EGB. Consequently, H3 is further validated, H4 is confirmed, while H5 is not supported.

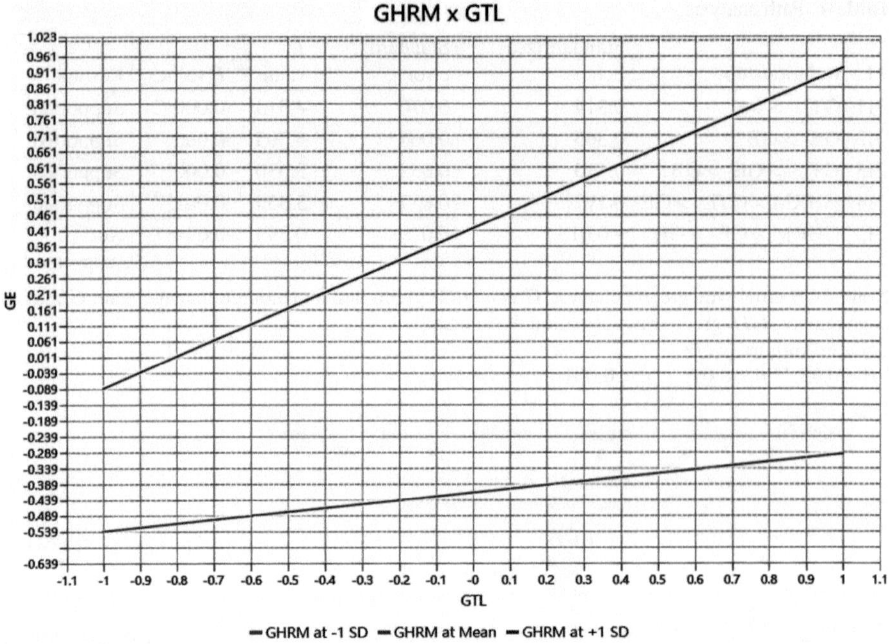

Fig. 3 Moderating effect of green HRM on GE (Source: Authors' work)

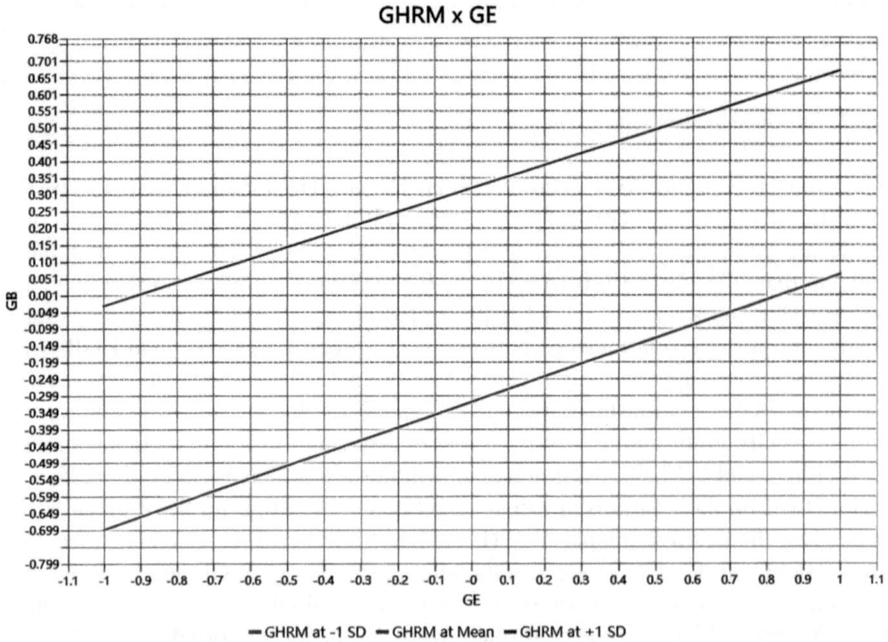

Fig. 4 Moderating effect of green HRM on EGB (Source: Authors' work)

5 Discussion and Implications

This study significantly contributes to both SCT and green HRM literature, particularly within the under-researched context of the Chinese private sector. By demonstrating that GTL directly influences employees' EGB and indirectly through the mediating roles of green HRM and GE initiatives, our research provides a robust empirical extension to SCT. SCT posits that human behavior is influenced by a reciprocal interaction between cognitive, behavioral, and environmental factors. Our findings show how GTL, as an environmental factor (leaders' actions and values), shapes employees' cognitive processes (environmental awareness, commitment) and subsequently their EGB. The mediating role of GE further illuminates the complex interplay, indicating that the organizational environment (green HRM practices) and individual GE are crucial pathways through which leadership influences the green behavior. Our study is among the first to integrate GTL, green HRM, GE, and EGB within Chinse private enterprises, thereby enriching the theoretical understanding of how social and organizational factors drive environmental behavior.

The practical implications of this study are substantial for HR managers and policymakers in China. First, the strong effects of GTL on EGB underscore the critical need for private enterprises to invest in developing green transformational leaders. HR managers should prioritize leadership training programs that equip managers with the skills to articulate a compelling environmental vision, motivate employees toward green goals, and foster a culture of environmental responsibility. This includes training on setting clear environmental objectives, providing encouragement for eco-friendly initiatives, and promoting enthusiasm for sustainable development. Second, the moderator role of green HRM highlights the imperative to strategically align HR practices with environmental objectives. Policymakers and HR managers in China should focus on implementing comprehensive green HRM practices. This means integrating environmental considerations into all facets of HRM, from eco-friendly recruitment and selection processes that prioritize candidates with green values and competencies, to training and development programs that enhance employees' environmental knowledge and skills. Furthermore, performance rating systems should explicitly incorporate environmental targets and reward environmentally responsible behaviors. Empowerment strategies should also be developed to encourage employees to take proactive ownership of green initiatives. For instance, creating platforms for employee-led green projects or establishing internal awards for environmental contributions can significantly boost GE. By fostering such an environment, Chinese private enterprises can not only enhance their environmental performance but also cultivate a more engaged and environmentally conscious workforce, contributing significantly to the nation's broader sustainability goals.

6 Future Directions

Future research could overcome these limitations by incorporating longitudinal approaches to explore the causal relationships between GTL, green HRM, GE, and EGB. A longitudinal design would enable researchers to track the dynamics of these relationships over time, providing a clearer understanding of how green leadership and green HRM practices influence EGB in the long run. Expanding the scope of the study to include a variety of sectors, industries, and cultural settings would further improve the generalizability of the findings, as different organizational contexts may yield distinct results. By examining how green leadership and green HRM practices are applied across different industries or cultural backgrounds, researchers could identify common patterns as well as sector-specific differences in how environmental practices are implemented and embraced.

Moreover, future studies could explore the impact of additional factors, such as organizational culture, external environmental policies, and industry-specific challenges, on EGB. Understanding the role of organizational culture is particularly important, as it can significantly shape employees' attitudes toward sustainability and influence how green initiatives are perceived and executed within the organization. External environmental regulations, such as government policies or industry standards, may also play a critical role in shaping EGB, as organizations often respond to external pressures to comply with environmental norms. Additionally, sector-specific challenges, such as the nature of the industry or the resources available, could affect how green leadership and green HRM practices are implemented and how effective they are in promoting EGB.

To further enhance the validity and depth of future research, adopting mixed-methods approaches could prove valuable. Combining quantitative surveys with qualitative interviews would not only help mitigate common method bias but also provide richer insights into the complex interactions between green leadership, green HRM, and employee behavior. Qualitative interviews could offer a more nuanced understanding of how employees perceive and respond to green leadership initiatives and green HRM practices, shedding light on underlying motivations, barriers, and facilitators of EGB. By integrating both quantitative and qualitative data, future studies could provide a more comprehensive picture of the factors that drive EGB in organizations.

7 Conclusion

Although this study offers important insights into the influence of GTL, green HRM, and GE on promoting EGB, it also has certain limitations that warrant attention in future research. First, the sample for this study was drawn exclusively from employees working in private-sector organizations in China. This focus on a single industry and cultural context may limit the generalizability of the findings to other industries or cultural settings. Different sectors or regions might exhibit unique organizational dynamics, and employees' responses to green leadership and green HRM practices could vary depending on the local culture, government regulations, and economic conditions. Therefore, future studies could benefit from a more diverse sample, including public-sector organizations or companies in other countries, to enhance the external validity of the results.

Second, the limitation of the study lies in its cross-sectional design. As the data were collected at a single time point, establishing causal relationships between the variables proved challenging. While our study suggests associations between GTL, green HRM, and EGB, it does not provide evidence of how these factors influence each other over time. Longitudinal studies could be useful in capturing the dynamic nature of these relationships and offering a clearer understanding of how green leadership and green HRM practices evolve and impact employee behavior over an extended period. Third, the reliance on self-reported data introduces the possibility of common method bias. Employees' responses may be influenced by social desirability, personal biases, or their desire to present themselves in a positive light, especially when it comes to environmental behavior. Future studies could utilize diverse data sources, including supervisor evaluations or objective performance metrics, to achieve a more precise and dependable evaluation of EGB.

Finally, while our study focuses on the relationship between GTL, green HRM, and GE, it may overlook other important factors that could influence these relationships. For example, organizational culture, external environmental pressures, or industry-specific challenges could play a significant role in shaping how employees respond to GTL and green HRM practices. Understanding these additional factors could offer a more comprehensive view of the forces that drive EGB in organizations. Future research could expand the scope by exploring these contextual elements and examining how they interact with GTL and green HRM to influence employee behavior. As previously said, EGB has emerged as an important and influential field of study for numerous academics. Multiple research studies are underway to investigate different factors that influence EGB. This research aims to bridge the existing knowledge gap by examining the correlation between GTL and EGB in private industries in China, while also considering the moderating effect of green HRM and the mediating effect of GE. Employing a measurement approach to examine the influence of values would contribute to the current body of research. Furthermore, the study concludes that managers can incorporate green ideals into HRM processes to improve EGB in the private sector. Furthermore, the study demonstrated that GE serves as a key mediator in enhancing environmental performance.

This study offers valuable suggestions to leaders and managers on promoting EGB and leveraging it to enhance environmental sustainability to surpass competitors in the market. Building on the findings of this study, it is necessary for companies to actively cultivate and foster green leadership qualities to effectively apply green HRM practices. The HR department can implement ecologically green HRM practices to attract, cultivate, and retain employees who are dedicated to the company's environmentally conscious business operations and products. Therefore, we suggest that the managers' transformational leadership provide an environment that supports individuals who are engaged in environmentally friendly work and exhibit EGB and provides them with opportunities to fulfill their potential in this area.

This study has demonstrated that GTL, green HRM, and GE are significant factors in fostering EGB among employees in private enterprises in China. The data presented in this study may be beneficial for organizational leaders or scholars interested in doing similar studies in different countries to assess the generalizability of our findings, including beyond China. In general, global business leaders have recognized the need for environmental conservation. Organizations must comprehend how to convert their social responsibilities into commercial advantages. This research mainly focuses on the importance of green HRM and GE in business practices. It examines how corporations can assume social responsibility for themselves and society by being ecologically and economically friendly. This study offers fresh perspectives on the growing movement to harmonize ecological sustainability with economic practices in business.

Ethics Statement Participants were informed that by completing the survey they provided their voluntary consent to participate in the study.

References

Aboramadan, M. (2022). The effect of green HRM on employee green behaviors in higher education: The mediating mechanism of green work engagement. *International Journal of Organizational Analysis, 30*(1), 7–23. https://doi.org/10.1108/IJOA-05-2020-2190

Aboramadan, M., Albashiti, B., Alharazin, H., & Dahleez, K. A. (2019). Human resources management practices and organizational commitment in higher education: The mediating role of work engagement. *International Journal of Educational Management, 34*(1), 154–174. https://doi.org/10.1108/IJEM-04-2019-0160

Aboramadan, M., Dahleez, K., & Hamad, M. H. (2020). Servant leadership and academics outcomes in higher education: The role of job satisfaction. *International Journal of Organizational Analysis, 29*(3), 562–584. https://doi.org/10.1108/IJOA-11-2019-1923

Agarwal, U. A., Datta, S., Blake-Beard, S., & Bhargava, S. (2012). Linking LMX, innovative work behavior and turnover intentions: The mediating role of work engagement. *Career Development International, 17*(3), 208–230. https://doi.org/10.1108/13620431211241063

Alfes, K., Truss, C., Soane, E. C., Rees, C., & Gatenby, M. (2013). The relationship between line manager behavior, perceived HRM practices and individual performance: Examining the mediating role of engagement. *Human Resource Management, 52*(6), 839–859. https://doi.org/10.1002/hrm.21512

Alshaabani, A., Naz, F., & Rudnák, I. (2021). Impact of green human resources practices on green work engagement in the renewable energy departments. *International Business Research, 14*(6), 44. https://doi.org/10.5539/ibr.v14n6p44

Al-Swidi, A. K., Gelaidan, H. M., & Saleh, R. M. (2021). The joint impact of green human resource management, leadership and organizational culture on employees' green behavior and organisational environmental performance. *Journal of Cleaner Production, 316*, 128112. https://doi.org/10.1016/j.jclepro.2021.12811

Amin, M., Khan, I., Shamim, A., Ting, D. H., Jan, A., & Abbasi, A. Z. (2024). Employee motivations in shaping customer value co-creation attitude and behavior: Job position as a moderator. *Journal of Retailing and Consumer Services, 79*, 103819. https://doi.org/10.1016/j.jretconser.2024.103819

Anwar, N., Mahmood, N. H. N., Yusliza, M. Y., Ramayah, T., Faezah, J. N., & Khalid, W. (2020). Green human resource management for organisational citizenship behavior towards the environment and environmental performance on a university campus. *Journal of Cleaner Production, 256*, 120401. https://doi.org/10.1016/j.jclepro.2020.120401

Arshad, M. (2020). The impact of green transformational leadership on environmental performance: Mediating effect of green human resource management and green innovation. *Journal of Organization and Business, 1*(1), 34–44. Retrieved from https://journals.internationalrasd.org/index.php/job/article/view/1156.

Arulrajah, A. A., & Opatha, H. H. D. N. P. (2016). Analytical and theoretical perspectives on green human resource management: A simplified underpinning. *International Business Research, 9*(12), 153–164. https://doi.org/10.5539/ibr.v9n12p153

Awan, F. H., Dunnan, L., Jamil, K., & Gul, R. F. (2023). Stimulating environmental performance via green human resource management, green transformational leadership, and green innovation: A mediation-moderation model. *Environmental Science and Pollution Research, 30*(2), 2958–2976. https://doi.org/10.1007/s11356-022-22424-y

Bandura, A. (1986). *Social foundations of thought and action: A social cognitive theory*. Prentice-Hall.

Bandura, A. (2001). Social cognitive theory: An agentic perspective. *Annual Review of Psychology, 52*(1), 1–26. https://doi.org/10.1146/annurev.psych.52.1.1

Begum, S., Ashfaq, M., Xia, E., & Awan, U. (2022). Does green transformational leadership lead to green innovation? The role of green thinking and creative process engagement. *Business Strategy and the Environment, 31*(1), 580–597. https://doi.org/10.1002/bse.2911

Berkanovic, E. (1980). The effect of inadequate language translation on Hispanics' responses to health surveys. *American Journal of Public Health, 70*(12), 1273–1276. https://doi.org/10.2105/AJPH.70.12.1273

Brislin, R. W. (1970). Back-translation for cross-cultural research. *Journal of Cross-Cultural Psychology, 1*(3), 185–216. https://doi.org/10.1177/135910457000100301

Cahyadi, A., Natalisa, D., Poór, J., Perizade, B., & Szabó, K. (2022). Predicting the relationship between green transformational leadership, green human resource management practices, and employees' green behavior. *Administrative Sciences, 13*(1), 5. https://doi.org/10.3390/admsci13010005

Chaudhary, R. (2019). Green human resource management in Indian automobile industry. *Journal of Global Responsibility, 10*(2), 161–175. https://doi.org/10.1108/JGR-12-2018-0084

Chen, Y. S., & Chang, C. H. (2013). The determinants of green product development performance: Green dynamic capabilities, green transformational leadership, and green creativity. *Journal of Business Ethics, 116*(1), 107–119. https://doi.org/10.1007/s10551-012-1452-x

Chen, T., & Wu, Z. (2022). How to facilitate employees' green behavior? The joint role of green human resource management practice and green transformational leadership. *Frontiers in Psychology, 13*, 906869. https://doi.org/10.3389/fpsyg.2022.906869

China Public Private Partnerships Center. (2020). *National PPP comprehensive information platform project management database 2019 annual report* (National PPP Comprehensive Information Platform Project Management Database Annual Report 2019). Retrieved December 11, 2020, from https://www.cpppc.org/jb/1781.jhtml.

Darvishmotevali, M., & Altinay, L. (2022). Toward pro-environmental performance in the hospitality industry: Empirical evidence on the mediating and interaction analysis. *Journal of Hospitality Marketing & Management, 31*(4), 431–457. https://doi.org/10.1080/19368623.2022.2019650

De Roeck, K., & Farooq, O. (2018). Corporate social responsibility and ethical leadership: Investigating their interactive effect on employees' socially responsible behaviors. *Journal of Business Ethics, 151*, 923–939. https://doi.org/10.1007/s10551-017-3656-6

Dumont, J., Shen, J., & Deng, X. (2017). Effects of green HRM practices on employee workplace green behavior: The role of psychological green climate and employee green values. *Human Resource Management, 56*(4), 613–627. https://doi.org/10.1002/hrm.21792

Farrukh, M., Ansari, N., Raza, A., Wu, Y., & Wang, H. (2022). Fostering employee's pro-environmental behavior through green transformational leadership, green human resource management and environmental knowledge. *Technological Forecasting and Social Change, 179*, 121643. https://doi.org/10.1016/j.techfore.2022.121643

Fawehinmi, O., Yusliza, M. Y., Mohamad, Z., Jamil, N. F., & Muhammad, Z. (2020). Assessing the green behavior of academics: The role of green human resource management and environmental knowledge. *International Journal of Manpower, 41*(7), 879–900. https://doi.org/10.1108/IJM-07-2019-0347

Fawehinmi, O., Yusliza, M. Y., Ogbeibu, S., Tanveer, M. I., & Chiappetta Jabbour, C. J. (2022). Academic employees' green behavior as praxis for bolstering environmental sustainable development: A linear moderated mediation evaluation. *Business Strategy and the Environment, 31*(7), 3470–3490. https://doi.org/10.1002/bse.3095

Fornell, C., & Larcker, D. F. (1981). Evaluating structural equation models with unobservable variables and measurement error. *Journal of Marketing Research, 18*(1), 39–50. https://doi.org/10.1177/002224378101800104

Francoeur, V., & Paillé, P. (2022). *Green behaviors in the workplace: Nature, complexity, and trends*. Springer International Publishing. https://doi.org/10.1007/978-3-030-94541-1

Frimpong, P. (2025). https://africachinacentre.org/private-economys-role-in-china-driving-stability-and-enhancingprosperity/#:~:text=China's%20private%20economy%20has%20emerged%20as%20a,the%20workforce%20across%20the%20country's%20urban%20cities

Furlan, L. A., Cassady, J. C., & Pérez, E. R. (2009). Adapting the cognitive test anxiety scale for use with Argentinean university students. *International Journal of Testing, 9*(1), 3–19. https://doi.org/10.1080/15305050902733448

Gilal, F. G., Ashraf, Z., Gilal, N. G., Gilal, R. G., & Channa, N. A. (2019). Promoting environmental performance through green human resource management practices in higher education institutions: A moderated mediation model. *Corporate Social Responsibility and Environmental Management, 26*(6), 1579–1590. https://doi.org/10.1002/csr.1835

Gustiah, I. P., & Nurhayati, M. (2022). The effect of green transformational leadership on green employee performance through green work engagement. *Scholars Journal of Economics, Business and Management, 9*, 159–168. https://doi.org/10.36347/sjebm.2022.v09i07.002

Hair, J. F., Gabriel, M., & Patel, V. (2014). AMOS covariance-based structural equation modeling (CB-SEM): Guidelines on its application as a marketing research tool. *Brazilian Journal of Marketing, 13*(2), 44–55. https://doi.org/10.1108/EBR-10-2013-0128

Hair, J. F., Hult, G. T. M., Ringle, C. M., Sarstedt, M., & Thiele, K. O. (2017). Mirror, mirror on the wall: A comparative evaluation of composite-based structural equation modeling methods. *Journal of the Academy of Marketing Science, 45*, 616–632. https://doi.org/10.1007/s11747-017-0517-x

Hameed, Z., Khan, I. U., Islam, T., Sheikh, Z., & Naeem, R. M. (2020). Do green HRM practices influence employees' environmental performance? *International Journal of Manpower, 41*(7), 1061–1079. https://doi.org/10.1108/IJM-08-2019-0407

Hameed, Z., Naeem, R. M., Hassan, M., Naeem, M., Nazim, M., & Maqbool, A. (2022). How GHRM is related to green creativity? A moderated mediation model of green transformational leadership and green perceived organizational support. *International Journal of Manpower, 43*(3), 595–613. https://doi.org/10.1108/IJM-05-2020-0244

Hartman, R. L., & Barber, E. G. (2020). Women in the workforce: The effect of gender on occupational self-efficacy, work engagement and career aspirations. *Gender in Management: An International Journal, 35*(1), 92–118. https://doi.org/10.1108/GM-04-2019-0062

Haynie, J. J., Mossholder, K. W., & Harris, S. G. (2016). Justice and job engagement: The role of senior management trust. *Journal of Organizational Behavior, 37*(6), 889–910. https://doi.org/10.1002/job.2082

Hu, M. K., & Kee, D. M. H. (2023). Transformational leadership for effective corporate governance in public sector enterprises: A Malaysian experience. In *Leadership and governance for sustainability* (pp. 95–116). IGI Global. https://doi.org/10.4018/978-1-6684-9711-1.ch006

Huang, S. Y., Ting, C. W., & Li, M. W. (2021). The effects of green transformational leadership on adoption of environmentally proactive strategies: The mediating role of green engagement. *Sustainability, 13*(6), 3366. https://doi.org/10.3390/su13063366

Irani, F., & Kilic, H. (2022). An assessment of implementing green HRM practices on environmental performance: The moderating role of green process innovation. *Journal of Global Hospitality and Tourism, 1*(1), 16–30. https://doi.org/10.5038/2771-5957.1.1.1001

Irfan, M., Elavarasan, R. M., Ahmad, M., Mohsin, M., Dagar, V., & Hao, Y. (2022). Prioritizing and overcoming biomass energy barriers: Application of AHP and G-TOPSIS approaches. *Technological Forecasting and Social Change, 177*, 121524. https://doi.org/10.1016/j.techfore.2022.121524

Jabeen, R., Mehmood, S., Ahmed, M., Ghani, T., Javaid, Z. K., & Popp, J. (2024). The role of green HRM on environmental performance: Mediating role of green ambidexterity and green behavior and moderating role of responsible leadership. *Journal of Chinese Human Resources Management, 15*(2), 70–90. https://doi.org/10.47297/wspchrmWSP2040-800504.20241502

Jia, F., & Kilic, H. (2022). An assessment of implementing green HRM practices on environmental performance: The moderating role of green process innovation. *Journal of Global Hospitality and Tourism, 1*(1), 16–30. https://doi.org/10.5038/2771-5957.1.1.1001

Kahn, W. A. (1990). Psychological conditions of personal engagement and disengagement at work. *Academy of Management Journal, 33*(4), 692–724. https://doi.org/10.5465/256287

Karatepe, O. M., & Olugbade, O. A. (2016). The mediating role of work engagement in the relationship between high-performance work practices and job outcomes of employees in Nigeria. *International Journal of Contemporary Hospitality Management, 28*(10), 2350–2371. https://doi.org/10.1108/IJCHM-03-2015-0145

Kee, D. M. H., Rahman, N. A., & Tan, A. W. (2020). The impact of transformational leadership and team innovation on team performance: Empirical evidence from Malaysia. *Journal of Management and Marketing Review, 5*(2), 99–106. https://doi.org/10.35609/jmmr.2020.5.2(2)

Khan, A. N., & Khan, N. A. (2022). The nexuses between transformational leadership and employee green organisational citizenship behavior: Role of environmental attitude and green dedication. *Business Strategy and the Environment, 31*(3), 921–933. https://doi.org/10.1002/bse.2926

Kim, Y. J., Kim, W. G., Choi, H. M., & Phetvaroon, K. (2019). The effect of green human resource management on hotel employees' eco-friendly behavior and environmental performance. *International Journal of Hospitality Management, 76*, 83–93. https://doi.org/10.1016/j.ijhm.2018.04.007

Kramar, R. (2014). Beyond strategic human resource management: Is sustainable human resource management the next approach? *The International Journal of Human Resource Management, 25*(8), 1069–1089. https://doi.org/10.1080/09585192.2013.816863

Luu, T. T. (2017). CSR and organizational citizenship behavior for the environment in hotel industry: The moderating roles of corporate entrepreneurship and employee attachment style. *International Journal of Contemporary Hospitality Management, 29*(11), 2867–2900. https://doi.org/10.1108/IJCHM-02-2016-0080

Machida, M., & Schaubroeck, J. (2011). The role of self-efficacy beliefs in leader development. *Journal of Leadership & Organizational Studies, 18*(4), 459–468. https://doi.org/10.1177/1548051811404419

Malik, S. Y., Cao, Y., Mughal, Y. H., Kundi, G. M., Mughal, M. H., & Ramayah, T. (2020). Pathways towards sustainability in organizations: Empirical evidence on the role of green human resource management practices and green intellectual capital. *Sustainability, 12*(8), 3228. https://doi.org/10.3390/su12083228

Ministry of Ecology and Environment, the People's Republic of China. (2020). *The 2019 report on the state of the ecology and environment in China is hereby announced in accordance with the*

Environmental Protection Law of the People's Republic of China. Retrieved from https://english.mee.gov.cn/Resources/Reports/soe/SOEE2019/202012/P020201215587453898053.pdf.

Mousa, S. K., & Othman, M. (2020). The impact of green human resource management practices on sustainable performance in healthcare organisations: A conceptual framework. *Journal of Cleaner Production, 243*, 118595. https://doi.org/10.1016/j.jclepro.2019.118595

Muisyo, P. K., & Qin, S. (2021). Enhancing the FIRM'S green performance through green HRM: The moderating role of green innovation culture. *Journal of Cleaner Production, 289*, 125720. https://doi.org/10.1016/j.jclepro.2020.125720

Muster, V., & Schrader, U. (2011). Green work-life balance: A new perspective for green HRM. *German Journal of Human Resource Management, 25*(2), 140–156. https://doi.org/10.1177/239700221102500205

Norton, T. A., Parker, S. L., Zacher, H., & Ashkanasy, N. M. (2015). Employee green behavior: A theoretical framework, multilevel review, and future research agenda. *Organization & Environment, 28*(1), 103–125. https://doi.org/10.1177/1086026615575773

Norton, T. A., Zacher, H., Parker, S. L., & Ashkanasy, N. M. (2017). Bridging the gap between green behavioral intentions and employee green behavior: The role of green psychological climate. *Journal of Organizational Behavior, 38*(7), 996–1015. https://doi.org/10.1002/job.2178

Patwa, N., & Kee, D. M. H. (2024). Sustainability starts with HR: Exploring green HRM practices in the UAE. In M. Y. Yusliza & D. Renwick (Eds.), *Green human resource management*. Springer. https://doi.org/10.1007/978-981-99-7104-6_12

Pham, N. T., Hoang, H. T., & Phan, Q. P. T. (2020). Green human resource management: A comprehensive review and future research agenda. *International Journal of Manpower, 41*(7), 845–878. https://doi.org/10.1108/IJM-07-2019-0350

Podsakoff, P. M., MacKenzie, S. B., Lee, J. Y., & Podsakoff, N. P. (2003). Common method biases in behavioral research: A critical review of the literature and recommended remedies. *Journal of Applied Psychology, 88*(5), 879. https://doi.org/10.1037/0021-9010.88.5.879

Preacher, K. J., & Hayes, A. F. (2004). SPSS and SAS procedures for estimating indirect effects in simple mediation models. *Behavior Research Methods, Instruments, & Computers, 36*, 717–731. https://doi.org/10.3758/BF03206553

Preacher, K. J., & Hayes, A. F. (2008). Asymptotic and resampling strategies for assessing and comparing indirect effects in multiple mediator models. *Behavior Research Methods, 40*(3), 879–891. https://doi.org/10.3758/BRM.40.3.879

Renwick, D. W., Redman, T., & Maguire, S. (2013). Green human resource management: A review and research agenda. *International Journal of Management Reviews, 15*(1), 1–14. https://doi.org/10.1111/j.1468-2370.2011.00328.x

Rich, B. L., Lepine, J. A., & Crawford, E. R. (2010). Job engagement: Antecedents and effects on job performance. *Academy of Management Journal, 53*(3), 617–635. https://doi.org/10.5465/amj.2010.51468988

Rodwell, J., McWilliams, J., & Gulyas, A. (2017). The impact of characteristics of nurses' relationships with their supervisor, engagement and trust, on performance behaviors and intent to quit. *Journal of Advanced Nursing, 73*(1), 190–200. https://doi.org/10.1111/jan.13102

Rubel, M. R. B., Kee, D. M. H., & Rimi, N. N. (2021a). Green HRM practices and supervisor pro-environmental behavior: The role of green work climate perceptions. *Journal of Cleaner Production, 313*(1), 1–13. https://doi.org/10.1016/j.jclepro.2021.127669

Rubel, M. R. B., Kee, D. M. H., & Rimi, N. N. (2021b). The influence of green HRM practices on green service behaviors: The mediating effect of green knowledge sharing. *Employee Relations, 43*(1), 996–1015. https://doi.org/10.1108/ER-04-2020-0163

Sabokro, M., Masud, M. M., & Kayedian, A. (2021). The effect of green human resources management on corporate social responsibility, green psychological climate and employees' green behavior. *Journal of Cleaner Production, 313*, 127963. https://doi.org/10.1016/j.jclepro.2021.127963

Sahan, F., Demir, E., & Zhou, Y. (2025). The impact of green human resource management on organizational sustainability. *Journal of Environmental Management, 350*, 119527. https://doi.org/10.1016/j.jenvman.2024.119527
Saks, A. M. (2006). Antecedents and consequences of employee engagement. *Journal of Managerial Psychology, 21*(7), 600–619. https://doi.org/10.1108/02683940610690169
Salama, W., Nor El Deen, M., Albakhit, A., & Zaki, K. (2022). Understanding the connection between sustainable human resource management and the hotel business outcomes: Evidence from the green-certified hotels of Egypt. *Sustainability, 14*(9), 5647. https://doi.org/10.3390/su14095647
Saleh, R. M., & Al-Swidi, A. (2019). The adoption of green building practices in construction projects in Qatar: A preliminary study. *Management of Environmental Quality: An International Journal, 30*(6), 1238–1255. https://doi.org/10.1108/MEQ-12-2018-0208
Schaufeli, W. B. (2017). Applying the job demands-resources model: A 'how to' guide to measuring and tackling work engagement and burnout. *Organizational Dynamics, 46*(2), 120–132. https://doi.org/10.1016/j.orgdyn.2017.04.008
Shahzad, M. A., Jianguo, D., & Junaid, M. (2023). Impact of green HRM practices on sustainable performance: Mediating role of green innovation, green culture, and green employees' behavior. *Environmental Science and Pollution Research, 30*(38), 88524–88547. https://doi.org/10.1007/s11356-023-28498-6
Shen, J., Dumont, J., & Deng, X. (2019). Green human resource management in Chinese Enterprises. In *Green human resource management in Chinese enterprises*. https://doi.org/10.4324/9780429286971.
Singh, S. K., Del Giudice, M., Chierici, R., & Graziano, D. (2020). Green innovation and environmental performance: The role of green transformational leadership and green human resource management. *Technological Forecasting and Social Change, 150*, 119762. https://doi.org/10.1016/j.techfore.2019.119762
Sobaih, A. E. E., Hasanein, A., Gharbi, H., & Abu Elnasr, A. E. (2022). Going green together: Effects of green transformational leadership on employee green behavior and environmental performance in the Saudi food industry. *Agriculture, 12*(8), 1100. https://doi.org/10.3390/agriculture12081100
Sulea, C., Virga, D., Maricutoiu, L. P., Schaufeli, W., Zaborila Dumitru, C., & Sava, F. A. (2012). Work engagement as mediator between job characteristics and positive and negative extra-role behaviors. *Career Development International, 17*(3), 188–207. https://doi.org/10.1108/13620431211241054
Suliman, M. A., Abdou, A. H., Ibrahim, M. F., Al-Khaldy, D. A. W., Anas, A. M., Alrefae, W. M. M., & Salama, W. (2023). Impact of green transformational leadership on employees' environmental performance in the hotel industry context: Does green work engagement matter? *Sustainability, 15*(3), 2690. https://doi.org/10.3390/su1503269
Sun, X., El Askary, A., Meo, M. S., & Hussain, B. (2022). Green transformational leadership and environmental performance in small and medium enterprises. *Economic Research-Ekonomska Istraživanja, 35*(1), 5273–5291. Retrieved from https://www.tandfonline.com/loi/rero20.
Synodinos, N. E. (2003). The "art" of questionnaire construction: Some important considerations for manufacturing studies. *Integrated Manufacturing Systems, 14*(3), 221–237. https://doi.org/10.1108/09576060310463172
Taamneh, M. M., Al-Okaily, M., Abudoleh, J. D., Albdareen, R., & Taamneh, A. M. (2025). Nexus between green human resource management practices and corporate social responsibility: Does transformational leadership make difference? *International Journal of Organizational Analysis, 33*(1), 1–24. https://doi.org/10.1108/IJOA-08-2023-3903
Tamimi, S., Arshad, R., Haq, S. U., & Chughtai, M. A. (2024). Application of green HRM and its impact on environmental performance: A case of construction projects. *International Journal of Business Environment, 15*(1), 41–62. https://doi.org/10.1504/IJBE.2024.135698

Tang, G., Chen, Y., Jiang, Y., Paillé, P., & Jia, J. (2018). Green human resource management practices: Scale development and validity. *Asia Pacific Journal of Human Resources, 56*(1), 31–55. https://doi.org/10.1111/1744-7941.12147

Tanova, C., & Bayighomog, S. W. (2022). Green human resource management in service industries: The construct, antecedents, consequences, and outlook. *The Service Industries Journal, 42*(5–6), 412–452. https://doi.org/10.1080/02642069.2022.2045279

Tran, H. M. N., Le, D. V., Que, A. P. N., Chau, T. N., & Thao, T. L. T. (2024). The relationship between transformational leadership style, job creativity and job performance outcomes of employees in Vietnamese startup enterprises. *Huaxia Human Resource Management (English Version), 15*(2), 55–69. https://doi.org/10.47297/wspchrmWSP2040-800503.20241502

Visamitanan, K., & Assarut, N. (2021). Impact of green supply chain management practices on employee engagement and organizational commitment: Mediating role of firm performance. *Global Business Review, 25*(5), 1336–1355. https://doi.org/10.1177/09721509211018569

Wang, Q., & Shao, Z. (2024). Linking transformational leadership and digital creativity from the lens of social cognitive theory. *Industrial Management & Data Systems, 124*(6), 2312–2332. https://doi.org/10.1108/IMDS-07-2023-0444

World Health Organization. (2021). *WHO global air quality guidelines: Particulate matter (PM2.5 and PM10), ozone, nitrogen dioxide, sulfur dioxide and carbon monoxide*. World Health Organization. Retrieved from https://iris.who.int/handle/10665/345329.

You, F., & Kee, D. M. H. (2024). Green HRM, green creativity, and sustainability: Evidence from Chinese private enterprises in China. In M. Y. Yusliza & D. Renwick (Eds.), *Green human resource management*. Springer. https://doi.org/10.1007/978-981-99-7104-6_18

Yusoff, Y. M., Nejati, M., Kee, D. M. H., & Amran, A. (2020). Linking green human resource management practices to environmental performance in hotel industry. *Global Business Review, 21*(3), 663–680. https://doi.org/10.1177/0972150918779294

Zhang, Y., Luo, Y., Zhang, X., & Zhao, J. (2019). How green human resource management can promote green employee behavior in China: A technology acceptance model perspective. *Sustainability, 11*(19), 5408. https://doi.org/10.3390/su11195408

Zhu, J., Tang, W., Wang, H., & Chen, Y. (2021). The influence of green human resource management on employee green behavior—A study on the mediating effect of environmental belief and green organizational identity. *Sustainability, 13*(8), 4544. https://doi.org/10.3390/su13084544

Zoogah, D. B. (2011). The dynamics of green HRM behaviors: A cognitive social information processing approach. *German Journal of Human Resource Management, 25*(2), 117–139. https://doi.org/10.1177/239700221102500204

Sustainable Development Goals as a Catalyst Between Ethical Leadership and Employee Green Behaviour

Woon Leong Lin, Nelvin XeChung Leow, Jing Yi Yong, Chuen Khee Pek, Kelvin Yong Ming Lee, and Siong Choy Chong

Abstract This study examines the relationship between ethical leadership and employee green behaviour, with a particular focus on the moderating role of commitment to the Sustainable Development Goals (SDGs). Employing a dynamic panel data approach using System Generalised Method of Moments (GMM) estimations, the analysis draws on data from 405 firms across multiple industries between 2012 and 2019. The findings reveal that while ethical leadership positively influences employee green behaviour, this effect is significantly amplified in firms demonstrating strong SDG commitment. Specifically, organisations with a high alignment to the SDGs exhibit a stronger relationship between ethical leadership and environmentally responsible employee actions. Moreover, SDG commitment independently exerts a strong positive influence on the green behaviour, regardless of the leadership style. This research contributes to the growing body of literature on sustainable business practices by demonstrating that integrating leadership strategies with institutional sustainability frameworks, such as the SDGs, enhances organisational effectiveness. Theoretically, the study is grounded in Social Learning Theory and Organisational Citizenship Behaviour, while practically, it suggests that firms should embed sustainability into their core strategies to fully leverage the benefits of ethical leadership. Limitations and future research directions are discussed, including the need for industry-specific analyses and deeper exploration of individual-level drivers of green behaviour.

W. L. Lin (✉) · N. X. Leow · J. Y. Yong · C. K. Pek
School of Management and Marketing, Faculty of Business and Law, Taylor's University, Subang Jaya, Selangor, Malaysia
e-mail: Woonleong.lin@taylors.edu.my; NelvinXeChung.Leow@taylors.edu.my; Serene.Yong@taylors.edu.my; Eugene.Pek@taylors.edu.my

K. Y. M. Lee
Faculty of Business and Law, School of Accounting and Finance, Taylor's University, Subang Jaya, Selangor, Malaysia
e-mail: KelvinYongMing.Lee@taylors.edu.my

S. C. Chong
Faculty of Business and Law, Taylor's University, Subang Jaya, Selangor, Malaysia
e-mail: eddy.chong@taylors.edu.my

M. Y. Yusliza et al. (eds.), *Employee Green Behavior*,
https://doi.org/10.1007/978-981-95-1422-9_13

1 Introduction

The global push for sustainability has become one of the most urgent challenges of the twenty-first century, driven by escalating environmental concerns such as climate change, biodiversity loss and the depletion of natural resources. Given their significant environmental footprint, businesses face increasing pressure to adopt sustainable practices, and there is growing pressure on them to adopt more sustainable practices (Hu et al., 2024). In response, the United Nations introduced the Sustainable Development Goals (SDGs) in 2015 as a comprehensive framework to address these global challenges. The SDGs comprise 17 goals, which include objectives such as combating climate change, promoting responsible consumption and fostering sustainable economic growth (Hu et al., 2024; Aftab et al., 2023). For organisations, the SDGs offer a strategic opportunity to integrate sustainability into long-term planning and corporate social responsibility (CSR) initiatives (Aftab et al., 2023). Many companies have adopted the SDGs not only as an ethical obligation but also as a means to enhance their reputations and build stakeholder trust. However, to realise the full potential of the SDGs, they must be operationalised within daily business practices—driven by ethical leadership that is aligned with sustainability principles (Ahmed et al., 2024).

This is where ethical leadership plays a critical role. Ethical leaders, who value fairness, responsibility and transparency, help their organisations align with the SDGs (Hu et al., 2024; Ahmed et al., 2024; Lin et al., 2020). Research shows that ethical leadership fosters a culture of environmental and social responsibility, which influences employee behaviour and decision-making (Patwary, 2023). By promoting sustainability values, ethical leaders create an environment where employees are encouraged to engage in green behaviour such as energy conservation, recycling and responsible resource use (Patwary, 2023; Hu et al., 2024).

Employee green behaviour (EGB) has become a crucial aspect of corporate sustainability. These behaviours, driven by individual employees, directly contribute to reducing the environmental footprint of organisations (Patwary, 2023). However, fostering these behaviours requires more than just policies; it necessitates leadership capable of inspiring and facilitating sustainable practices (Ahmed et al., 2021; Hu et al., 2024). The SDGs serve as a bridge between ethical leadership and EGB by providing a structured framework for leaders to align their sustainability vision with employee actions (Aftab et al., 2023).

By embracing the SDGs, ethical leaders can inspire employees to adopt behaviours that support long-term environmental goals, ensuring that sustainability becomes a fundamental part of the organisational culture (Hu et al., 2024; Patwary, 2023). This connection between ethical leadership, SDGs and EGB is essential for achieving broader organisational sustainability goals (Aftab et al., 2023; Bhardwaj et al., 2025).

However, many organisations face significant challenges in translating ethical leadership into tangible EGBs that align with the broader sustainability goals (i.e. the SDGs). Although ethical leadership is widely acknowledged as vital for fostering a positive organisational climate for sustainability, the mechanisms by which it influences employee behaviour remain underexplored (Iqbal et al., 2025; Wood et al., 2024). While existing studies indicate that ethical leadership enhances organisational trust, responsibility-taking and green citizenship behaviours, the link between these behaviours and the SDGs at an operational level remains unclear (Hasan et al., 2024; Younas et al., 2023).

Moreover, the SDGs, with their global mandate to address environmental, social and economic challenges, have yet to be fully utilised as a framework for bridging the gap between ethical leadership and employee-driven sustainability outcomes. Current research has primarily focused on leadership styles and green organisational climates but has not sufficiently explored how ethical leadership can leverage the SDGs to drive EGBs (Asghar et al., 2023; Meirun et al., 2024). This disconnect hinders organisations from fully capitalising on the potential of their leadership to mobilise employees towards sustainability initiatives.

The lack of a clear framework linking ethical leadership to EGB through the lens of the SDGs presents a significant challenge (Liu et al., 2024). Organisations risk missing out on critical opportunities to foster employee engagement in sustainability initiatives, which is essential for achieving long-term environmental objectives. To address this gap, it is imperative to explore how the SDGs can function as a mediating framework that aligns ethical leadership with pro-environmental behaviours, ensuring that employee actions contribute meaningfully to both organisational and global sustainability targets (Zafar & Jafar, 2024; Xiao et al., 2021).

This research aims to explore the impact of ethical leadership on EGBs within global car manufacturing firms, focusing on both in-role and extra-role green behaviours. Ethical leadership, which promotes ethical conduct and decision-making, plays a critical role in encouraging employees to engage in behaviours that support sustainability and environmental objectives. However, the mechanisms by which ethical leadership influences these green behaviours, particularly in large-scale manufacturing industries like the automotive sector, remain underexplored.

Additionally, this research examines the moderating role of SDGs in this relationship by assessing how the presence of SDGs influences the strength of the relationship between ethical leadership and EGB in the automotive industry. By understanding this moderating effect, the research will offer firms with a clearer roadmap for integrating global sustainability goals with internal leadership practices, thus promoting more effective employee engagement in environmental initiatives.

To analyse these relationships, the study employs panel data from global car manufacturers from 2012 to 2019, prior to the COVID-19 pandemic. The System Generalised Method of Moments (GMM) serves as the primary analytical tool. This method is well-suited for addressing dynamic relationships and controlling for unobserved heterogeneity, simultaneity and endogeneity issues in panel data. System GMM is particularly effective in handling panel datasets with multiple time periods, making it ideal for analysing how ethical leadership influences green behaviours across multiple global firms over time.

The study deliberately excludes data from the COVID-19 period due to the unprecedented disruptions caused by the pandemic. These disruptions led to significant changes in business operations and employee work environments, which could introduce biases that do not reflect normal business conditions. By focusing on the pre-COVID period, the research aims to provide a more accurate understanding of the baseline dynamics between ethical leadership and green behaviour in car manufacturers without the confounding effects of pandemic-related disruptions.

This research is significant for several reasons. First, it addresses a critical gap in the literature by focusing on how ethical leadership influences EGBs within the global car manufacturing industry, a sector known for its environmental impact. Given the contributions of the automotive industry to carbon emissions and resource consumption, fostering green behaviour within organisations is crucial for meeting regulatory and societal expectations around sustainability.

Second, by employing System GMM on panel data from 2012 to 2019, the study provides robust empirical insights into the dynamic relationship between ethical leadership and EGB over time. System GMM addresses endogeneity and allows for precise estimates, ensuring that the findings are reliable and generalisable across global car manufacturers.

Furthermore, the research adds value by examining the SDGs as a moderating factor. The SDGs provide a universally recognised framework for addressing sustainability challenges. Understanding how the SDGs moderate the relationship between ethical leadership and EGB enables firms to align their leadership practices with global sustainability goals. This is becoming increasingly important as businesses face growing pressure from stakeholders to demonstrate environmental responsibility.

The chapter begins with an introduction outlining the research aims and significance. The theoretical background delves into concepts like social learning theory and institutional theory. The methodology outlines the use of CSRHub data and the System GMM approach to address endogeneity. In the Results section, the findings highlight the positive interaction between ethical leadership and SDG commitment. The chapter concludes with a discussion of implications, along with limitations and future research directions.

2 Literature Review and Hypothesis Development

2.1 Theoretical Background

The relationship between ethical leadership and EGB is underpinned by key theories that explain how leaders influence pro-environmental actions within organisations. These theories provide insights into the mechanisms by which ethical leadership drives sustainability-focused behaviours and integrates environmental goals into the core values of an organisation. This relationship is informed by two major theories: Social Learning Theory (SLT) and Organisational Citizenship Behaviour (OCB).

The SLT, introduced by Bandura (1977), describes how individuals learn behaviours by observing and imitating others, particularly leaders. Employees observe ethical leaders who prioritise sustainability and model their actions accordingly. Ethical leadership, characterised by fairness, responsibility and a commitment to sustainability, sets a powerful example for promoting pro-environmental behaviours (Aftab et al., 2023). When leaders engage in green practices, employees internalise and incorporate these behaviours into their daily work routines, contributing to the sustainability goals of the company (Hu et al., 2024).

OCB, especially towards the environment (OCBE), refers to voluntary, sustainability-focused actions that benefit the organisation. Ethical leaders foster OCBE by creating an ethical climate that encourages employees to engage in green behaviours beyond their formal roles, such as reducing energy use and supporting sustainability initiatives (Afsar et al., 2020; Patwary, 2023). Ethical leadership thus plays a vital role in driving voluntary green behaviours critical for achieving organisational sustainability.

2.2 *Ethical Leadership and Its Impact on Green Behaviour*

Ethical leadership has emerged as a key factor in promoting responsible and sustainable business practices. Defined as demonstrating normatively appropriate conduct through personal actions, communication and decision-making (Brown et al., 2005), ethical leadership fosters a culture where ethical standards guide behaviour. In the context of environmental sustainability, ethical leadership drives EGB, which includes actions to reduce environmental harm and enhance organisational sustainability (Robertson & Barling, 2013). Studies show that ethical leaders influence employee attitudes and behaviours by embedding sustainability in decision-making and fostering organisational citizenship behaviour toward the environment (Afsar et al., 2020; Aftab et al., 2023). One way ethical leadership promotes green behaviour is by creating an ethical climate where employees align their actions with environmental values (Zhang et al., 2021). This climate fosters voluntary pro-environmental engagement, as seen by employees reducing waste and conserving energy (Patwary, 2023; Ahmed et al., 2021). Ethical leadership also enhances the psychological ownership of sustainability initiatives, motivating employees to take responsibility and act proactively, such as recycling or reducing paper use (Kim et al., 2022; Aftab et al., 2023). Green transformational leadership, which overlaps with ethical leadership, also encourages innovation in eco-friendly practices (Zhang et al., 2021; Patwary, 2023). Leaders inspire employees to seek out new ways to minimise environmental impact (Aftab et al., 2023).

2.3 Ethical Leadership and Role Modelling in Green Employee Behaviour

According to SLT, ethical leadership significantly influences EGB through role modelling. Employees tend to observe and imitate the behaviours of leaders who prioritise sustainability, adopting similar eco-friendly practices like energy conservation and waste reduction (Bandura, 1977). Ethical leaders set a clear example for employees to follow by integrating sustainable practices into their decision-making processes, fostering a culture in which pro-environmental behaviours become the norm (Hu et al., 2024; Kim et al., 2022).

In addition, OCBE further supports the role of ethical leadership in driving green behaviours. Ethical leaders create a positive ethical climate that encourages voluntary employee actions beyond formal job responsibilities, such as reducing resource usage and participating in environmental initiatives (Afsar et al., 2020; Patwary, 2023). By creating this ethical climate, leaders help cultivate employee ownership of sustainability goals, motivating them to engage in eco-friendly behaviours that align with the broader environmental objectives of the organisation (Aftab et al., 2023; Ahmed et al., 2021).

Ethical leadership also strengthens employee commitment to sustainability through psychological ownership, which occurs when employees feel personally responsible for environmental outcomes (Kim et al., 2022). Leaders who visibly support sustainability efforts inspire employees to internalise these goals, making them more likely to voluntarily engage in actions like recycling or energy conservation without the need for direct supervision (Zhang et al., 2021). Thus, ethical leadership not only fosters compliance with green policies but also encourages discretionary, voluntary behaviours that contribute significantly to the overall environmental performance of the organisation.

In this context, ethical leadership is a key driver of green behaviour, as employees learn from the visible commitment of their leaders to sustainability. This is reinforced by a workplace culture that values voluntary contributions to environmental efforts, ensuring that sustainability becomes a shared responsibility across all levels of the organisation (Afsar et al., 2020; Patwary, 2023). This leads to the following hypothesis:

Hypothesis 1 Ethical leadership positively influences EGB through role modelling. (Role modelling positively mediates the relationship between ethical leadership and EGB.)

2.4 Sustainable Development Goals as a Moderator

The SDGs provide a framework for aligning organisational practices with global sustainability objectives. The SDGs act as a powerful moderator by enhancing the relationship between leadership and pro-environmental employee engagement. By embedding SDG principles into organisational strategies, ethical leaders create clear pathways for employees to engage in green behaviours such as waste reduction and energy conservation (Hu et al., 2024). Accordingly, leaders who emphasise SDG 12 (responsible consumption) and SDG 13 (climate action) reinforce the importance of sustainability in achieving organisational and societal goals (Aftab et al., 2023). This alignment increases employee buy-in because they understand how their actions contribute to global goals. The SDGs also provide measurable benchmarks, such as reducing carbon footprints, further motivating employees to engage in sustainability initiatives (Ahmed et al., 2021). By framing green behaviours within the SDGs, ethical leaders cultivate a stronger organisational culture of sustainability and commitment (Patwary, 2023; Kim et al., 2022). The following hypothesis thus ensues:

Hypothesis 2 The presence of SDG commitment strengthens the positive relationship between ethical leadership and EGB.

3 Methodology

3.1 Data Collection

The data for this study was sourced from CSRHub, covering 405 firms in the Motor Vehicle Manufacturing industry from 2012 to 2019. CSRHub aggregates Environmental, Social and Governance (ESG) data from over 500 sources, including ESG rating agencies and sustainability reports, providing reliable measures of the sustainability practices of companies (Lin, 2023; Rupp & Limpaphayom, 2024). Key variables include ethical leadership, as measured by sustainability and governance practices and EGB, as inferred from CSR ratings. SDG commitment data was also included to explore its moderating effect. Data is publicly available on CSRHub (Lin, 2023).

3.2 Generalised Method of Moments (GMM)

The Generalised Method of Moments (GMM), developed by Hansen (1982), is ideal for addressing endogeneity, serial correlation and unobserved heterogeneity in dynamic panel data models. GMM is used in this study due to its ability to handle the dynamic relationships between ethical leadership and EGB, with SDG commitment as a moderator. By leveraging lagged variables as instruments, GMM corrects for endogeneity, ensuring consistent estimates (Blundell & Bond, 1998). This technique is particularly suitable as it captures the evolving nature of EGB while controlling for unobserved heterogeneity and organisational factors.

3.3 Application of GMM in the Empirical Model

Following the recommendations from Roodman (2009) for implementing GMM in panel data analysis, we use System GMM to address issues related to endogeneity, serial correlation and unobserved heterogeneity. The approach focuses on controlling for the proliferation of instruments, ensuring instrument validity and addressing common pitfalls in GMM estimation.

The general specification of the model is as follows:

$$\mathrm{GreenBehaviour}_{it} = \alpha + \beta_1 \mathrm{EthicalLeadership}_{it-1} + \beta_2 \mathrm{SDGCommitment}_{it} + \left(\beta_3 \mathrm{EthicalLeadership}_{it-1} \times \mathrm{SDGCommitment}_{it}\right) + X_{it}\gamma + \mu_i + o_{it}$$

Where:

$\mathrm{GreenBehaviour}_{it}$ represents the green behaviour of employee i at time t.

$\mathrm{EthicalLeadership}_{it-1}$

$\mathrm{SDGCommitment}_{it}$ refers to the organisational commitment to the SDGs.

X_{it} represents control variables such as:

- Total Assets: A measure of firm size, as larger firms may have more resources to implement green initiatives.
- Free Cash Flow: Indicates the liquidity of the firm and its ability to fund sustainability projects.
- R&D Investment: Represents the focus of the firm on innovation, which could impact its environmental strategies and employee behaviours.
- Leverage: Debt level of the firm, as highly leveraged firms might face constraints that impact their ability to invest in green practices.

μ_i captures unobserved individual-specific effects.

ϵ_{it} is the error term.

This model is estimated using System GMM, which combines equations in both levels and first differences. The lagged variables of endogenous regressors, such as ethical leadership and SDG commitment, are used as instruments to address endogeneity. This approach ensures that the estimates are unbiased and consistent, providing reliable insights into how ethical leadership influences green behaviours, and moderated by SDG commitment.

3.4 Diagnostics and Tests

To ensure the robustness of the GMM estimates, several diagnostic tests are conducted:

1. Hansen J-test for Overidentifying Restrictions: This test assesses the validity of the instruments used in the GMM model. A non-significant result indicates that the instruments are valid and uncorrelated with the error term.
2. Arellano–Bond Test for Autocorrelation: This test checks for serial correlation in the first-differenced residuals. The absence of second-order serial correlation is necessary for the GMM estimates to be reliable.
3. Robust Standard Errors: To address potential issues of heteroscedasticity, robust standard errors will be computed in the two-step System GMM, ensuring the reliability of the results.

In summary, System GMM is an ideal estimator for this study due to its ability to handle the dynamic nature of EGBs, address endogeneity concerns and control for unobserved heterogeneity. By including key financial control variables such as total assets, free cash flow, research and development (R&D) investment and leverage, the model accounts for factors that may also influence green behaviours within firms. This approach ensures consistent and efficient estimates of the relationship between ethical leadership and green behaviours, which are moderated by SDG commitment. Given the complexities of these relationships, GMM offers the most robust and appropriate approach for analysing data.

3.5 Ethical Leadership Score

The ethical leadership score, ranging from 0 to 100, measures how well a company integrates ethical principles into its governance and decision-making processes. This score was sourced from CSRHub, a leading provider of ESG data, which compiles ratings from over 500 sources, including non-governmental organisations (NGOs) and sustainability reports (CSRHub, n.d.). A higher score indicates strong ethical leadership, reflecting the commitment of a company towards ethical decision-making, transparency and alignment with global standards such as the SDGs (Patwary, 2023). Companies with high ethical leadership scores are more likely to integrate sustainability into their operations, fostering strong stakeholder relationships and improved environmental performance. A lower score, on the other hand, signals weak ethical leadership, which is associated with poor governance and insufficient commitment to sustainable practices, potentially leading to increased reputational risks and difficulties in meeting sustainability goals. CSRHub provides a robust framework for evaluating ethical leadership by aggregating data from trusted sources like MSCI ESG Research, Sustainalytics and CDP (Huang et al., 2022).

Firms with high ethical leadership scores are often linked to better employee engagement in environmental initiatives and resource conservation, as ethical leaders inspire sustainable corporate cultures. This study also uses CSRHub scores to assess how ethical leadership influences organisational sustainability and employee behaviours (Lin et al., 2020; Park et al., 2024).

3.6 Employee Green Behaviour Score

The EGB score, ranging from 0 to 100, measures how actively employees engage in environmentally friendly practices within their organisation. This score was sourced from CSRHub, which aggregates the ESG data from multiple sources, including NGOs and sustainability reports (CSRHub, n.d.). EGB encompasses activities such as waste reduction, resource conservation and participation in sustainability initiatives. A high score (near 100) indicates strong employee engagement in green practices, such as recycling and energy conservation, reflecting an organisational culture that supports environmental stewardship. Companies with high scores often provide the necessary incentives and tools for employees to contribute to sustainability goals (Patwary, 2023). In contrast, a low score (near 0) suggests weak employee participation, possibly due to poor leadership communication or a lack of organisational commitment to sustainability (Lin et al., 2020; Park et al., 2024).

This score is valuable as it quantifies how employee actions align with organisational sustainability objectives. High EGB helps reduce environmental impact and enhance organisational reputation. Research shows that active employee engagement in green initiatives leads to increased job satisfaction and productivity (Huang et al., 2022; Aftab et al., 2023), helping companies meet broader SDGs (Patwary, 2023).

3.7 SDG Commitment Score

The SDG commitment variable (scored from 0 to 100), which serves as a moderator in this study, examines how the alignment of a company with the SDGs affects the relationship between ethical leadership and EGB. A higher score (closer to 100) indicates a stronger integration of SDG principles into company strategies, more transparent sustainability reporting and active stakeholder engagement. In organisations with high SDG commitment, ethical leadership is more likely to influence employee behaviours, encouraging environmentally responsible actions such as energy conservation and waste reduction (Patwary, 2023). SDG commitment enhances the ethical message from leadership, ensuring employees understand that their green behaviours contribute not only to organisational goals but also to global sustainability objectives. Conversely, a low SDG commitment score (closer to 0) suggests weak integration of SDGs, limiting the moderating effect and reducing leadership's ability to inspire the green behaviour (Lin et al., 2020; Park et al., 2024).

This variable evaluates factors such as compliance with the Global Reporting Initiative (GRI), transparency in sustainability efforts and engagement in global frameworks like the United Nations Global Compact. Organisations with high SDG commitment often involve employees in sustainability training and green initiatives, fostering a culture where ethical leadership and sustainability goals are aligned (Huang et al., 2022).

4 Results and Finding

The descriptive statistics in Table 1 provide key insights into the variables used in the analysis, including EGB (EMP_Green), ethical leadership (Ethical_Leader) and SDG commitment (SDG), alongside control variables such as leverage, free cash flow, R&D investment and total assets. The mean score of 5.734 for EGB, with a standard deviation of 5.84, reflects moderate engagement in green practices across firms, but with variability from 0 to 10, indicating some firms thrive while others lag in promoting sustainability. Ethical leadership has a mean score of 1.235, with a high standard deviation (3.322), suggesting significant variation in how firms implement ethical leadership. The SDG commitment variable shows a mean of 6.555 and a standard deviation of 1.83, indicating moderate commitment to the SDGs, with some firms showing stronger alignment than others. For the control variables, leverage has a wide range (mean of 7.267, standard deviation of 18.59), and free cash flow ranges from −214.36 to 241.11, reflecting major differences in liquidity across firms. Diagnostic tests indicate that the model passed the Variance Inflation Factor (VIF) test (2.89), confirming no multicollinearity concerns. Additionally, the model passed both normality and heteroscedasticity tests, ensuring unbiased and efficient estimates with normally distributed residuals and consistent variance.

Table 1 Descriptive statistics

Variable	Unit of measurement	Obs	Mean	Std. dev	Min	Max
EMP_Green	Scores from 0 to 10	3240	5.734	5.84	19	100
Ethical_ Leader	Scores from 0 to 10	3240	1.235	3.322	24	100
SDG	Scores from 0 to 10	3240	6.555	1.83	45	100
Leverage	Long-term debt of firms to their total equity	3240	7.267	18.59	2.720	91.21
Free cash flow	Free cash flow to the total of sales	3240	6.510	26.55	−214.36	241.11
ln R&D	R&D investment to the total net sales of the firm	3240	11.284	12.10	0.86	108.48
ln Total assets	Log of total asset of the firm	3240	11.470	1.26	6.38	12.63

Notes: All statistics are based on original data values. The correlation matrix in Table 2 highlights the key relationships between variables. EGB is positively correlated with ethical leadership (0.5158), implying that stronger ethical leadership enhances the green behaviour among employees. The correlation with SDG commitment is weaker (0.1429), suggesting a moderate link between SDG alignment and green behaviour. Leverage correlates positively with both EGB (0.2546) and SDG commitment (0.1841), showing that higher-leverage firms may prioritise sustainability for risk management. Interestingly, free cash flow negatively correlates with the green behaviour (−0.0414) but shows a strong positive relationship with ethical leadership (0.5407). These findings emphasise the importance of ethical leadership, supported by SDG commitment, while financial variables provide context

Source: Author's own work

Table 2 Correlation

	1	2	3	4	5	6	7
EMP_Green	1						
Ethical_Leader	0.1158	1					
SDG	0.1429	0.1391	1				
Leverage	0.2646	0.0869	0.4541	1			
Free cash flow	0.2214	0.5407	0.0643	0.0117	1		
ln R&D	0.2615	0.0540	0.0208	0.1137	0.0301	1	
ln Total assets	0.6487	0.2382	0.2384	−0.2246	0.4503	0.2138	1

Notes: All statistics are based on original data values
Source: Author's own work

4.1 Ethical Leadership and Employee Green Behaviour

The results in Table 3 provide key insights into the influence of ethical leadership on EGB and the importance of control variables. The System GMM model shows a positive and significant coefficient of 0.278 for lagged EGB, highlighting that past green behaviour strongly influences future actions. This underscores the importance of maintaining early sustainability efforts to create a lasting green culture within organisations. Although ethical leadership shows a positive relationship with EGB in the Pooled OLS model (0.458), it becomes insignificant in the Fixed Effects and System GMM models, indicating that unobserved factors and past behaviours reduce its direct impact. This suggests that other factors, such as financial resources or previous sustainability efforts, may be more influential over time.

Interestingly, SDG commitment has a negative impact on green behaviour in the Fixed Effects model (−0.178) and remains negative, though insignificant, in the System GMM model, indicating the challenges in translating strategic sustainability goals into employee actions. The strong positive effect of R&D investment across models, particularly in System GMM (4.153), emphasises the critical role of innovation in fostering sustainable employee practices. The diagnostic tests, including the VIF (2.89), Hansen test and AR tests, confirm the robustness and reliability of the models.

4.2 The Moderating Effect of SDG on Ethical Leadership and Employee Green Behaviour

Table 4 presents the results of the System GMM models, examining the moderating effect of SDG commitment on the relationship between ethical leadership and EGB. Across all models, lagged EGB (EMP_Green_{t-1}) remains significant with coefficients of 0.247, 0.187 and 0.212 in Models 1, 2 and 3, respectively, at the 1% level. This confirms that past green behaviour strongly influences future actions, stressing the importance of early and sustained sustainability efforts within

Table 3 Impact of ethical leadership on EGB performance

	Static		Dynamic		
Variables	Pooled OLS	Fixed effect	Pooled OLS	Fixed effect	System GMM
EMP_Green_{t-1}			0.563*** (0.0255)	0.245*** (0.0269)	0.278*** (0.1280)
Ethical_Leader	0.459*** (0.0629)	0.078 (0.887)	0.234 (0.478)	−0.098 (0.687)	0.109 (0.486)
Leverage	−0.00878* (0.00386)	−0.00471 (0.00531)	−0.00247 (0.00516)	−0.00413 (0.00279)	−0.000831 (0.0000879)
Free Cash	0.846*** (0.00869)	0.184** (0.00787)	0.234*** (0.00817)	0.212 (0.872)	0.1032 (0.221)
SDG	0.176 (0.771)	0.178*** (0.065)	0.553*** (0.00897)	0.231*** (0.0534)	0.267*** (0.0066)
ln Total assets	3.752*** (0.213)	1.124*** (0.548)	0.765*** (0.243)	−5.864*** (0.789)	−1.245*** (1.458)
ln R&D	1.205*** (0.291)	2.899*** (0.641)	0.356** (0.245)	6.709*** (0.763)	4.153** (1.745)
Year dummy	Yes	Yes	Yes	Yes	Yes
Constant	4.891** (1.924)	12.879** (5.789)	1.874 (1.991)	−11.453* (6.324)	58.642 (10.841)
Observation	3240	3240	3240	3240	3240
Number of firms		405		405	405
Number of instruments					22
AR (1)					5.17 (0.001)
AR (2)					3.14 (0.756)
Hansen test					28.37 (0.278)
Difference-in-Hansen test					7.45 (0.165)

Notes: The standard errors are reported in parentheses, except for the Hansen test, AR (1), AR (2) and Difference-in-Hansen which are *p*-values. ***, ** and * indicate significance at 1%, 5% and 10% levels, respectively. Time dummies are included in the model specification, but the results are not reported to save space. The System GMM model is estimated by using the Blundell and Bond (1998) dynamic panel System GMM estimations and Roodman (2009)—Stata xtabond2 command
Source: Author's own work

organisations. In Model 1, ethical leadership alone is not statistically significant (coefficient: −0.245), indicating a weak direct influence on green behaviour. However, when SDG commitment is introduced in Model 2, ethical leadership becomes negative and significant (−5.591, 5% level), suggesting that without SDG alignment, ethical leadership may not effectively promote green behaviour.

The interaction term (SDG * Ethical Leader) in Model 3 is −0.899 (significant at the 5% level), indicating that in highly SDG-aligned organisations, the role of ethical leadership might diminish as sustainability practices become institutionalised. SDG commitment alone has a positive and significant impact (7.598) on green behaviour, showing that clear sustainability goals independently drive employee engagement. Control variables show that R&D investment has a strong positive

Table 4 Moderating effect of SDG on ethical leadership and EGB performance

	Dynamic		
Variables	Model 1	Model 2	Model 3
EMP_Green_{t-1}	0.247*** (0.0015)	0.187*** (0.0029)	0.212*** (0.0024)
Ethical Leader		0.245 (0.077)	5.591** (2.315)
SDG*Ethical Leader			0.889** (0.579)
SDG	0.135 (0.263)	0.474* (0.344)	7.597** (3.862)
Leverage	−0.0047 (0.00906)	−0.000413 (0.00689)	−0.0000831 (0.000789)
Free cash flow	−0.000897 (0.224)	−0.000458 (0.243)	−0.000558 (0.431)
ln Total assets	−2.863*** (0.143)	−2.254*** (0.369)	−5.265*** (0.478)
ln R&D	4.356*** (0.175)	5.709*** (0.333)	4.153*** (0.885)
Year dummy	Yes	Yes	Yes
Constant	4.878 (5.981)	8.455 (9.366)	0.00645 (0.551)
Observation	3240	3240	3240
Number of firms	405	405	405
Number of instruments	126	126	126
AR (1)	−4.45 (0.001)	−5.15 (0.003)	−6.18 (0.001)
AR (2)	0.13 (0.879)	0.85 (0.468)	−0.74 (0.856)
Hansen test	23.08 (0.325)	40.33 (0.423)	34.37 (0.658)
Difference-in-Hansen test	5.37 (0.166)	4.74 (0.196)	2.45 (0.145)

Notes: The standard errors are reported in parentheses, except for the Hansen test, AR (1), AR (2) and Difference-in-Hansen, which are *p*-values. ***, ** and * indicate significance at 1%, 5% and 10% levels, respectively. Time dummies are included in the model specification, but the results are not reported to save space. The System GMM model is estimated by using the Blundell and Bond (1998) dynamic panel System GMM estimations and Roodman (2009)—Stata xtabond2 command
Source: Author's own work

effect, while total assets have a negative relationship with the green behaviour. The diagnostic tests, including AR (1), AR (2) and the Hansen test, confirm the robustness of the models and the validity of the instruments used.

5 Discussion of Findings and Hypothesis Support

5.1 Ethical Leadership and Employee Green Behaviour (H1)

Hypothesis 1 (H1) proposed that ethical leadership positively influences EGB through role modelling. The System GMM model from Table 3 shows that the relationship between ethical leadership and EGB becomes positive but insignificant (0.209). This suggests that, after controlling for past behaviour and unobserved firm-level characteristics, ethical leadership continues to have a positive but not significant impact on EGB.

This finding partially supports the hypothesis that ethical leadership fosters environmentally friendly practices, as employees tend to emulate the ethical examples set by their leaders. However, the lack of statistical significance suggests that other organisational factors, such as financial structure or established sustainability practices, may play a greater role in shaping the employee behaviour. Nevertheless, the positive direction of the coefficient suggests that ethical leadership remains aligned with promoting green behaviour, even if its independent effect is not as strong in the dynamic context (Afsar et al., 2020). This is consistent with Graves and Sarkis (2018), who discovered that leadership plays a role in shaping environmental outcomes but that organisational processes and policies significantly mediate this relationship.

Furthermore, Zafar and Jafar (2024) found that ethical leadership can promote corporate social responsibility (CSR) and environmental behaviours, but only when there is strong organisational support, indicating that leadership influence alone is often insufficient without broader structural and cultural backing.

Thus, H1 is partially supported in the System GMM model, highlighting that ethical leadership may contribute to promoting green behaviour, though its effect is potentially mediated or moderated by other factors.

5.2 The Moderating Role of SDG Commitment (H2)

Hypothesis 2 (H2) posited that SDG commitment enhances the positive relationship between ethical leadership and EGB. The results of the System GMM model in Table 4 support this hypothesis, showing that SDG commitment significantly moderates this relationship. The interaction term (SDG * Ethical Leader) is positive and significant (0.899, significant at the 5% level), indicating that when organisations align with the SDGs, ethical leadership becomes more effective at promoting the environmentally friendly behaviour. This indicates that employees are influenced not only by leadership but also by the organisation's overarching commitment to sustainability. (Huang et al., 2022). Moreover, SDG commitment alone has a positive and significant effect on green behaviour (7.598), indicating that organisations with stronger SDG alignment are more successful at fostering green practices, even irrespective of leadership (Patwary, 2023).

These findings corroborate the work of Fernando et al. (2020) and Paillé et al. (2019), who argue that SDG alignment provides a roadmap for sustainability, reducing the need for direct leadership intervention. Thus, H2 is fully supported where SDG commitment not only strengthens the impact of ethical leadership but also enhances overall sustainability efforts. This underscores the importance of aligning leadership with SDG frameworks to foster an environment conducive to green behaviour.

6 Conclusion

This study examined the impact of ethical leadership on EGB and the moderating role of SDG commitment. Using System GMM estimations, the findings suggest that while ethical leadership fosters environmentally friendly practices, its influence is nuanced and often mediated by broader organisational factors. Ethical leadership showed a positive but not always significant effect on green behaviour, indicating that other elements, such as corporate culture and established sustainability practices, might play a more critical role. The study also highlights the crucial role of SDG commitment as a moderator. Organisations strongly aligned with the SDGs see an enhanced effect of ethical leadership on EGB. When sustainability goals are embedded within the framework of organisations, ethical leaders are better positioned to inspire pro-environmental actions. Additionally, SDG commitment alone had a strong positive influence, underscoring the importance of formal sustainability strategies in fostering a culture of environmental responsibility (Huang et al., 2022; Patwary, 2023).

6.1 Theoretical Implications

From a theoretical standpoint, this study contributes to the literature on leadership and environmental behaviour by highlighting the conditional nature of the ethical leadership–green behaviour relationship. The findings indicate that the role of leadership in shaping employee behaviours can be significantly moderated by broader organisational frameworks such as SDG alignment. The study extends the SLT by showing that while leaders serve as role models, their impact is amplified when organisational structures provide clear sustainability guidelines (Graves & Sarkis, 2018). Furthermore, the study bridges the gap between ethical leadership and sustainability practices by integrating institutional theory to emphasise how formal commitments, like the SDGs, shape leadership effectiveness in driving the environmental behaviour.

6.2 Practical Implications

The findings have significant practical implications for organisations seeking to enhance sustainability practices. First, companies should institutionalise sustainability goals, such as those aligned with the SDGs, to establish a solid framework that supports green behaviour. This involves embedding sustainability into corporate strategies, operations and culture, rather than relying solely on leadership. Second, while ethical leadership remains essential, firms should ensure that leaders are equipped with the resources and policies that allow them to foster environmental responsibility more effectively. Organisations can invest in leadership development programmes that prioritise sustainability as a core value, helping leaders to integrate green practices into their daily operations (Paillé et al., 2019). The results also suggest that fostering green behaviour requires a long-term commitment and persistence in sustainability efforts, highlighting the necessity of continuity in leadership and environmental strategies.

6.3 Limitations

While this study offers valuable insights, several limitations must be acknowledged. First, the study relies on secondary data from CSRHub, which may restrict the depth of insight into firm-specific practices or leadership behaviours. Future research could incorporate primary data such as surveys or interviews with employees to gain a more nuanced understanding of how ethical leadership influences individual-level green behaviours. Second, the study focuses on organisations from a single industry, which may affect the generalisability of the findings. Industry-specific dynamics may play a role in shaping how leadership and SDG commitment influence behaviour. Third, the reliance on System GMM for estimation, while addressing endogeneity, may not fully capture all possible confounding factors. Future research could explore alternative estimates (Roodman, 2009).

6.4 Future Research Directions

Future research should explore the cross-industry applicability of these findings by examining how ethical leadership and SDG alignment influence green behaviour in different sectors. Additionally, more research is needed to investigate the individual-level mechanisms through which ethical leadership shapes green behaviour, focusing on the role of employee motivation, engagement and environmental attitudes. Another area for future study is to examine the long-term effects of SDG commitment, assessing whether organisations that maintain high levels of SDG alignment over extended periods see greater gains in EGB and overall sustainability performance. Finally, exploring the role of external factors, such as regulatory pressure or market competition, may provide a more comprehensive understanding of how external forces interact with leadership and sustainability initiatives to shape green behaviour.

Acknowledgements This work was supported by the Fundamental Research Grant Scheme (FRGS) from the Ministry of Higher Education, Malaysia FRGS/1/2022/SS01/TAYLOR/02/3.

References

Afsar, B., Maqsoom, A., Shahjehan, A., Afridi, S. A., Nawaz, A., & Fazliani, H. (2020). Responsible leadership and employee's proenvironmental behavior: The role of organizational commitment, green shared vision, and internal environmental locus of control. *Corporate Social Responsibility and Environmental Management, 27*(1), 297–312.

Aftab, J., Abid, N., Cucari, N., & Savastano, M. (2023). Green human resource management and environmental performance: The role of green innovation and environmental strategy in a developing country. *Business Strategy and the Environment, 32*(4), 1782–1798.

Ahmed, Z., Nathaniel, S. P., & Shahbaz, M. (2021). The criticality of information and communication technology and human capital in environmental sustainability: Evidence from Latin American and Caribbean countries. *Journal of Cleaner Production, 286*, 125529.

Ahmed, T., Yousaf, A., Clavijo, R. C., & Sanders, K. (2024). Entrepreneurial pathways to sustainability: A theoretical paper on green human resource management, green supply chain management, and entrepreneurial orientation. *Sustainability, 16*(15), 6357.

Asghar, N., Khattak, A., Danish, M. H., Bokhari, I. H., & Waris, M. (2023). Managerial skills, technology adaptation and firm performance: Mediating role of process innovation and product innovation. *Cogent Business & Management, 10*(3), 228148.

Bandura, A. (1977). Self-efficacy: Toward a unifying theory of behavioral change. *Psychological Review, 84*(2), 191.

Bhardwaj, B., Sharma, D., & Chand, M. (2025). Ensuring sustainability through green HRM practices: A review, synthesis and research avenues. *Journal of Organizational Effectiveness: People and Performance, 12*(1), 57–90.

Blundell, R., & Bond, S. (1998). Initial conditions and moment restrictions in dynamic panel data models. *Journal of Econometrics, 87*(1), 115–143.

Brown, M. E., Treviño, L. K., & Harrison, D. A. (2005). Ethical leadership: A social learning perspective for construct development and testing. *Organizational Behavior and Human Decision Processes, 97*(2), 117–134.

CSRHub. (n.d.). *Corporate social responsibility ratings and information*. Retrieved from https://www.csrhub.com/.

Fernando, G. D., Jain, S. S., & Tripathy, A. (2020). This cloud has a silver lining: Gender diversity, managerial ability, and firm performance. *Journal of Business Research, 117*, 484–496.

Graves, L. M., & Sarkis, J. (2018). The role of employees' leadership perceptions, values, and motivation in employees' provenvironmental behaviors. *Journal of Cleaner Production, 196*, 576–587.

Hansen, L. P. (1982). Large sample properties of generalized method of moments estimators. *Econometrica: Journal of the Econometric Society, 50*, 1029–1054.

Hasan, S., Isık, Z., & Demirdögen, G. (2024). Evaluating the contribution of lean construction to achieving sustainable development goals. *Sustainability, 16*(8), 3502.

Hu, B., Peng, N., & Zhu, G. (2024). Sustainable finance in natural resource extraction: Navigating green recovery pathways. *Resources Policy, 96*, 105197.

Huang, Q., Li, Y., Lin, M., & McBrayer, G. A. (2022). Natural disasters, risk salience, and corporate ESG disclosure. *Journal of Corporate Finance, 72*, 102152.

Iqbal, I., Zafar, H., & e Alam, S. M. F. (2025). Achieving human resource management sustainability in banking sector. *Review Journal of Social Psychology & Social Works, 3*(1), 119–145.

Kim, S., Jeong, S. H., & Seo, M. H. (2022). Nurses' ethical leadership and related outcome variables: Systematic review and meta-analysis. *Journal of Nursing Management, 30*(7), 2308–2323.

Lin, W. L. (2023). Corporate social responsibility and irresponsibility: Effects on supply chain performance in the automotive industry under environmental turbulence. *Journal of Cleaner Production, 428*, 139033.

Lin, W. L., Yip, N., Ho, J. A., & Sambasivan, M. (2020). The adoption of technological innovations in a B2B context and its impact on firm performance: An ethical leadership perspective. *Industrial Marketing Management, 89*, 61–71.

Liu, Y., Shao, Y., & Yan, J. (2024). The double-edged sword of ethical leadership: Investigating when and why ethical leadership promotes versus inhibits team performance. *Applied Psychology, 73*(4), 2129–2152.

Meirun, T., Ahmed, Z., Alzoubi, R. H., Khosa, M., & Nguyen, N. T. (2024). The road to eco-excellence: How does environmentally specific empowering leadership foster hospitality employees' green creativity through green creative self-efficacy and green learning orientation. *International Journal of Hospitality Management, 120*, 103790.

Paillé, P., Mejía Morelos, J. H., Raineri, N., & Stinglhamber, F. (2019). The influence of the immediate manager on the avoidance of non-green behaviors in the workplace: A three-wave moderated-mediation model. *Journal of Business Ethics, 155*, 723–740.

Park, S., Kim, B., & Song, S. (2024). Generalist vs. specialist CEOs: Towards an understanding of CSR scheme in the restaurant industry. *International Journal of Hospitality Management, 121*, 103797.

Patwary, A. K. (2023). Examining environmentally responsible behaviour, environmental beliefs and conservation commitment of tourists: A path towards responsible consumption and production in tourism. *Environmental Science and Pollution Research, 30*(3), 5815–5824.

Robertson, J. L., & Barling, J. (2013). Greening organizations through leaders' influence on employees' pro-environmental behaviors. *Journal of Organizational Behavior, 34*(2), 176–119.

Roodman, D. (2009). How to do xtabond2: An introduction to difference and system GMM in Stata. *The Stata Journal, 9*(1), 86–136.

Rupp, A., & Limpaphayom, P. (2024). Benefits of corporate social responsibility during a pandemic: Evidence from stock price reaction to COVID-19 related news. *Research in International Business and Finance, 68*, 102169.

Wood, G., Cooke, F. L., Brou, D., Wang, J., & Ghauri, P. (2024). Rethinking contexts and institutions for research on human resource management in multinational enterprises in an age of polycrisis: Reflections and suggestions. *The International Journal of Human Resource Management, 35*(19), 3173–3206.

Xiao, X., Zhou, Z., Yang, F., & Qi, H. (2021). Embracing responsible leadership and enhancing organizational citizenship behavior for the environment: A social identity perspective. *Frontiers in Psychology, 12*, 632629.

Younas, S., Shoukat, S., Awan, A., & Arslan, S. M. (2023). Comparing effects of green innovation and renewable energy on green economy: The metrics of green economy as nucleus of SDGs. *Pakistan Journal of Humanities and Social Sciences, 11*(2), 1035–1051.

Zafar, M. B., & Jafar, A. (2024). Human capital and Islamic banking: A systematic literature review. *Journal of Islamic Accounting and Business Research.*

Zhang, N., Bu, X., Xu, Z., Gong, Z., & Gilal, F. G. (2021). Effect of ethical leadership on moral sensitivity in Chinese nurses: A multilevel structural equation model. *Advances in Nursing Science, 44*(3), E78–E92.

Harnessing Employee Engagement for Environmental Sustainability and CSR

Nitin Patwa, Roshni Paul, S. Vittal, Veena Jadhav, and Bhanu Ranjan

Abstract This research delves into the pivotal role of employees in advancing environmental sustainability and Corporate Social Responsibility (CSR) within organizations. Employees play a crucial role in the implementation and enforcement of sustainability policies, as well as in identifying environmental risks and opportunities. Environmental initiatives require collaborative efforts between organizations and employees in today's landscape. Leadership Commitment to Environmental Sustainability sets the organization's tone; frequent training and education programs help enhance knowledge and create more employee awareness about environmental issues; ultimately, employee engagement and attitudes help create a green culture.

1 Introduction

Environmental sustainability and Corporate Social Responsibility (CSR) are paramount aspects of modern business operations. "Environmental sustainability refers to the ability to maintain an ecological balance in our planet's natural environment and conserve natural resources to support the well-being of current and future generations (Sen, 2016). This necessitates the adoption of sustainable practices, responsible production and consumption, and proactive measures to address climate concerns. Similarly, CSR underscores the obligation of businesses to contribute to sustainable development by delivering economic, social, and environmental

N. Patwa (✉) · R. Paul
S P Jain School of Global Management | Dubai, Dubai International Academic City, Dubai, UAE
e-mail: nitin.patwa@spjain.org

S. Vittal
S P Jain School of Global Management | Dubai, Dubai, UAE

V. Jadhav · B. Ranjan
S P Jain School of Global Management | Singapore, Singapore, Singapore

M. Y. Yusliza et al. (eds.), *Employee Green Behavior*,
https://doi.org/10.1007/978-981-95-1422-9_14

benefits to stakeholders (ElAlfy et al., 2020). There is a growing emphasis on integrating green innovation practices into CSR initiatives, reflecting an organizational commitment to enhance sustainability efforts (Le, 2022).

Employees play a central role in advancing the goals of environmental sustainability and CSR within organizations (Stahl et al., 2020). They are responsible for implementing and executing these practices daily, affecting positive change, reducing environmental hazards, and fostering sustainable behavior (Chams & García-Blandón, 2019). Social learning and identification theories elucidate how social responsibility and employees' green commitments are intertwined and mutually reinforcing. Their attitudes bolster green Human Resources Management (HRM) practices and promote pro-environmental behavior, which, in turn, influence employee engagement in environmentally conscious work practices through environmental congruence (Dono et al., 2010).

This research explores the essential role employees play in driving environmental sustainability and Corporate Social Responsibility (CSR) within organizations. As key agents of change, employees are instrumental in executing sustainability initiatives, upholding environmental policies, and recognizing both risks and opportunities related to environmental impact. In today's evolving business landscape, meaningful progress in sustainability increasingly depends on active collaboration between organizations and their workforce (Margerum, 2008). Recognizing the importance of such partnerships, organizations actively engage in environmental and social initiatives. These efforts are often driven by internal priorities and external societal pressures, collectively defined as corporate social responsibility.

These initiatives significantly influence consumers' perceptions of an organization's environmental commitments. Despite their significance, the role and contributions of employees in these initiatives are often undervalued. This research explores employees' multifaceted roles, potential, and contributions to such initiatives. Additionally, it seeks to identify employees' challenges in this context and propose strategies for moving forward with greater precision and efficacy.

2 Research Problem and Objectives

Despite the growing recognition of environmental sustainability and Corporate Social Responsibility (CSR), why many organizations struggle to harness employee engagement to achieve these goals effectively?

Specific objectives include:

1. Identifying key factors influencing employee engagement in sustainability and CSR initiatives
2. Evaluating the role of leadership, organizational culture, and communication in promoting employee participation
3. Providing actionable recommendations for organizations to nurture a sustainable and socially responsible workforce

3 Literature Review

Employee engagement is important in successfully practicing environmental sustainability practices and Corporate Social Responsibility (CSR) initiatives within organizations (Casey & Sieber, 2016). Various theoretical frameworks connect the dynamics of employee attitudes, behaviors, and organizational strategies to adopting sustainable practices. These theories provide insights into the mechanisms that drive employee participation in environmental initiatives, nurturing a deeper understanding of harnessing their potential for sustainability and CSR.

4 Theoretical Development

Social exchange theory helps to understand employee involvement in environmental initiatives (Cropanzano et al., 2017). According to it, employees are motivated to reciprocate positive organizational actions, such as training programs or visible organizational commitment to sustainability. When organizations invest in initiatives prioritizing environmental sustainability, employees perceive these efforts as part of a mutual relationship (Mishra, 2017). This perception nurtures a sense of obligation and encourages active participation and contribution toward sustainability goals. By aligning organizational efforts with the employee expectations of reciprocity, social exchange theory underscores the importance of cultivating a relational framework for driving sustainability (Jerab & Mabrouk, 2023). Another theory, i.e., the Theory of Planned Behavior (TPB), provides a behavioral perspective, emphasizing that an employee's attitudes, subjective norms, and perceived behavioral control significantly influence their intention to engage in sustainable practices (Bouarar, 2021). Positive attitudes toward environmental sustainability, reinforced by awareness and knowledge of environmental issues, are critical in adopting environmentally responsible actions (Shah & Asghar, 2024). The theory suggests that organizations can create an intrinsically motivated workforce to adopt sustainable practices by shaping employee attitudes and perceptions through education and awareness programs (Holdsworth et al., 2019).

Additionally, subjective norms—such as peer influence and organizational culture—can further enhance engagement in sustainability initiatives. Understanding perceived organizational support's critical role in shaping sustainable employee behaviors is also crucial. Employees who perceive that their organization is committed to environmental responsibility—through leadership endorsement, training programs, and resource allocation—are more likely to align their actions with these priorities (Paillé & Meija-Morelos, 2019). OST (Organizational Support Theory) postulates that such perceptions enhance employees' sense of value within the organization, promoting their engagement in environmentally sustainable practices (Eisenberger et al., 2020). This theory underlines the importance of institutional commitment as a precursor to individual and collective contributions to sustainability goals.

Self-Determination Theory (SDT) focuses on the intrinsic motivations that drive the employee behavior. It suggests that employees are likelier to engage wholeheartedly in sustainable practices when they feel competent, autonomous, and meaningfully involved (Deci et al., 2017). Training and education programs play a crucial role in advancing these feelings by equipping employees with the necessary skills and knowledge. Additionally, participation in environmental initiatives enhances a sense of ownership and purpose, strengthening their commitment to sustainability. SDT underscores the value of empowering employees through meaningful engagement to achieve long-term sustainability objectives.

Stakeholder Theory offers a complete perspective, emphasizing the interconnectedness of organizational and employee interests with broader societal goals. By addressing the expectations of all stakeholders, including employees, organizations can create a culture that aligns internal behaviors with external sustainability objectives. Employees are instrumental in achieving sustainability goals as key stakeholders when actively contributing to these initiatives. Stakeholder Theory highlights the importance of encouraging a collaborative environment where employees feel valued and integral to the organization's CSR and sustainability efforts. This alignment ensures that organizational practices resonate with societal demands for environmental responsibility (Kalra, 2024).

Integrating these theoretical frameworks provides a comprehensive understanding of harnessing employee engagement for environmental sustainability and CSR. Social Exchange Theory emphasizes the importance of reciprocity in driving engagement, while the Theory of Planned Behavior focuses on shaping attitudes and intentions. Organizational Support Theory underscores the role of perceived support, Self-Determination Theory highlights intrinsic motivation, and Stakeholder Theory connects employee engagement with broader organizational and societal objectives. Together, these theories offer a robust foundation for developing strategies that leverage employee participation to achieve sustainability goals, ensuring that environmental responsibility becomes an integral part of organizational culture and practice.

These theories help to understand the following five as antecedents to achieving environmental sustainability practices in the organization and help in structuring our research model and research framework as referred in Fig. 1

1. Employee Knowledge and Awareness of Environmental Issues (EKAA)
2. Training and Education Programs on Environmental Sustainability for Employees (T&EP)
3. Leadership Commitment to Environmental Sustainability (LCES)
4. Employee Attitudes Toward Sustainability (EATS)
5. Employee Engagement in Environmental Initiatives (EEIEI)

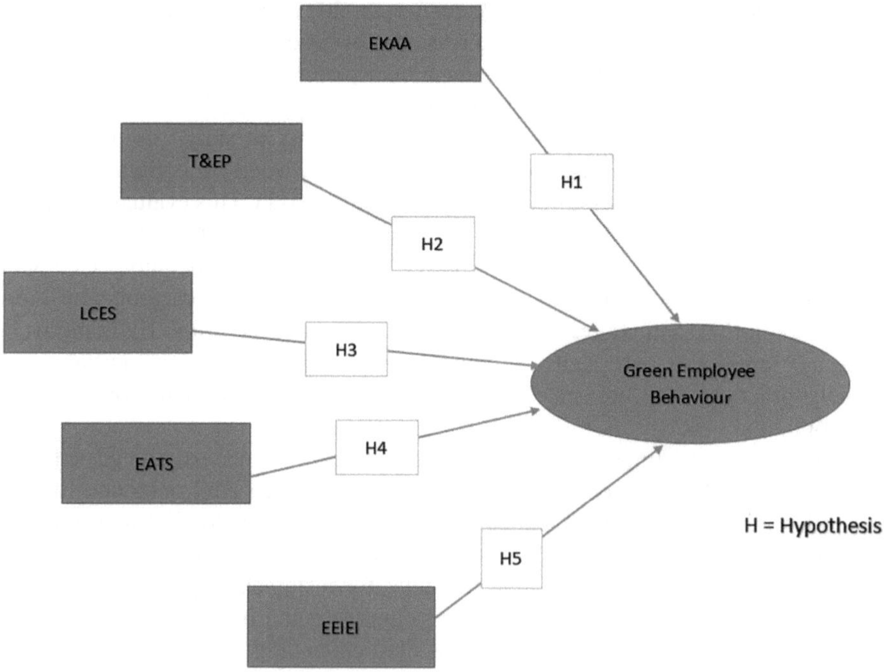

Fig. 1 Research framework (Source: Author's own work)

5 Research Model

1. Employee Knowledge and Awareness of Environmental Issues (EKAA)
2. Training and Education Programs on Environmental Sustainability for Employees (T&EP)
3. Leadership Commitment to Environmental Sustainability (LCES)
4. Employee Attitudes Toward Sustainability (EATS)
5. Employee Engagement in Environmental Initiatives (EEIEI)

6 Employee Knowledge and Awareness of Environmental Issues

This variable assesses employees' knowledge, awareness, and problem-solving capabilities regarding environmental issues within the workplace. It includes training on environmental awareness related to the organization's operations, including pollution, climate change, and resource conservation topics. A culture of sustainability gets strengthened through employees who are better educated, more aware of environmental issues, and capable of innovating and advocating for eco-friendly practices among their peers (Neruja & Arulrajah, 2021). Recent research suggests that the presence of green practices, such as employees using public transport, when embedded in the human resources management policies, makes employees better aware of environmental issues and their impact on the planet (Faeni, 2024).

Research suggests that environmental literacy refers to staff's comprehensive education and training on vital environmental issues, such as pollution, climate change, and resource conservation (Scholz & Binder, 2011). This evaluates employees' understanding of fundamental environmental concepts and the influence of this knowledge on the work environment (Fryxell & Lo, 2003). It encompasses a wide range of topics, including the causes and consequences of climate change, different types of pollution, and strategies for natural resource conservation. By being well-informed, employees are better equipped to recognize problems, initiate eco-friendly practices, and advocate for sustainable policies within the organization (Viterouli et al., 2023).

Moreover, the findings revealed the psychological distance associated with climate change and its influence on the relationship between environmental literacy and collaborative consumption tendencies (Aktan & Kethüda, 2024). Environmental literacy is invaluable to a business, as it empowers employees to adapt to environmental challenges and create positive strategies for sustainable operations. An environmentally literate workforce is likelier to innovate and implement technologies that reduce environmental impacts while enhancing corporate efficiency. Organizations can foster environmental literacy through employee training programs, knowledge-sharing initiatives, and cultivating continuous learning and sustainability awareness (Ahsan, 2024).

Employees should understand the environmental impact of their organization's activities, such as industrial processes, supply chain management, waste generation, and energy consumption. Employees become aware of how their organization contributes to environmental damage, such as greenhouse gas emissions, pollution, habitat destruction, and resource depletion. Bilderback (2024) suggests that integrating the UN's sustainability development goals (SDGs) into training, developmental programs, and incentives can promote sustainable practices and overall organizational sustainability. This will also guide the organization to set long-term goals toward sustainability, which will lead to a more comprehensive awareness of organizational impact among the employees (Bilderback, 2024).

Raising the awareness of organizational environmental impact is essential for nurturing accountability. Employees with a deeper understanding of their company's ecological footprint are more likely to explore ways to reduce negative impacts and amplify positive contributions. Engaging employees in sustainability efforts—such as recycling, waste reduction, energy efficiency initiatives, and adopting green technologies—can significantly improve the environmental performance. This collective effort, supported by open communication and access to environmental data, creates a cohesive workforce focused on sustainability goals (Lin & Zhao, 2023).

Staying informed about current environmental trends, regulations, and developments is crucial for employees to perform effectively. This includes understanding the latest environmental policies, scientific discoveries, and technological advancements that are relevant to their jobs (Watson et al., 2010). Employees knowledgeable about current environmental issues are better equipped to predict regulatory changes, adjust their operations accordingly, and implement the most efficient methods to meet environmental requirements (Sharma, 2000). Even new theories such as "humanistic organizing" have suggested a tripartite framework of mindful organizational communication whereby organizations can keep their workforce better informed. This will include rethinking organizational communication in a more humanistic manner to understand better the constant disruption faced by the workforce across social, economic, and environmental domains and also considering the employees' stress and disengagement with such issues (Town et al., 2024).

Providing employees with access to up-to-date information on environmental developments—through news sources, information sessions, and regular updates—enables them to stay at the forefront of environmental performance. This continuous learning culture improves organizational compliance and energy efficiency and empowers employees to contribute meaningfully to the organization's sustainability objectives. Adopting a proactive workforce that addresses environmental challenges and opportunities creates a competitive advantage for the organization.

Organizations can significantly enhance their sustainability efforts and achieve long-term environmental goals by promoting environmental literacy, raising awareness of organizational impacts, and ensuring employees stay informed about current environmental events.

7 Training and Education Programs on Environmental Sustainability for Employees

Training and education programs are crucial in fostering employee engagement in environmental sustainability. These programs empower workers to contribute actively toward greening the workplace and ensuring a more sustainable future (Saeed et al., 2019). By educating employees on sustainable practices and ecological issues, businesses can promote environmental conservation, economic efficiency, and resource optimization in the workplace (Mishra, 2017). The effectiveness of these training programs depends mainly on how well they align with the organization's sustainability goals and how relevant they are to the employees' specific roles. Training should equip employees with the knowledge and skills to address sustainability-related issues within their departments (Lacy et al., 2009). For instance, the HR department may focus on reducing energy consumption, while marketing teams may implement eco-friendly initiatives. Practical activities such as promoting a paperless workday or encouraging water-saving practices ensure that employees can apply their knowledge to everyday tasks (Willard, 2012).

These programs must go beyond raising general awareness about environmental sustainability and involve employees in concrete actions that improve their work environment (Boiral, 2009). The emphasis should be on optimizing performance, developing sustainable processes, and contributing to the company's energy conservation efforts (Omer, 2008). Equally vital is ensuring the accessibility of the training programs. Various delivery methods, such as online courses, workplace seminars, videos, printed materials, and site visits, can cater to different learning styles and preferences (Driscoll, 2010). All training materials must be accessible to employees with disabilities, and the option for captioning videos, providing transcripts for audio, or supplying image descriptions for visual aids should be considered. Employees should also receive training in their native language to enhance understanding and engagement (Noe et al., 2010).

Additionally, offering training on multiple devices with minimal Internet requirements helps include employees with limited technological access. Bilderback's (2024) research study shows the impact of integrating sustainability training with SDGs on organizational sustainability, promotes sustainable practices, improves employee satisfaction and productivity, reduces environmental impact, and enhances the organization's reputation. The study also found that regular progress reviews and long-term objectives will be crucial for organizations to remain competitive and adaptable to the changes (Bilderback, 2024).

Employee involvement can be boosted by clearly explaining the benefits of sustainability efforts and by implementing recognition strategies to reward contributions to green projects (Marrucci et al., 2024). The success of these training programs must be measured through comprehensive evaluations. Pre- and post-training assessments allow organizations to gauge the employees' understanding of sustainability concepts and practices before and after training. Interactive events, such as simulations or role-playing exercises, help to confirm that employees have absorbed the lessons and can exhibit sustainable behavior in the workplace. Long-term impact assessments should be conducted several months after the initial training to see whether positive behavior changes are sustained. A cost-benefit analysis is also essential to evaluate the financial effects of the training program. By comparing training costs with the savings realized through reduced resource usage, companies can determine the economic value of the initiative. Moreover, analyzing the return on investment (ROI) allows businesses to see both the tangible and intangible benefits of sustainability training. By making these training programs accessible, relevant, and engaging, businesses can inspire their employees to contribute to sustainability goals and societal benefits, resulting in lasting social, environmental, and economic advantages (Epstein, 2018). Regular evaluations ensure that these programs continue to evolve and remain aligned with organizational objectives while motivating employees to stay committed to sustainability initiatives. Furthermore, besides an intrinsic organizational interest in sustainability, regulations and stakeholder pressure can also encourage organizations to prioritize sustainability training. Also, corporate sustainability training should not be performed from the perspective of greenwashing but should be taken as a token for creating meaningful social impact in alignment with the organizational strategy and long-term performance goals (Sult et al., 2024).

8 Leadership Commitment to Environmental Sustainability

It plays a pivotal role in shaping the direction and culture of organizations today. Information technology leaders can influence organizational culture and priorities and guide, train, and practice sustainability initiatives (Roscoe et al., 2019). Their commitment signals the importance of green practices, such as energy-efficient data centers, reduced carbon footprints, and responsible electronic waste management (Abdullah & Lim, 2023). This commitment can inspire employees and stakeholders to align with sustainable goals, creating a culture of environmental responsibility (Galpin et al., 2015).

The Information Technology (IT) sector can potentially lead sustainability efforts significantly. Elliot points out that the IT industry has become an influential force capable of driving change, while Gholami et al. (2013) suggest that IT can initiate movements against pollution and contribute to environmental conservation. The sector is lauded for enhancing efficiency, reducing resource consumption, and improving operational sustainability. Dedrick also highlights how IT processes save energy and offer companies competitive advantages by enabling green operations.

However, the environmental impact of the IT sector is not without its challenges. Concerns have been raised by Ruth (2009) regarding the increasing energy consumption and the rise in electronic waste. This paradox reveals the downside of rapid technological advancement—while IT improves efficiency, it also contributes to growing environmental degradation. Vidal et al. (2023) cautions that the unchecked growth of the sector could lead to the depletion of resources and the generation of excessive pollutants, further straining the environment.

The frequency and transparency with which organizational leaders communicate sustainability goals also significantly foster commitment. According to Aguinis and Glavas (2019), organizations that openly share their eco-friendly initiatives tend to be perceived as more trustworthy and reputable. This approach cultivates a sense of responsibility among employees and clients, helping reinforce sustainable practices. Gonsalves et al. further argue that regular updates on progress toward sustainability goals keep employees engaged and motivated. Leaders who prioritize open communication and actively listen to feedback can address challenges in real time, reinforcing their commitment to environmental stewardship. As leaders become more transparent about their green actions, they increase their chances of success in contributing to a healthier environment. Leaders who engage in open dialogues with their teams create a sense of belonging, essential for building a unified effort toward sustainability.

The extent to which leaders demonstrate their commitment to sustainability through walk-a-talk is another crucial aspect. Research by Lehman and Geller (2004), and Steg et al. (2014) shows that environmental problems often stem from harmful human actions. Leaders who promote environmentally friendly behaviors and discourage harmful practices are better equipped to adapt to sustainability challenges. They value employees who actively contribute to environmental protection in their personal and workplace lives. Since employees spend a significant portion of their time at work, leaders must nurture a culture of sustainability within the workplace (Lülfs & Hahn, 2013).

One way to measure the pro-environmental behavior is through the "Organizational Citizenship Behavior for the Environment," which focuses on employees' actions to benefit the environment, such as volunteering for green projects or supporting sustainable initiatives (Paillé & Boiral, 2013). Research into pro-environmental behaviors has examined everyday actions, such as reducing water and electricity usage, and more substantial efforts, such as supporting eco-friendly policies (Stern, 2000; Larson, 2010). Leaders within companies that adopt eco-friendly strategies, such as reducing waste and efficiently using natural resources, are better positioned to achieve their sustainability goals. Many leaders also integrate environmental objectives into their human resources strategies by educating employees about sustainability and recognizing those who make significant contributions (Renwick et al., 2013; Daily & Huang, 2001).

Moreover, nurturing green behavior from the start of an employee's journey—during recruitment, orientation, and training—can help build a strong foundation for sustainability within the organization. Leaders should also create a work environment that encourages innovation and supports green initiatives (Renwick et al., 2013; Milliman & Clair, 2017). Despite the growing focus on green practices in the workplace, there is still much to learn about the drivers of pro-environmental behavior and the role green HRM practices play in promoting it (Anwar et al., 2020; Levy & Marans, 2012).

Leadership, management, and organizational culture are vital in shaping employees' green behavior and ecological performance. Leaders who demonstrate a strong commitment to sustainability through their actions, communication, and resource allocation are likelier to succeed in their environmental efforts and create a lasting positive impact on the planet.

9 Employee Attitudes Toward Sustainability

Employee attitudes toward sustainability vary greatly depending on the work culture, business culture, and individual beliefs. These factors can significantly influence how employees perceive and engage with sustainability efforts. Each employee may have a unique perspective, but specific behavior patterns have been observed across different organizations. Some employees are enthusiastic sustainability proponents and are deeply passionate about protecting the environment. They take pride in their roles as stewards of the environment and often take the lead in driving sustainability initiatives. Their commitment makes them highly effective in engaging others and pushing for positive environmental change within their organizations. Research investigations into whether organizational sustainability policies affect the attitude of employees by Dahiya (2020) showed that psychological mechanisms are prominent, such as work climate perceptions, which become the link between the policies and the green behavior of the employees (Dahiya, 2020).

Other employees, while not as actively involved, still care about sustainability. These concerned citizens may not participate directly in environmental campaigns or assume leadership roles, but they support the overall goals of sustainability. They contribute to environmental impact assessments and maintain an awareness of their organization's sustainability efforts. Some employees, however, are indifferent to sustainability, viewing it as a secondary concern or a task they are encouraged to follow through with incentives or bonuses. Their lack of awareness about the broader impact of sustainability might lead them to participate only when prompted by rewards, and they may not fully grasp the importance of their contributions to environmental efforts (Wang, 2018).

A key factor influencing employees' attitudes toward sustainability is the leadership and culture within the organization. When sustainability is promoted from the top down, it becomes a more integrated part of the organizational mindset, influencing employees to engage more actively in sustainability programs. However, if leadership does not prioritize sustainability, the goals and efforts around environmental responsibility can be jeopardized (Fu et al., 2023).

Personal values also play a significant role in shaping the employee behavior toward sustainability. Employees with solid ethical beliefs are more likely to engage in practices that reduce waste, preserve resources, and consider the broader societal impacts of their actions. Their knowledge of the environmental and social consequences of unsustainable behavior drives them to do what is suitable for the environment. Individuals becoming more socially and environmentally conscious are more inclined to participate in sustainability programs, often unconsciously incorporating these values into their daily routines (Yasir et al., 2022).

The company culture and commitment to sustainability initiatives also influence employees' attitudes. The collective beliefs and values shared within an organization shape how employees view environmental efforts. A company nurturing a culture of sustainability and environmental responsibility is likelier to see its employees work collaboratively toward achieving sustainability goals. Company culture is crucial in molding employees' attitudes and beliefs on these issues, ensuring sustainability becomes a core value. Therefore, culture is crucial in shaping how employees approach sustainability, and a strong cultural foundation can motivate groups of employees to work together toward common environmental goals (Epstein et al., 2010; Ketprapakorn & Kantabutra, 2022).

Leaders also influence employees' attitudes toward driving sustainability within organizations. Leaders who are actively involved in sustainability efforts communicate their goals and encourage employee participation, which plays a pivotal role in the success of these initiatives. Leaders who personally engage in environmental activities inspire their employees and demonstrate their commitment to sustainability through their actions. This visibility helps to build trust and credibility, both within the organization and in the broader community. Leadership's involvement also attracts media attention, highlighting the company's commitment to environmental responsibility. Leaders must align their goals with their actions, making sustainability an integral part of the organizational mission (Marshall et al., 2017).

Recognizing and incentivizing employees to contribute to sustainability can also significantly boost engagement. Whether through awards, bonuses, or performance-based incentives, employees who are recognized for their contributions are more likely to remain committed to the organization's sustainability goals. Incentivizing sustainability motivates employees to participate and defines the company's environmental initiatives' success and progress. When sustainability becomes a measurable part of an employee's performance, it reinforces the importance of environmental responsibility at every level of the organization (Burnett, 2022).

Nurturing a sustainability culture requires strong leadership, organizational commitment, and the integration of personal values into the business's daily operations. Employees need to feel supported and motivated to participate in sustainability efforts, and organizations must prioritize environmental responsibility as a core value. By doing so, companies can create a work environment where sustainability is a shared goal, leading to better organizational outcomes and contributing positively to the broader environmental landscape.

10 Employee Engagement in Environmental Initiatives

Employee engagement in environmental initiatives is a critical area of focus for organizational sustainability. The degree of employee involvement, organizational culture, leadership support, and communication tactics all play vital roles in determining how engaged employees are with sustainability efforts (Kota & Solomon, 2024). Research highlights that management commitment, from top-down to middle management, is a crucial factor in employees aligning their sense of purpose with organizational values (Heyden et al., 2017). A supportive culture that promotes green practices and effective communication strategies can significantly enhance employee motivation and participation in environmental sustainability programs (Saeed et al., 2019).

Employee involvement in environmental issues often depends on the level of participation and voluntary engagement in campaigns. Successful environmental programs empower employees across all levels, fostering a sense of ownership and self-motivation toward sustainability (Farooq et al., 2023). This ownership creates a work environment encouraging curiosity about professionally and personally environmentally friendly practices. When employee participation is low, it can lead to feelings of disengagement and powerlessness, which further reduce involvement in sustainability initiatives (Karatepe et al., 2020). Designing proper structures for managing employee sustainability programs effectively is essential to avoid such outcomes (Daily & Huang, 2001).

Key elements that promote employee involvement include award systems, communication channels, leadership, and organizational support. Employee engagement in sustainable initiatives is critical to achieving the organization's environmental goals (Saeed et al., 2019). When employees come together for environmental causes, they become part of a broader effort toward conservation and sustainable practices. This engagement is strongly linked to knowledge and awareness (Stringer, 2010). Employees with a deep understanding of ecological issues and sustainable practices are more likely to contribute actively to their organization's environmental initiatives. Awareness encompasses knowledge and an understanding of the importance of environmental protection and sustainability (Zacher et al., 2023).

Highly educated employees are more motivated and equipped to handle green projects. When workers know how their daily activities impact the environment, they are more likely to integrate sustainable practices into their work (Wehrmeyer, 2017). Employees aware of environmental degradation's consequences are often more motivated to act, primarily when their company clearly communicates its environmental policies and goals (Ones & Dilchert, 2012). Workshops and training sessions can improve understanding and awareness, providing employees with hands-on experience and empowering them to take charge of sustainability projects. Leadership plays a crucial role in fostering a culture of environmental consciousness. Employees are more likely to follow suit when leaders demonstrate their commitment to sustainability and integrate it into decision-making. Leaders can inspire employees by engaging in sustainability efforts and prioritizing environmental issues in the organization's overall strategy (Woo & Kang, 2020).

Raising awareness of environmental issues among employees is critical to fostering environmental literacy and encouraging active participation in sustainability programs (Nehal & Gupta, 2024). Employers must provide employees with the tools and resources necessary to engage with these initiatives. The role of knowledge and awareness in employee engagement cannot be overstated. Further research is needed to understand how awareness and knowledge translate into concrete behaviors and outcomes in organizational settings (Jones & Volpe, 2011).

Developing a more conscious attitude toward sustainability and corporate social responsibility has significantly impacted employee participation in sustainability activities. Behavioral theory, such as Transtheoretical Model (TTM), offer valuable insights into how individuals adopt new behaviors. These theories suggest that various factors influence behavior change, including individual values, opinions, group attitudes, and societal norms. Social norms within the organization also play a crucial role in shaping behavior. Employees are more likely to engage in environmental initiatives if they see their leaders and peers actively participating in these efforts (Robertson & Barling, 2013). Creating a social atmosphere based on peer support and social influence can further boost employee engagement in sustainability activities.

Equally important is the support and resources provided by the organization, which can motivate employees to adopt sustainable behaviors. Companies that invest in new facilities, staff awareness programs, and resources for sustainability initiatives demonstrate their commitment to environmental responsibility (Babiak & Trendafilova, 2011). This type of organizational support not only boosts employee morale but also empowers teams to achieve significant accomplishments in sustainability. Additionally, weaving environmental objectives into performance metrics and organizational culture reinforces the importance of sustainability and ensures long-term development. When employees understand that environmental responsibility aligns with the company's values, they are more likely to embrace behavior change and engage in environmental projects.

11 Contribution of Research

Harnessing employee engagement for environmental sustainability and CSR is crucial in understanding the integration of employee-driven initiatives within organizational sustainability goals. Theories like social exchange theory, theory of planned behavior, stakeholder theory, and more help in understanding multidimensional frameworks for enhancing sustainability practices. Employees will likely engage more deeply in environmental initiatives when they perceive organizational support and mutual benefits; this gives a belief in the importance of nurturing a culture of reciprocal commitment through incentives and recognition for sustainability efforts, driving higher engagement in environmental initiatives.

The Theory of Planned Behavior emphasizes that attitudes, subjective norms, and perceived behavioral control drive employee behavior. Employee attitudes toward sustainability influence their intentions and actions toward environmental initiatives. Education and training programs that shift attitudes and enhance perceived behavioral control over environmental actions are designed. Organizational Support Theory underlines the importance of perceived organizational support in nurturing employee commitment and discretionary efforts.

Organizational commitment to environmental sustainability is like organizational support, a key factor that motivates employees to engage in voluntary environmental behaviors, helps to visibly prioritize and resource environmental sustainability, and reinforces employees' sense of support and alignment with corporate values.

Research also highlights the role of intrinsic motivation and psychological needs (autonomy, competence, and relatedness) in driving behavior in line with the self-determination theory. It explains the critical role of training and education programs in adopting intrinsic motivation by enhancing the competence and autonomy in sustainability practices. It demonstrates how empowering employees with knowledge and self-directed initiatives leads to sustained engagement in environmental initiatives.

12 Conclusion and Scope of Future Research

Boosting employee engagement in sustainability initiatives involves encouraging environmentally friendly actions, creating a leadership culture supporting environmental management, and identifying the factors influencing behavior change. These factors include personal beliefs, sociocultural influences, organizational support, and proactive communication. Future research should focus on long-term observations and empirical studies to better understand the relationship between behavior dynamics and employee engagement in sustainability. Through these efforts, organizations can continue to enhance employee participation in environmental initiatives and work toward a more sustainable future.

References

Abdullah, N., & Lim, A. (2023). The incorporating sustainable and green IT practices in modern IT service operations for an environmentally conscious future. *Journal of Sustainable Technologies and Infrastructure Planning, 7*(3), 17–47.

Aguinis, H., & Glavas, A. (2019). Corporate social responsibility, sensemaking, and the search for meaningfulness through work. *Journal of Management, 45*(3), 1057–1086.

Ahsan, M. J. (2024). Green leadership and innovation: Catalysts for environmental performance in Italian manufacturing. *International Journal of Organizational Analysis, 33*, 1646.

Aktan, M., & Kethüda, Ö. (2024). The role of environmental literacy, psychological distance of climate change, and collectivism on generation Z's collaborative consumption tendency. *Journal of Consumer Behaviour, 23*(1), 126–140.

Anwar, N., Mahmood, N. H. N., Yusliza, M. Y., Ramayah, T., Faezah, J. N., & Khalid, W. (2020). Green human resource management for organizational citizenship behavior towards the environment and environmental performance on a university campus. *Journal of Cleaner Production, 256*, 120401.

Babiak, K., & Trendafilova, S. (2011). CSR and environmental responsibility: Motives and pressures to adopt green management practices. *Corporate Social Responsibility and Environmental Management, 18*(1), 11–24.

Bilderback, S. (2024). Integrating training for organizational sustainability: The application of sustainable development goals globally. *European Journal of Training and Development, 48*(7/8), 730–748.

Boiral, O. (2009). Greening the corporation through organizational citizenship behaviors. *Journal of Business Ethics, 87*, 221–236.

Bouarar, A. C. (2021). *Using the planned behavior theory to explore employees' intentions to implement green practices.*

Burnett, A. (2022). *Incentivising 'regenerative value' to improve sustainability outcomes.*

Casey, D., & Sieber, S. (2016). Employees, sustainability, and motivation: Increasing employee engagement by addressing sustainability and corporate social responsibility. *Research in Hospitality Management, 6*(1), 69–76.

Chams, N., & García-Blandón, J. (2019). On the importance of sustainable human resource management for the adoption of sustainable development goals. *Resources, Conservation and Recycling, 141*, 109–122.

Cropanzano, R., Anthony, E. L., Daniels, S. R., & Hall, A. V. (2017). Social exchange theory: A critical review with theoretical remedies. *Academy of Management Annals, 11*(1), 479–516.

Dahiya, R. (2020). Does organizational sustainability policies affect environmental attitude of employees? The missing link of green work climate perceptions. *Business Strategy & Development, 3*(3), 395–403.

Daily, B. F., & Huang, S. C. (2001). Achieving sustainability through attention to human resource factors in environmental management. *International Journal of Operations & Production Management, 21*(12), 1539–1552.

Deci, E. L., Olafsen, A. H., & Ryan, R. M. (2017). Self-determination theory in work organizations: The state of a science. *Annual Review of Organizational Psychology and Organizational Behavior, 4*(1), 19–43.

Driscoll, M. (2010). *Web-based training: Creating e-learning experiences.* John Wiley & Sons.

Eisenberger, R., Rhoades Shanock, L., & Wen, X. (2020). Perceived organizational support: Why caring about employees counts. *Annual Review of Organizational Psychology and Organizational Behavior, 7*(1), 101–124.

ElAlfy, A., Palaschuk, N., El-Bassiouny, D., Wilson, J., & Weber, O. (2020). Scoping corporate social responsibility (CSR) research evolution in the sustainable development goals (SDGs) era. *Sustainability, 12*(14), 5544.

Epstein, M. J. (2018). *Making sustainability work: Best practices in managing and measuring corporate social, environmental and economic impacts*. Routledge.

Epstein, M. J., Buhovac, A. R., & Yuthas, K. (2010). Implementing sustainability: The role of leadership and organizational culture. *Strategic Finance, 91*(10).

Faeni, D. P. (2024). Green practices and employees performance: The mediating roles of green human resources management policies and knowledge development. *Journal of Infrastructure, Policy, and Development, 8*(8), 4924.

Farooq, K., Yusliza, M. Y., Muhammad, Z., Omar, M. K., & Nik Mat, N. H. (2023). Employee ecological behavior among academicians at the workplace. *Social Responsibility Journal, 19*(4), 713–740.

Fryxell, G. E., & Lo, C. W. (2003). The influence of environmental knowledge and values on managerial behaviors on behalf of the environment: An empirical examination of managers in China. *Journal of Business Ethics, 46*, 45–69.

Fu, H., Guo, W., Sun, Z., & Xia, T. (2023). Asymmetric impact of natural resources rent, monetary and fiscal policies on environmental sustainability in BRICS countries. *Resources Policy, 82*, 103444.

Galpin, T., Whitttington, J. L., & Bell, G. (2015). Is your sustainability strategy sustainable? Creating a culture of sustainability. *Corporate Governance, 15*(1), 1–17.

Gholami, R., Sulaiman, A. B., Ramayah, T., & Molla, A. (2013). Senior managers perception on green information systems (IS) adoption and environmental performance: Results from a field survey. *Information & Management, 50*(7), 431–438.

Heyden, M. L., Fourné, S. P., Koene, B. A., Werkman, R., & Ansari, S. (2017). Rethinking 'top-down' and 'bottom-up' roles of top and middle managers in organizational change: Implications for employee support. *Journal of Management Studies, 54*(7), 961–985.

Holdsworth, S., Sandri, O., Thomas, I., Wong, P., Chester, A., & McLaughlin, P. (2019). The assessment of graduate sustainability attributes in the workplace: Potential advantages of using the theory of planned behavior (TPB). *Journal of Cleaner Production, 238*, 117929.

Jerab, D., & Mabrouk, T. (2023). *The role of leadership in changing organizational culture*. Retrieved from SSRN 4574324.

Jones, C., & Volpe, E. H. (2011). Organizational identification: Extending our understanding of social identities through social networks. *Journal of Organizational Behavior, 32*(3), 413–434.

Kalra, P. (2024). Corporate social responsibility and stakeholder theory: An integrated review. *Journal of Management & Entrepreneurship, 10*.

Karatepe, O. M., Rezapouraghdam, H., & Hassannia, R. (2020). Job insecurity, work engagement and their effects on hotel employees' non-green and nonattendance behaviors. *International Journal of Hospitality Management, 87*, 102472.

Ketprapakorn, N., & Kantabutra, S. (2022). Toward an organizational theory of sustainability culture. *Sustainable Production and Consumption, 32*, 638–654.

Kota, V., & Solomon, C. (2024). *Exploring the impact of organizational culture and leadership support on employee engagement in achieving green sustainability—A manufacturing sector perspective*.

Lacy, P., Arnott, J., & Lowitt, E. (2009). The challenge of integrating sustainability into talent and organization strategies: Investing in the knowledge, skills and attitudes to achieve high performance. *Corporate Governance: The International Journal of Business in Society, 9*(4), 484–494.

Larson, K. L. (2010). An integrated theoretical approach to understanding the sociocultural basis of multidimensional environmental attitudes. *Society and Natural Resources, 23*(9), 898–907.

Le, T. T. (2022). How do corporate social responsibility and green innovation transform corporate green strategy into sustainable firm performance? *Journal of Cleaner Production, 362*, 132228.

Lehman, P. K., & Geller, E. S. (2004). Behavior analysis and environmental protection: Accomplishments and potential for more. *Behavior and Social Issues, 13*, 13–33.

Levy, B. L., & Marans, R. W. (2012). Towards a campus culture of environmental sustainability: Recommendations for a large university. *International Journal of Sustainability in Higher Education, 13*(4), 365–377.

Lin, D., & Zhao, Y. (2023). The impact of environmental regulations on enterprises' green innovation: The mediating effect of managers' environmental awareness. *Sustainability, 15*(14), 10906.

Lülfs, R., & Hahn, R. (2013). Corporate greening beyond formal programs, initiatives, and systems: A conceptual model for voluntary pro-environmental behavior of employees. *European Management Review, 10*(2), 83–98.

Margerum, R. D. (2008). A typology of collaboration efforts in environmental management. *Environmental Management, 41*, 487–500.

Marrucci, L., Daddi, T., & Iraldo, F. (2024). Creating environmental performance indicators to assess corporate sustainability and reward employees. *Ecological Indicators, 158*, 111489.

Marshall, J., Coleman, G., & Reason, P. (2017). *Leadership for sustainability: An action research approach*. Routledge.

Milliman, J., & Clair, J. (2017). Best environmental HRM practices in the US. In *Greening people* (pp. 49–73). Routledge.

Mishra, P. (2017). Green human resource management: A framework for sustainable organizational development in an emerging economy. *International Journal of Organizational Analysis, 25*(5), 762–788.

Nehal, N. B., & Gupta, A. (2024). The role of environmental education in fostering climate change awareness and action. *Remittances Review, 9*(2), 3254–3267.

Neruja, S., & Arulrajah, A. A. (2021). The impact of environmental knowledge and awareness on sustainability performance of organizations: The mediating role of employee green behaviour. *International Business Research, 14*(9), 1–68.

Noe, R. A., Tews, M. J., & McConnell Dachner, A. (2010). Learner engagement: A new perspective for enhancing our understanding of learner motivation and workplace learning. *The Academy of Management Annals, 4*(1), 279–315.

Omer, A. M. (2008). Energy, environment and sustainable development. *Renewable and Sustainable Energy Reviews, 12*(9), 2265–2300.

Ones, D. S., & Dilchert, S. (2012). Environmental sustainability at work: A call to action. *Industrial and Organizational Psychology, 5*(4), 444–466.

Paillé, P., & Boiral, O. (2013). Pro-environmental behavior at work: Construct validity and determinants. *Journal of Environmental Psychology, 36*, 118–128.

Paillé, P., & Meija-Morelos, J. H. (2019). Organizational support is not always enough to encourage employee environmental performance. The moderating role of exchange ideology. *Journal of Cleaner Production, 220*, 1061–1070.

Renwick, D. W., Redman, T., & Maguire, S. (2013). Green human resource management: A review and research agenda. *International Journal of Management Reviews, 15*(1), 1–14.

Robertson, J. L., & Barling, J. (2013). Greening organizations through leaders' influence on employees' pro-environmental behaviors. *Journal of Organizational Behavior, 34*(2), 176–194.

Roscoe, S., Subramanian, N., Jabbour, C. J., & Chong, T. (2019). Green human resource management and the enablers of green organizational culture: Enhancing a firm's environmental performance for sustainable development. *Business Strategy and the Environment, 28*(5), 737–749.

Ruth, S. (2009). Green it more than a three percent solution? *IEEE Internet Computing, 13*(4), 74–78.

Saeed, B. B., Afsar, B., Hafeez, S., Khan, I., Tahir, M., & Afridi, M. A. (2019). Promoting employee's proenvironmental behavior through green human resource management practices. *Corporate Social Responsibility and Environmental Management, 26*(2), 424–438.

Scholz, R. W., & Binder, C. R. (2011). *Environmental literacy in science and society: From knowledge to decisions.*
Sen, A. (2016). The ends and means of sustainability. In *The capability approach and sustainability* (pp. 5–19). Routledge.
Shah, S. S., & Asghar, Z. (2024). Individual attitudes towards environmentally friendly choices: A comprehensive analysis of the role of legal rules, religion, and confidence in government. *Journal of Environmental Studies and Sciences, 14*, 629–651.
Sharma, S. (2000). Managerial interpretations and organizational context as predictors of corporate choice of environmental strategy. *Academy of Management Journal, 43*(4), 681–697.
Stahl, G. K., Brewster, C. J., Collings, D. G., & Hajro, A. (2020). Enhancing the role of human resource management in corporate sustainability and social responsibility: A multi-stakeholder, multidimensional approach to HRM. *Human Resource Management Review, 30*(3), 100708.
Steg, L., Bolderdijk, J. W., Keizer, K., & Perlaviciute, G. (2014). An integrated framework for encouraging pro-environmental behaviour: The role of values, situational factors and goals. *Journal of Environmental Psychology, 38*, 104–115.
Stern, P. C. (2000). New environmental theories: Toward a coherent theory of environmentally significant behavior. *Journal of Social Issues, 56*(3), 407–424.
Stringer, L. (2010). *The green workplace: Sustainable strategies that benefit employees, the environment, and the bottom line*. St. Martin's Press.
Sult, A., Wobst, J., & Lueg, R. (2024). The role of training in implementing corporate sustainability: A systematic literature review. *Corporate Social Responsibility and Environmental Management, 31*(1), 1–30.
Town, S., Reina, C. S., Brummans, B. H., & Pirson, M. (2024). Humanistic organizing: The transformative force of mindful organizational communication. *Academy of Management Review, 49*, 824.
Vidal, I., Sucozhañay, G., Haupt, M., & Vanegas, P. (2023). Towards the definition of WEEE recycling targets in Ecuador. A case of study for mobile phones. *Waste Management & Research, 41*(10), 1539–1548.
Viterouli, M., Belias, D., Koustelios, A., Tsigilis, N., & Bakogiannis, D. (2023). Fostering sustainability through the integration of green human resource management and change management: Nurturing eco-conscious organizational practices. In *Managing successful and ethical organizational change* (pp. 241–278). IGI Global.
Wang, Y. (2018). Commitment to sustainable development: Exploring the factors affecting employee attitudes towards corporate social responsibility-oriented management. *Corporate Social Responsibility and Environmental Management, 25*(6), 1284–1292.
Watson, R. T., Boudreau, M. C., & Chen, A. J. (2010). Information systems and environmentally sustainable development: Energy informatics and new directions for the IS community. *MIS Quarterly, 34*, 23–38.
Wehrmeyer, W. (Ed.). (2017). *Greening people: Human resources and environmental management*. Routledge.
Willard, B. (2012). *The new sustainability advantage: Seven business case benefits of a triple bottom line*. New Society Publishers.
Woo, E. J., & Kang, E. (2020). Environmental issues as an indispensable aspect of sustainable leadership. *Sustainability, 12*(17), 7014.
Yasir, N., Xie, R., & Zhang, J. (2022). The impact of personal values and attitude toward sustainable entrepreneurship on entrepreneurial intention to enhance sustainable development: Empirical evidence from Pakistan. *Sustainability, 14*(11), 6792.
Zacher, H., Rudolph, C. W., & Katz, I. M. (2023). Employee green behavior as the core of environmentally sustainable organizations. *Annual Review of Organizational Psychology and Organizational Behavior, 10*(1), 465–494.

Exploring the Impact of Green HRM Practices on Employee Well-being: The Role of Environmental Knowledge and Consciousness

Khalid Farooq, M. Y. Yusliza, Alaa Amin Abdalla, and Ratri Wahyuningtyas

Abstract Integrating sustainability goals into HR practices has become increasingly critical amid growing global challenges like climate change and social inequalities. This study examines the intricate relationship between green human resource management (HRM) practices, employee well-being (EWB), and two essential mediating factors: Environmental Knowledge and Consciousness. This study conceptualizes whether green HRM positively influences environmental knowledge, environmental consciousness, and EWB. The research seeks to advance theoretical understanding, providing insights into how sustainable HRM practices may conceptually impact EWB and environmental consciousness. This chapter enhances the body of knowledge in green HRM by offering a theoretical foundation rooted in social exchange theory for organizations aiming to align HR strategies with sustainability goals, paving the way for future empirical research in this vital field.

1 Introduction

In the context of extraordinary global issues, including climate change, resource scarcity, and social disparities, sustainability has emerged as a central theme for organizations across the globe (Bahmani et al., 2023). Sustainable methodologies

K. Farooq (✉) · M. Y. Yusliza
Faculty of Business, Economics and Social Development, Universiti Malaysia Terengganu, Kuala Nerus, Terengganu, Malaysia
e-mail: khalid.farooq@umt.edu.my; yusliza@umt.edu.my

A. A. Abdalla
College of Business, Abu Dhabi University, Abu Dhabi, United Arab Emirates
e-mail: alaa.abdalla@adu.ac.ae

R. Wahyuningtyas
Faculty of Economics and Business, Universitas Telkom, Bandung, Indonesia
e-mail: ratriwahyu@telkomuniversity.ac.id

M. Y. Yusliza et al. (eds.), *Employee Green Behavior*,
https://doi.org/10.1007/978-981-95-1422-9_15

transcend the conventional boundaries of corporate accountability and penetrate the fundamental essence of human resource management (HRM) (Jackson et al., 2011). This transformation has given rise to green HRM, a strategic approach that harmonizes HR practices with environmental sustainability objectives. Green HRM encompasses a spectrum of policies, procedures, and initiatives to foster a sustainable work environment while enhancing employee well-being (EWB) (Jain et al., 2023).

The global landscape of HRM has evolved significantly in recent years, shaped by the imperatives of sustainability, environmental stewardship, and corporate social responsibility (Madero Gómez et al., 2022). Organizations progressively acknowledge the significance of incorporating ecological considerations into their HRM strategies, leading to Green HRM (Farooq et al., 2022a). This paradigm shift in HRM approaches is driven by the realization that the well-being of employees and the environment are inextricably linked and that fostering a sustainable work environment can yield numerous benefits (Yusliza et al., 2021).

EWB is paramount in organizations for several reasons (Kuriakose & Sreejesh, 2023). First, it significantly influences productivity within the workplace. When employees are physically and mentally healthy, they are more likely to be focused and efficient in their roles (Kleszewski & Otto, 2023). They can perform tasks more efficiently, make well-informed decisions, and enhance the organization's overall performance. This boost in productivity translates to better outcomes for the company. Moreover, an emphasis on EWB helps reduce absenteeism. Employees who feel that their employers value their health and well-being are less likely to take sick leave due to stress-related issues or physical ailments (Kuriakose & Sreejesh, 2023). This ensures that work processes remain uninterrupted and tasks are completed on time, preventing delays and disruptions.

In the twenty-first century, as society grapples with the consequences of unsustainable practices, the importance of understanding the intricate relationships between Green HRM, EWB, and the broader environmental context cannot be overstated (Jain et al., 2023). As organizations adopt sustainable practices, the need for comprehensive research in Green HRM becomes evident (Zhang et al., 2021). Green HRM encompasses strategies and policies that prioritize environmental sustainability while enhancing EWB (Zhang et al., 2021). Nevertheless, despite its increasing significance, a dearth of scholarly inquiry investigates the relationship between green HRM, employee well-being, and two crucial mediating factors: Environmental Awareness and Ecological Consciousness. In order to fill this void in the literature, the present investigation aims to examine the interconnections among green HRM practices, employee well-being, and the mediating variables of environmental awareness and environmental knowledge. In light of the aforementioned considerations, the following research questions guide this study:

1. What is the relationship between green HRM, Environmental Knowledge, Environmental Consciousness, and EWB?
2. What role do Environmental Knowledge and Consciousness play in mediating the relationship between green HRM and EWB?

This conceptual chapter offers significant contributions to the existing body of knowledge on green HRM and employee well-being. First, it seeks to advance our theoretical understanding of the potential positive relationship between green HRM and EWB. While prior research has primarily focused on the operational and environmental benefits of green HRM (Yusliza et al., 2021), a notable gap exists in exploring its impact on employee-centric outcomes, such as well-being (Farooq & Yusliza, 2023). Through an examination of the conceptual frameworks, this research elucidates the theoretical basis for utilizing social exchange theory to articulate green HRM practices' impact on employees' well-being.

Second, it explores the conceptual mediating relationships between green HRM, environmental knowledge, and environmental consciousness. Existing literature often overlooks the psychological and cognitive pathways that translate sustainable HRM practices into enhanced employee outcomes (Farooq & Yusliza, 2023). By tackling this deficiency, the research endeavors to furnish theoretical perspectives regarding the processes by which green HRM practices conceptually affect EWB, thereby presenting a more refined comprehension of the indirect implications of green HRM. These conceptual frameworks provide significant directions for entities endeavoring to harmonize their human resource strategies with objectives about environmental sustainability, thereby establishing a foundation for the forthcoming empirical research.

2 Literature Review

2.1 Theoretical Foundation

The Social Exchange Theory provides an insightful framework for examining the complex relationships between green HRM practices and EWB. This theoretical perspective posits that individuals participate in social interactions and relationships guided by a principle of reciprocity, wherein they anticipate that their efforts and contributions will elicit affirmative reactions (Islam et al., 2022). In organizations adopting green HRM, employees perceive these sustainability practices as investments made by the organization in their well-being and the environment (Cooke et al., 2020). Employees who recognize the organization's commitment to fostering a sustainable workplace culture through eco-friendly policies, opportunities for environmentally responsible activities, and clear communication of sustainability goals are more likely to reciprocate positively (Afsar & Umrani, 2019). This reciprocation manifests as heightened job satisfaction, improved morale, and an enhanced sense of purpose in their work (Kuriakose & Sreejesh, 2023). Employees interpret green HRM practices as a mechanism of support, concluding that green HRM is consistent with ecological sustainability and significantly contributes to fostering EWB within the organizational context (Bahmani et al., 2023).

Social Exchange Theory further illuminates the conceptual relationships between green HRM, environmental knowledge, and environmental consciousness. As organizations embrace green HRM strategies and demonstrate their commitment to sustainability, employees perceive the organization as investing in their development (Anwar et al., 2020). Employees, in return, respond by actively seeking and acquiring environmental knowledge, becoming more aware of the organization's environmental policies, and gaining a deeper understanding of broader sustainability issues (Farooq & Yusliza, 2023). Simultaneously, the theory suggests that these employees will likely exhibit a heightened environmental consciousness. In essence, by valuing and encouraging environmentally responsible behaviors, green HRM catalyzes the growth of environmental knowledge and consciousness among employees (Okumus et al., 2019). This theoretical perspective provides insights into how green HRM practices may influence EWB not only directly but also indirectly by promoting greater environmental awareness and consciousness among the workforce, ultimately contributing to a more sustainable organizational culture.

In summary, Social Exchange Theory offers a valuable lens for understanding how employees perceive and respond to green HRM practices. By framing these initiatives as reciprocal exchanges, SET helps explain how organizational investments in sustainability not only foster enhanced environmental awareness and knowledge but also promote employee well-being. This theoretical foundation thus reinforces the idea that green HRM strategies can cultivate a mutually beneficial relationship between the organization and its employees, leading to a more engaged, informed, and sustainable workforce.

2.2 *Proposition Development*

2.2.1 Green HRM and EWB

Research into the relationship between green HRM and EWB reveals a compelling synergy between these two dimensions of organizational functioning. Green HRM initiatives, exemplified by environmentally conscious policies, sustainable operational practices, and fostering employee involvement in ecological responsibility, are progressively acknowledged for their contribution to the enhancement of EWB (Järlström et al., 2018). Scholars such as Renwick et al. (2016) emphasize that organizations prioritizing sustainability within their HRM strategies tend to cultivate a workforce with higher job satisfaction and engagement, causative to EWB. This alignment between green HRM and EWB is rooted in the sense of purpose- and value-driven work environment that sustainability initiatives create (Jain et al., 2023). Employee well-being encompasses employees' physical, psychological, and emotional health within an organization. It is influenced by various factors, including workplace environment, job satisfaction, work–life balance, and access to resources that support personal and professional growth (Zhang et al., 2021).

Employees who perceive their organization as environmentally responsible experience greater job satisfaction and a heightened sense of well-being.

Green HRM and EWB extend beyond job satisfaction to encompass psychological empowerment and health benefits. Green HRM practices empower employees to actively participate in sustainability efforts actively, fostering a sense of ownership and empowerment (Jain et al., 2023). Research by Rahi (2023) demonstrates that this empowerment positively influences well-being. Furthermore, green HRM contributes indirectly to well-being by cultivating healthier and less stressful workplace environments. Sustainable workplaces, resulting from green HRM practices, are conducive to improved physical and mental health among employees (Shen et al., 2018). The reduction in stress levels and the promotion of healthier behaviors underscore how green HRM not only supports sustainability goals but also plays a pivotal role in enhancing the overall well-being of the workforce (Jain et al., 2023). Therefore, we proposed that;

Proposition 1 *Green HRM has a positive effect on employee well-being.*

2.2.2 Green HRM and Environmental Knowledge

Organizations that prioritize green HRM practices actively engage employees in sustainability efforts, making them more aware of environmental issues, policies, and the organization's commitment to sustainability (Saeed et al., 2021). Environmental knowledge refers to a person's comprehension and cognizance of ecological challenges, biological systems, and sustainable methodologies. It includes knowledge about the causes and consequences of environmental problems and strategies and actions required to mitigate them (Okumus et al., 2019). Green HRM initiatives, characterized by eco-friendly policies, sustainability training, and the promotion of environmentally conscious behaviors among employees, have significantly contributed to the development of environmental knowledge within the workforce (Farooq & Yusliza, 2023). This heightened awareness extends beyond the workplace, influencing employees' understanding of broader ecological challenges. Research by Paillé et al. (2020) reveals that when employees recognize their organization as being environmentally conscious, their understanding of environmental issues increases, thereby exemplifying the constructive correlation between green HRM and the enhancement of environmental knowledge.

Green HRM not only enhances the individual environmental knowledge of employees but also cultivates a culture of ecological awareness (Darvishmotevali & Altinay, 2022). It encourages employees to incorporate sustainability into daily routines and decision-making processes (Norton et al., 2014). Employees who experience green HRM practices are more likely to adopt environmentally responsible behaviors at work and in their personal lives. This broader understanding and adoption of eco-friendly practices contribute to increased environmental knowledge among employees. Thus, the positive relationship between green HRM and environmental knowledge extends beyond the workplace, creating a workforce that is well-versed in sustainability practices and committed to environmental consciousness in their daily lives (Farooq et al., 2022b). In essence, green HRM serves as a catalyst for elevating professional and personal environmental knowledge among employees. Hence, we proposed that.

Proposition 2 *Green HRM has a positive effect on environmental knowledge.*

2.2.3 Green HRM and Environmental Consciousness

Green HRM practices encompass initiatives that serve as catalysts for developing environmental consciousness within the workforce. Environmental consciousness is an individual's awareness, attitudes, and values regarding environmental protection and sustainability. It reflects a person's commitment to addressing environmental challenges and willingness to engage in behaviors that minimize ecological harm. Organizations that prioritize green HRM actively engage employees in sustainability efforts, making them more aware of environmental issues, their sustainability goals, and their role in contributing to a more sustainable world. Research highlights that when employees are empowered to participate in environmental sustainability efforts, they are more likely to develop a sense of ownership and commitment to ecological values (Sudin & Saad, 2018). This underscores the positive relationship between green HRM and the cultivation of environmental consciousness.

Employees who experience green HRM practices are more likely to internalize sustainability values and adopt them as part of their core belief system. This transformation extends beyond the workplace, influencing employees' lifestyles and daily decision-making processes (Tuan, 2022). Employees who are part of organizations with strong green HRM practices are more inclined to embrace sustainability in their professional roles and personal lives (Al-Swidi et al., 2022). Consequently, green HRM plays a pivotal role in creating a workforce that is both environmentally knowledgeable and deeply committed to ecological consciousness in both their professional and personal spheres. In essence, green HRM acts as a catalyst for elevating environmental consciousness among employees, contributing to a more sustainable and responsible organizational culture; therefore,

Proposition 3 *Green HRM has a positive effect on environmental consciousness.*

2.2.4 Mediating Role of Environmental Consciousness

The mediating factor of environmental awareness significantly influences the association between green HRM practices and EWB. Initiatives related to green HRM enhance consciousness regarding environmental issues among the workforce. Research conducted by Li et al. (2023) highlights that employees who actively participate in sustainability initiatives cultivate a deeper commitment to ecological principles and adopt a purpose-oriented perspective in their professional roles. In this framework, environmental awareness serves as the conduit that magnifies the beneficial effects of green HRM on EWB. Its mediating influence on the connection between green HRM and EWB transcends mere awareness and extends into the domain of proactive measures (Darvishmotevali & Altinay, 2022). Employees who possess environmental awareness are likely to engage in sustainability efforts (Safari et al., 2018). Such proactive involvement not only promotes their well-being but also enhances the collective well-being of the organization. Environmentally aware employees experience a heightened sense of purpose and fulfilment in their roles as their contributions resonate with their values and the organization's sustainability objectives (Darvishmotevali & Altinay, 2022). Thus, environmental consciousness acts as a catalyst that strengthens the link between green HRM practices and EWB and elevates the workforce's overall well-being, creating a workplace culture where sustainability and personal well-being are intertwined. In essence, environmental consciousness is a key mediator that strengthens the positive effects of green HRM on EWB.

Proposition 4 *Environmental consciousness positively mediates the relationship between green HRM and employee well-being.*

2.2.5 Mediating Role of Environmental Knowledge

The mediating role of environmental knowledge is pivotal in understanding how green HRM practices positively influence EWB. Employees exposed to green HRM initiatives adopt sustainable practices and enhance their knowledge of environmental issues and the organization's commitment to sustainability (Liu et al., 2020). Research by Boiral and Paillé (2012) indicates that as employees perceive their organization as environmentally responsible, their environmental knowledge expands. This augmented environmental knowledge, in turn, strengthens the link between green HRM practices and EWB. It equips employees with the information and awareness to engage with sustainability initiatives, fostering a sense of purpose and well-being. Well-informed employees are more likely to actively participate in sustainability efforts at work and in their personal lives (Liu et al., 2020). By actively engaging, they boost their own well-being while also strengthening the organization as a whole. Employees who possess a deep understanding of sustainability are more satisfied and engaged in their roles, as their efforts align with their values and the organization's sustainability goals (Durmaz & Fidanoğlu, 2023). Consequently, environmental knowledge serves as a catalyst that strengthens the positive effects of green HRM on EWB. In essence, it is a key mediator that bridges the relationship between green HRM practices and well-being, fostering a workplace culture where knowledge and personal well-being are intertwined, contributing to a more sustainable and fulfilling work environment (Fig. 1). Therefore,

Proposition 5 *Environmental knowledge positively mediates the relationship between green HRM practices and employee well-being.*

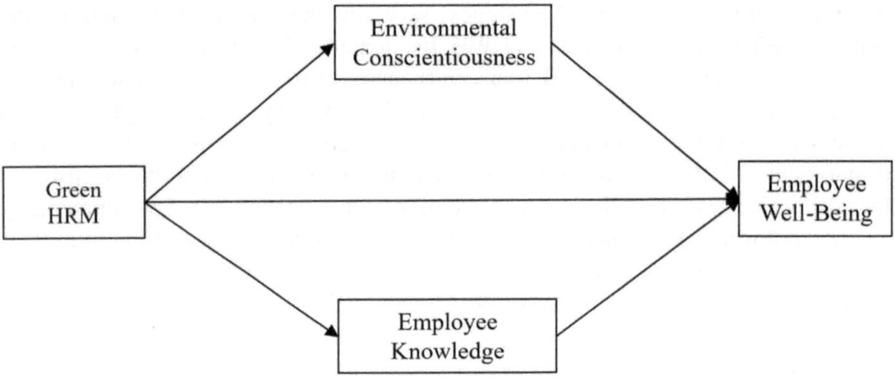

Fig. 1 Proposed conceptual model for employee well-being (Source: Author's own work)

3 Conclusion

This conceptual chapter significantly contributes to the growing body of knowledge in green HRM and EWB. It enhances theoretical understanding by proposing that green HRM practices are not only operational strategies but also mechanisms for promoting employee well-being, grounded in Social Exchange Theory. By applying this theoretical lens, the chapter offers a novel perspective on how green HR practices are perceived by employees as reciprocal investments, thereby reinforcing positive behavioral and psychological outcomes.

A key theoretical contribution lies in the integration of green HRM with constructs such as environmental knowledge and environmental consciousness, which are conceptualized as mediators in the relationship between green HRM and EWB. This multidimensional framework broadens the scope of existing HRM literature by linking individual-level psychological outcomes to organizational-level sustainability initiatives.

Furthermore, the chapter advances the theory by identifying green HRM as a foundational element in cultivating a sustainable organizational culture. It underscores the importance of aligning HRM practices with ecological values, thereby extending the boundaries of traditional HRM theory into the domain of environmental ethics and corporate social responsibility.

These insights provide a robust foundation for future empirical research and for organizations seeking to align their HR strategies with environmental sustainability objectives. As organizations navigate the dynamic landscape of sustainability, recognizing the dual role of green HRM—in promoting environmental responsibility and enhancing employee well-being—will be vital. Future empirical studies should test and refine the proposed theoretical model, ultimately contributing to more sustainable, ethically responsible, and fulfilling workplaces.

3.1 Practical Implications

The insights and recommendations from this study offer valuable guidance for organizations aiming to integrate sustainability into HRM practices while simultaneously enhancing EWB. The following areas outline key practical applications:

- Organizations should embed sustainability principles within their HR strategies by adopting eco-friendly policies, sustainability-focused training programs, and initiatives that encourage environmentally responsible behaviors among employees.
- Fostering a culture of ecological responsibility can contribute to long-term sustainability while simultaneously enhancing EWB. A sustainability-oriented workplace culture supports both environmental and human resource objectives.
- Leaders and HR professionals must work collaboratively to align HRM practices with environmental sustainability goals. Such alignment underscores the strategic value of sustainability in driving employee engagement and satisfaction.
- Empowering employees to participate actively in sustainability initiatives can foster a greater sense of purpose and fulfilment, directly supporting well-being. Participation also nurtures a sense of ownership and collective responsibility.
- Recognizing and rewarding employee involvement in environmental activities reinforces shared accountability and promotes sustained engagement in sustainability goals.
- Investing in educational initiatives such as training sessions, workshops, and internal communication campaigns can raise environmental awareness and consciousness. These efforts also signal the organization's commitment to ecological responsibility, thereby strengthening EWB.

3.2 Future Directions

Building upon the theoretical foundations of this study, future research should focus on empirically testing the proposed framework to deepen the understanding of the relationship between green HRM and EWB. Quantitative methods, including surveys, interviews, and longitudinal studies, can provide critical evidence on the interplay between green HRM, EWB, and the mediating roles of environmental knowledge and consciousness. Exploring the influence of cultural factors on the effectiveness of green HRM practices presents another promising avenue for research. Diverse cultural contexts may shape perceptions and outcomes of green HRM strategies, offering valuable insights into their global applicability. Longitudinal studies are also necessary to evaluate the sustainability of green HRM practices and their enduring effects on EWB. Tracking changes over time will clarify whether such practices lead to sustained improvements in employee well-being and environmental awareness. Additionally, future research should investigate the role of technology in facilitating green HRM initiatives and enhancing environmental consciousness. Digital platforms, data analytics, and virtual tools could be crucial in optimizing sustainability efforts and improving employee engagement. By addressing these future directions, researchers can enrich the understanding of green HRM, EWB, and environmental consciousness, thereby advancing organizational sustainability and employee satisfaction on a global scale.

Acknowledgments This work was supported by the Talent and Publications Enhancement Research Grant (TAPERG) from the Universiti Malaysia Terengganu, Malaysia (TAPERG/2023/UMT/2682).

References

Afsar, B., & Umrani, W. A. (2019). Corporate social responsibility and pro-environmental behavior at workplace: The role of moral reflectiveness, coworker advocacy, and environmental commitment. *Corporate Social Responsibility and Environmental Management, 27*(1), 109–125. https://doi.org/10.1002/csr.1777

Al-Swidi, A. K., Al-Hakimi, M. A., Gelaidan, H. M., & Al-Temimi, S. K. A. J. (2022). How does consumer pressure affect green innovation of manufacturing SMEs in the presence of green human resource management and green values? A moderated mediation analysis. *Business Ethics, Environment and Responsibility, 31*(4), 1157–1173. https://doi.org/10.1111/beer.12459

Anwar, N., Nik Mahmood, N. H., Yusliza, M. Y., Ramayah, T., Noor Faezah, J., & Khalid, W. (2020). Green human resource management for organisational citizenship behaviour towards the environment and environmental performance on a university campus. *Journal of Cleaner Production, 256*, 120401. https://doi.org/10.1016/j.jclepro.2020.120401

Bahmani, S., Farmanesh, P., & Khademolomoom, A. H. (2023). Effects of green human resource management on innovation performance through green innovation: Evidence from northern Cyprus on small island universities. *Sustainability, 15*(5), 4158. https://doi.org/10.3390/su15054158

Boiral, O., & Paillé, P. (2012). Organizational citizenship behaviour for the environment: Measurement and validation. *Journal of Business Ethics, 109*(4), 431–445. https://doi.org/10.1007/s10551-011-1138-9

Cooke, F. L., Schuler, R., & Varma, A. (2020). Human resource management research and practice in Asia: Past, present and future. *Human Resource Management Review, 30*(4), 100778. https://doi.org/10.1016/j.hrmr.2020.100778

Darvishmotevali, M., & Altinay, L. (2022). Green HRM, environmental awareness and green behaviors: The moderating role of servant leadership. *Tourism Management, 88*, 104401. https://doi.org/10.1016/j.tourman.2021.104401

Durmaz, Y., & Fidanoğlu, A. (2023). The mediating role of environmental knowledge in the effect of environmentally sensitive thoughts and behaviors on business performance: Practice in Turkey. *Journal of Cleaner Production, 422*, 138491. https://doi.org/10.1016/j.jclepro.2023.138491

Farooq, K., & Yusliza, M. Y. (2023). Two decades of workplace ecological behaviour: a systematic literature review. *Benchmarking: An International Journal, 30*(10), 4681–4716. https://doi.org/10.1108/BIJ-02-2022-0079

Farooq, K., Yusliza, M. Y., & Muhammad, Z. (2022a). Make it their decisions, not your directives: Exploring required green competencies for employee ecological behaviour. *Organizacija, 55*(2), 128–141. https://doi.org/10.2478/orga-2022-0009

Farooq, K., Yusliza, M. Y., Muhammad, Z., & Saputra, J. (2022b). Developing a conceptual model of employee ecological behavior using an integrative approach. *Journal of Environmental Management and Tourism, 131*(57), 29–38. https://doi.org/10.14505/jemt.v13.1(57).03

Islam, M. A., Mendy, J., Haque, A. K. M. A., & Rahman, M. (2022). Green human resource management practices and millennial employees' retention in small and medium enterprises: The moderating impact of creativity climate from a developing country perspective. *Business Strategy and Development, 5*(4), 335–349. https://doi.org/10.1002/bsd2.202

Jackson, S. E., Renwick, D. W., Jabbour, C. J., & Muller-Camen, M. (2011). State-of-the-art and future directions for green human resource management: Introduction to the special issue. *German Journal of Human Resource Management, 25*, 99–116. https://doi.org/10.1177/239700221102500203

Jain, D., Vyas, P., & Rawat, G. (2023). The psychological role of green HRM in improving employee well-being and environmental sustainability. *Journal for ReAttach Therapy and Developmental Diversities, 6*(7), 224–237. Retrieved from https://jrtdd.com/index.php/journal/article/view/786/552.

Järlström, M., Saru, E., & Vanhala, S. (2018). Sustainable human resource management with salience of stakeholders: A top management perspective. *Journal of Business Ethics, 152*(3), 703–724. https://doi.org/10.1007/s10551-016-3310-8

Kleszewski, E., & Otto, K. (2023). A matter of needs: Basic need satisfaction as an underlying mechanism between perfectionism and employee well-being. *Motivation and Emotion, 47*(5), 761–780. https://doi.org/10.1007/s11031-023-10029-y

Kuriakose, V., & Sreejesh, S. (2023). Co-worker and customer incivility on employee well-being: Roles of helplessness, social support at work and psychological detachment—A study among frontline hotel employees. *Journal of Hospitality and Tourism Management, 56*, 443–453. https://doi.org/10.1016/j.jhtm.2023.07.009

Li, Z., Yu, H., & Xing, L. (2023). The impact of green culture on employees' green behavior: The mediation role of environmental awareness. *Corporate Social Responsibility and Environmental Management, 30*(3), 1325–1335. https://doi.org/10.1002/csr.2422

Liu, P., Teng, M., & Han, C. (2020). How does environmental knowledge translate into pro-environmental behaviors? The mediating role of environmental attitudes and behavioral intentions. *Science of the Total Environment, 728*, 138126. https://doi.org/10.1016/j.scitotenv.2020.138126

Madero Gómez, S., Olivas-Lujan, M. R., Rubio Leal, Y. L., & Yusliza, M. Y. (2022). Sustainable human resource management: Practitioners' perspectives. *Management Research, 20*(2), 111–128. https://doi.org/10.1108/MRJIAM-07-2021-1202

Norton, T. A., Zacher, H., & Ashkanasy, N. M. (2014). Organisational sustainability policies and employee green behaviour: The mediating role of work climate perceptions. *Journal of Environmental Psychology, 38*, 49–54. https://doi.org/10.1016/j.jenvp.2013.12.008

Okumus, F., Köseoglu, M. A., Chan, E., Hon, A., & Avci, U. (2019). How do hotel employees' environmental attitudes and intentions to implement green practices relate to their ecological behavior? *Journal of Hospitality and Tourism Management, 39*, 193–200. https://doi.org/10.1016/j.jhtm.2019.04.008

Paillé, P., Valéau, P., & Renwick, D. W. (2020). Leveraging green human resource practices to achieve environmental sustainability. *Journal of Cleaner Production, 260*, 121137. https://doi.org/10.1016/j.jclepro.2020.121137

Rahi, S. (2023). Fostering employee work engagement and sustainable employment during COVID-19 crisis through HR practices, employee psychological well-being and psychological empowerment. *Industrial and Commercial Training, 55*(3), 324–345. https://doi.org/10.1108/ICT-04-2022-0023

Renwick, D. W. S., Jabbour, C. J. C., Muller-Camen, M., Redman, T., & Wilkinson, A. (2016). Contemporary developments in Green (environmental) HRM scholarship. *International Journal of Human Resource Management, 27*(2), 114–128. https://doi.org/10.1080/09585192.2015.1105844

Saeed, A., Rasheed, F., Waseem, M., & Tabash, M. I. (2021). Green human resource management and environmental performance: The role of green supply chain management practices. *Benchmarking, 29*(9), 2881–2899. https://doi.org/10.1108/BIJ-05-2021-0297

Safari, A., Salehzadeh, R., Panahi, R., & Abolghasemian, S. (2018). Multiple pathways linking environmental knowledge and awareness to employees' green behavior. *Corporate Governance, 18*(1), 81–103. https://doi.org/10.1108/CG-08-2016-0168

Shen, J., Dumont, J., & Deng, X. (2018). Employees' perceptions of green HRM and non-green employee work outcomes: The social identity and stakeholder perspectives. *Group and Organization Management, 43*(4), 594–622. https://doi.org/10.1177/1059601116664610

Sudin, S., & Saad, Z. M. (2018). Linking Green HRM and green intellectual capital with corporate environmental citizenship behaviour. *European Proceedings of Social and Behavioural Sciences, 44*. https://doi.org/10.15405/epsbs.2018.07.02.45.

Tuan, L. T. (2022). Promoting employee green behavior in the Chinese and Vietnamese hospitality contexts: The roles of green human resource management practices and responsible leadership. *International Journal of Hospitality Management, 105*, 103253. https://doi.org/10.1016/j.ijhm.2022.103253

Yusliza, M. Y., Muhammad, Z., Farooq, K., Kassim, W. Z. W., & Omar, M. K. (2021). A qualitative study of factors influencing employee ecological behavior of academics in higher education. In M. H. Bilgin, H. Danis, E. Demir, & G. Karabulut (Eds.), *Eurasian business and economics perspectives* (pp. 3–21). Springer International Publishing. https://doi.org/10.1007/978-3-030-85304-4_1

Zhang, B., Yang, L., Cheng, X., & Chen, F. (2021). How does employee green behavior impact employee well-being? An empirical analysis. *International Journal of Environmental Research and Public Health, 18*(4), 1669. https://doi.org/10.3390/ijerph18041669

Index

M. Y. Yusliza et al. (eds.), *Employee Green Behavior*,
https://doi.org/10.1007/978-981-95-1422-9

MIX
Papier aus verantwortungsvollen Quellen
Paper from responsible sources
FSC® C105338

If you have any concerns about our products,
you can contact us on
ProductSafety@springernature.com

In case Publisher is established outside the EU,
the EU authorized representative is:
Springer Nature Customer Service Center GmbH
Europaplatz 3, 69115 Heidelberg, Germany

Printed by Libri Plureos GmbH
in Hamburg, Germany